SEX
& BROADCASTING
A Handbook on Starting a
Radio Station for the Community

LORENZO W. MILAM

Introduction by
Thomas J. Thomas

Dover Publications, Inc.
Mineola, New York

Copyright

Copyright © 1988 by Lorenzo W. Milam
All rights reserved.

Bibliographical Note

This Dover edition, first published in 2017, is an unabridged republication of the fourth edition of the work, published as *The Original Sex and Broadcasting: A Handbook on Starting a Radio Station for the Community* by Mho & Mho Works, San Diego, in 1988. [First publication: 1971]

Library of Congress Cataloging-in-Publication Data

Names: Milam, Lorenzo W., author.
Title: Sex and broadcasting : a handbook on starting a radio station for the community / Lorenzo W. Milam ; introduction by Thomas J. Thomas.
Description: Mineola, New York : Dover Publications, Inc., 2017.
Identifiers: LCCN 2016054224| ISBN 9780486814490 | ISBN 0486814491
Subjects: LCSH: Radio broadcasting—United States. | Radio programs—United States. | Radio stations—United States.
Classification: LCC HE8698 .M53 2017 | DDC 384.54068/1—dc23
LC record available at https://lccn.loc.gov/2016054224

Manufactured in the United States by LSC Communications
81449101   2017
www.doverpublications.com

A PREFACE
TO THE NEW VARIORUM EDITION
OF *SEX AND BROADCASTING*

To put out a fully revised Fourth Edition of *Sex and Broadcasting* requires that we include the immense technical changes that have occurred in the past ten years: the invasion of cable radio and television; the now common home video recorder; the plethora of disks and tapes (both audio and visual); the bastard step-brother of community radio, CB; three-and-a-half million satellite dishes sprouting in backyards, great aluminum psilocybin mushrooms of knowledge and communication. The things we tried to do in 1965 or 1975 (haltingly, with minimal budget) being done on satellite and cable channels — channels that scarcely existed back then: C-SPAN, CNN, the Discovery channel, Z, Bravo, Nickelodeon, A&E, and PBS for video; National Public Radio and Pacifica network programs and the great sound services of the Canadian Broadcasting Corporation on Anik-D for audio.

Radio and radio law and radio practice seemed so much more direct in the old days. Information about the Federal Communications Commission was the privilege of the few — mostly broadcasters and their attorneys. The outsiders like us who bothered to study and learn the truth could be well rewarded with results. The FCC was slow and stupid, a great slobbering beast there at the edge of the Potomac. With certain goads and craftiness, one could move the monster into a spasm of action, preferably in the right direction. This was one of the lessons of *Sex and Broadcasting*, editions One, Two, and Three.

Nowadays, under the rubric of "deregulation," the FCC has emerged as an earlier incarnation: a clone of the

Federal Radio Commission. The FRC was the immediate predecesor to the FCC, and following John L. O'Sullivan's dicta that "The best government is that which governs least," managed, singlehandedly, to shove American radio in to a Media Dark Age which lasted well into our own time. The regulators were owned by the industries they were created to regulate — AT&T ran the Common Carrier Bureau, the Broadcast Bureau was the baliwick of the commercial broadcasters, and the Chairman and most of the FCC commissioners were content to live in tiny, dark holes — namely, the back pockets of ABC, CBS, and NBC, and the other networks. Like some mythic southern family, all of them revelled in their incest, but it was deadly for the lifeblood of American free speech. In other countries, it would be known as Socialism for the rich and powerful; in the United States, it was called "Good Business Practice."

I am happy to report that a natural antipathy to work, as well as a premature dotage, has spared me the hard job of rewriting *Sex and Broadcasting*. Thus, what you have in your hands is mostly a reprint of the Third Edition. In fairness to my lassitudinous self — I should point out that a great deal of this present edition is not ancient history. So much of what I wrote — and what people seem to remember with fondness (as I do myself) — is concerned with the *passion* of broadcasting. This, indeed, is the source of its ambivalent and somewhat garish title.

As well, there are still all those valid ideas for improving radio programming with minimum budget and primitive tools. In fact, these may be more important than ever with the proliferation of public broadcast outlets which have embarked on a slavish aping of commercial radio stations (even to the point of being lead by the ratings — rather than by ideals of great and good radio). The humble, rough-around-the-edges community radio station — a model for freedom of speech, and freedom for all of us to be involved — still deserves our love and our support.

Nabokov once claimed that each of us, as humans, are Double Monsters, joined at the hip of time. As such, we are remiss if we attempt to exorcise, or even deny, the passion-clogged, foolish and embarrassing parts of our previous lives. Except for a few typographical emendations, and the omission of one or two paragraphs (not my own) that might be grievously embarrassing to their previous owners — *Sex and Broadcasting* 1975 stands as it was. Readers will quickly spot the anarchisms: telephone numbers, addresses, names, directions, forms, foundations, stations, friends, enemies, piques, cholers and flatulent passions that no longer parse, much less exist. The ideas born out of those heady days shall remain, uncensored by me, their loving and deranged father. They are part of our set back when we felt we could so easily burn a change into the fundament of American communications.

God grant our friends, relatives, loves, and gods a chance to experience such rich forgiveness.

Lorenzo W. Milam
Ulan-Bator
January, 1988

INTRODUCTION

"What a blabby night I made of it.... I painted one of those lurid, dramatic pictures which, nevertheless, was true. And I told them the exact how-to-do-it. I told them how to get money for it; I told them about the army of people we knew in engineering and law and other existing radio stations who could share with them the knowledge.... I told them about frequency searches, and how to find the old, used equipment, and where to put their antenna. I gave them do-it-yourself radio lecture #1, complete with jokes, dramatic memories, and stories about busts and bombings....

"I waggled my head around, got that light in my eyes when I was talking about real radio alternatives, that mad-light of the radio crazies."

— Lorenzo Milam

    Sex and Broadcasting is a special telling of the old exact how-to-do-it: the full tilt, Do-It-Yourself Radio Lecture Number One. Circa 1974.

    The book is, in many respects, an artifact of an era—for community radio and the country itself—that now is past. Perusing its pages is a bit like leafing through an old copy of *The Whole Earth Catalogue*, or, for those of us who lived these stories, perhaps more like a school yearbook. Now a decade old, most of the hard-core practical information is seriously out-of-date. And truth be told, a lot of it never worked all that well, anyway.

But Milam's Radio Lecture Number One, babbly, dramatic and all the rest, was never *really* about how-to-do-it and neither is this book. His genius was and is the passion and excitement and vitality of radio pushed well beyond the limits of practicality, the tumbling transmission of ideas, musics and emotions that can be driven by the power of people's need to communicate and the energy that is captured when a channel of true connection is created within a community. *Sex and Broadcasting* virtually glows with the mad-light of a true radio crazy.

It is through this vision and emotion, along with a healthy dose of humor and a lot of plain good writing, that this "Handbook" transcends its time.

These essays are a delicious kick in the pants— for the Washington bureaucracies (how little has changed); for the serious public radio crowd who little realize how much of their success rests on these early experiments; and, perhaps most important, for today's community broadcasters who share this rich legacy but are increasingly distant from the spirit and sensibilities that are captured in these pages.

My first job in radio was at KDNA, a St. Louis station founded by Lorenzo Milam and his long-time partner, Jeremy Lansman. I labored day and night, every day of the week, for the glorious wage of $66 a month— and so did everybody else. We had broken down equipment, never enough money to pay the bills, and every day seemed an exercise in improvisation. But as Milam once wrote, "KDNA was fearless and magical...radio reversing itself: asking that the people who lived in a city bring the city into the radio station, and cascade it out to the far reaches of men's minds...." We touched people in a powerful way.

KDNA was sold to a commercial broadcaster in

1973 and became an easy listening station. The city tore down the studios not long after and today there's just pavement, a street, where our station once stood. But years later, I still meet people who, with a startling fondness and clarity, remember those days, those broadcasts. Because we touched people in a powerful way.

I did then and do now consider it to have been a privilege to have been part of that experience. Radio is such an ephemeral and immediate medium that such moments pass all too swiftly. This book preserves for us some of the adventure, the wonder, and the power.

Thomas J. Thomas

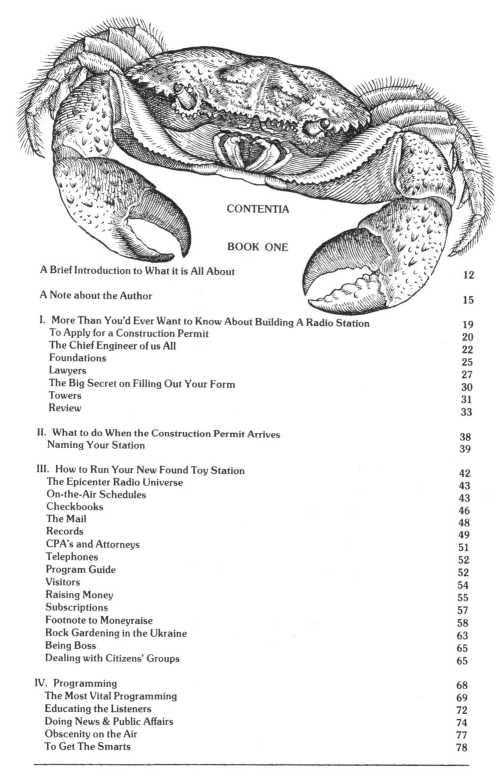

CONTENTIA

BOOK ONE

A Brief Introduction to What it is All About	12
A Note about the Author	15
I. More Than You'd Ever Want to Know About Building A Radio Station	19
To Apply for a Construction Permit	20
The Chief Engineer of us All	22
Foundations	25
Lawyers	27
The Big Secret on Filling Out Your Form	30
Towers	31
Review	33
II. What to do When the Construction Permit Arrives	38
Naming Your Station	39
III. How to Run Your New Found Toy Station	42
The Epicenter Radio Universe	43
On-the-Air Schedules	43
Checkbooks	46
The Mail	48
Records	49
CPA's and Attorneys	51
Telephones	52
Program Guide	52
Visitors	54
Raising Money	55
Subscriptions	57
Footnote to Moneyraise	58
Rock Gardening in the Ukraine	63
Being Boss	65
Dealing with Citizens' Groups	65
IV. Programming	68
The Most Vital Programming	69
Educating the Listeners	72
Doing News & Public Affairs	74
Obscenity on the Air	77
To Get The Smarts	78

V. Some Other Media Games	82
Short Wave	83
UHF Television	84
Cable Radio and Cable Television	88
Mental Communication	90
Communications with Space	92
Mau-Mauing Your Existing Local Schlock Station	94
Buying an Existing Station	100
Religious Radio	102
Illegal Transmitting	108
VI. Some Essays on Radio, Music, Programming, et al	114
External Transmission	115
Secretary's Notes on the Meeting of Tomentose	117
The Terrible Waste of Educational Stations	119
Sex and Broadcasting	122
Un-Collective Radio	123
Classical Muzak	125
Picketing the California Broadcasters Association	128
The Last Days of our Discontent in Estes Park	130
The Eeny-Weenies	134
A Small Hiccup for Humanity	137
Creep Critics	138
Sex and A Typical Day at Your Atypical Community Station	139
TV Past	142
The Roots of Militance	143
The National Public Radio Conference	144
A Scam on WCAM	145
Sex and the American Banking System	151
The KRAB Media Conference	154
Is There Anyone Can Show Me the Way to San Jose?	159
A Brief History of Listener-Supported Radio	162

BOOK TWO

VII. The Dixie Songbird and Other Forms of Silliness	172
The Dixie Songbird	172
Dear Bob and John	177
Dear Heart: The Poets' Weekly	179
Dear Draft Board	182
Whips & Spurs: A Family Magazine of Pain & Brutality	185
Dogmouth: A Magazine of the Arts	188
The Slaver	192
The Daily Handmaiden, A Journal of Onanism	198
The Early Guides	201

VIII. Appendix	215
References: Books & Magazines	216
The Reference Room at the FCC	222
Lawyers & Engineers	225
Equipment (New & Used)	226
Towers	227
Sympathetic Broadcasters	229
Programming Sources	231
The 100: KTAO's Greatest Hits!	238
Filling Out the Technical Part of Form 340	256
by Benj. Franklin Dawson III	
What to Do When the Construction Permit Arrives	258
by Benj. Franklin Dawson III	
Turning on Your New Transmitter	264
by The Aether Sprites	
Health, Education & Welfare, Corporation for Public Broadcasting Grants	265
by Tom Thomas	
PBS Go Boom!	273
by John Ross	
The Tax Man Will Gladly Fund You	276
by Alfred L. Blackhead	
Local Broadcasting in Remote Communities	280
by Douglas Ward	
A Case Study on Starting a Community Station	288
by Holt Maness	
Three More Paper Games	293
The KRAB Obscenity Trial	305
Running for Political Office	328
A Final Essay About Radio Politics	344
A Note on the Third Edition of SEX AND BROADCASTING	351

11

12

A Brief Introduction to What It's All About

Somewhere in the shadows of the early history of radio looms the mysterious figure of Nathan B. Stubblefield. Nathan B. Stubblefield? Nora Blatch? Reginald A. Fessenden? Professor Amos Dolbear? Where do they get those names?

Nathan B. Stubblefield was born in, grew up in, lived in, and died in Murray, Kentucky. The citizens of that miniscule town were affectionate towards their mad radio genius, and erected a monument to Stubblefield in 1930. They called him The Father of Radio.

Stubblefield was poor, and a mystic. He was a mendicant and a martyr to his invention. Everyone wanted to steal his invention from him. Jim Lucas said that his home was so wired "that if a stranger approached within a half-mile, it set off a battery of bells." And Stubblefield, stubby mystic that he was, said

> I have solved the problem of telephoning without wires through the earth as Signor Marconi has of sending signals through space. But, I can also telephone without wires through space as well as through the earth, because my medium is everywhere.

My medium is everywhere. Nathan B. Stubblefield, the self-taught inventor of Murray Kentucky, who would later tell people that he would turn whole hillsides light with 'mysterious beams.' Stubblefield, the mystic of the mystic transmission of waves everywhere, through air and land and water, to the nether reaches of the stars.

Everybody knew about Stubblefield's Black Box. The Black Box made the light, and the voice, out of the air. In 1892 (14 years before Fessenden's experiment from Brant Rock) he handed his friend Rainey T Wells a box, and told him to walk away from the shack. Stubblefield always lived in a shack. Wells said later

> I had hardly reached my post...when I heard HELLO RAINEY come booming out of the receiver. I jumped a foot and said to myself THIS FELLOW IS FOOLING ME. HE HAS WIRES SOMEPLACE. [Wells moved a few feet further on]. All the while he kept talking to me but there were no wires I tell you.

This fellow is fooling me...there were no wires, I tell you. Early radio, radio magic, the magic of sending the voice through *nothing*. Nathan B Stubblefield, the magician with the black box and all the lights, the man who could make the voice travel through *thin air*.

They stole his invention. Of course: they always do. The Wireless Telephone Company of America, set up by 'promoters' and 'speculators.' Smooth talkers (unlike unverbal mystic Stubblefield) who jacked up the price of the stock and disappeared. Stubblefield wrote for the prospectus:

> I can telephone without wires a mile or more now, and when the more powerful apparatus on which I am working is finished, combined with further development, the distance will be unlimited...

The apparatus on which I am working...distance...unlimited. Nathan B Stubblefield died in 1928 in a shack in Murray Kentucky. He died of starvation.

Stubblefield called the New York promoters a bunch of "damned rascals." He said they were "defrauding the public." What he meant was that they were defrauding his dream of unlimited voices, for unlimited distances, and unlimited lights. The mystic of radio with his loops and coils and magic was being defrauded; and all he wanted was to make the aether speak.

Nathan B. Stubblefield. Defrauded by the promoters. They wanted to take his loops and coils and Make Money. And Stubblefield was hurt wrenched torn by these animals from the city, these damned rascals. He went back home to his shack in 1913. And for fifteen years was barely seen. Sometimes the neighbors saw him 'from a distance.' For fifteen years, nothing, except:

Some observers reported seeing mysterious lights MYSTERIOUS LIGHTS and hearing weird sounds WEIRD SOUNDS in the vicinity of Stubblefield's home.

...Two weeks before his death, Stubblefield visited with a neighbor, Mrs. L. E. Owen. He asked her to write his story. He said

> *I've lived fifty years before my time. The past is nothing.*
> *I have perfected now the greatest invention the world*
> *has ever known. I've taken light from the air and the*
> *earth as I did with sound...I want you to know about*
> *making a whole hillside blossom with light...*

Nathan B. Stubblefield. Locked in his shack. Starved to death. The man who took the Black Box and made words travel through the space around us. The man who created strange noises and weird lights. The man who would make a whole hillside *blossom with light*. Nathan B. Stubblefield. Of Murray Kentucky. Dead at seventy of starvation and too many visions...

> Drawn from an article by Thos. W. Hoffer in THE JOURNAL OF BROADCASTING, Summer 1971

15

A note about the author

Lorenzo W. Milam was born near Idaho, Potato, in a log cabin on the edge of Tapioca State Pudding. As a child he was an imaginary playmate.

The University of Pineapple is his alma papaya, he was graduated *mango cum laude.* Early in his career, he revolutionized the broadcasting business, but he is not to be blamed: The San Andreas is nobody's fault.

Once Lorenzo said to me, "Tom, who's gonna chop your suey when I'm gone?" And it's true. Noone has chopped my suey for a won, won ton.

Tom Robbins

SEX & BROADCASTING

16

17

Book One

> "I have come to another conclusion respecting the earth: namely, that it is not round as they describe, but of the form of a pear...upon one part of which is a prominence like a woman's nipple...
> I believe it is impossible to ascend thither, because I am convinced that it is the spot of terrestrial paradise, whither no one can go but by God's permission."
> —Christopher Columbus, after his Third Voyage (1498). Quoted in Miller's NEW HISTORY OF THE U.S.

18

Part One

> How did they ever get a permit to build the world?
>
> —Leo Gugliocciello

MORE THAN YOU'D EVER WANT TO KNOW ABOUT BUILDING A RADIO STATION

19

Oh! Radio! Radio as it is...

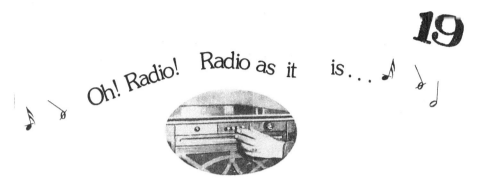

Broadcasting as it exists now in the United States is a pitiful, unmitigated whore. At some stage in its history, there was a chance to turn it to a creative, artful, *caring* medium; but then all the toads came along, realizing the power of radio and television to hawk their awful wares. The saga of broadcasting in America is littered with the bodies of those who wanted to do something *significant*---and who were driven out [or more correctly, sold out] by the pimps and thieves who now run the media.

Broadcasting does not have to be so vile and boorish. The Canadians best of all have shown that it is possible to have a superb blending of commercial and non-commercial radio and television: and Canadian communications are alive and alert and funny and meaningful. They do not have to bore people to death (as the 'educational' broadcasters in this country so obviously need to do); nor do they view the listener as some sort of dumb animal to be fed acres of pap---solely for the purpose of prying money from him. The art of radio can be used for artistic means; the radio-soul does not have to be made into a strumpet for soap and politicians.

The dismal state of American broadcasting is perpetuated by nitwits who should know better. BROADCASTING MAGAZINE---the memento mori of the whole 19th century robber baron tradition of commercial broadcasting, babbles about censorship every time the Federal Communications Commission moves timidly into the area of consumer (listener and viewer) protection. The turnips at the National Association of Broadcasters have millions of dollars to bang on every congressman or Senator who may dare to try to change the milking of the golden goose aether. And the FCC itself is harassed and badgered on all sides by an industry which has enormous power.

But the spectrum is as big as all outdoors---and there is a nitch here, a crack there, for those who care to try to squeeze some of the art back into radio. There are even ways for the poor and the dispossessed to get on the air, to have a chance to speak and be heard outside the next room, the next block. Although most of this vital natural resource has gone into the hands of the speculator-ruinators, there is a portion of the FM band which has been set aside for commercial-free operations. Due to some easings in the restrictions on those who may apply for these frequencies, it is possible for small groups to have their own broadcast outlets...even though they are independent of schools, colleges, and the big moneybag radio combines.

We are primarily concerned with the educational portion of the FM band, from 88.1-91.9

megahertz. Traditionally, this has been set aside for the big bores who run Schools of Communication at various colleges and universities and even public schools. But in 1954, the FCC permitted (suggested, in fact) that Pacifica Foundation in Berkeley make use of the educational band for a repeater operation—KPFB. Since then, other non-school groups have utilized the lower end of the FM band for community stations---educating listeners in the widest sense. KPFK in Los Angeles, KPFT in Houston Texas (both Pacifica stations) and, as well, KPOO (Poor People's Radio) of San Francisco, KBDY of St. Louis, KUSP (the 'Pataphysical Broadcasting Foundation) of Santa Cruz, and two or three other groups have asked for and received permits for non-commercial broadcast stations.

KUSP, KBDY, KCHU, and WYEP are the most interesting stations for the purposes of this booklet. For the first time in history of the FCC law and practice, non-school groups have been recognized as being legally qualified to broadcast on the non-commercial part of the broadcast band. 'Pataphysical Broadcasting Foundation, especially, was granted a permit to construct a low power station specifically using suggestions outlined in the first edition of SEX AND BROADCASTING. Sometimes the only way to test new directions in FCC policy is through the application process. It is not unlike the exasperating method of determining changes in the Russian government through obscure pictures on the back page of PRAVDA.

So, for the first time since the great wild early days of AM radio, back in the 1920s before the ogres took over our precious Aether---for the first time since those wild experimental days of free and loving transmission---radio operations have become available to anyone who might have that dreadful need to communicate. A need which some of us feel to be as strong and as vibrant as the need to love, and to eat, and to sleep.

A disease. Maybe we can even think of the art of transmission as a need of purblind sickness: a habit as hard and driving as the very shriek of the blower which cools the white-hot tubes in the broadcast transmitter. You and I, caught in the transmission of generations of words--- cascading them to the edges of our visible horizon: and perhaps even sending them to the outer edges of the universe to puzzle strange minds behind strange brows. I know that someday I may be able to explain to you my views on the blinding habit of radio...which has to do with self-image, and the needs for minorities (us) to see and hear themselves on a million screens and in a million speakers in a million homes...

...But that's for us to talk about at some time when we have a few beers, and the sun is stretching to die on the Santa Cruz Mountains, and I can titillate you with my image of the sensual nature of broadcasting, the fascinating tingle of control rooms, and rack panels, and the fine hard mesh of microphones, and the dizzying amplification of a Collins transmitter. That's for later: now I want to give you some hope on the how-to-do it—because you may be able to do it.

And then again you may not. For what I am going to start to try to do is to lead you into the maze of bureaucracy called Federal Communications Commission and Form 340 and site availability. And you and I can never NEVER project the strange meanderings of the governmental process of cotton and delay. You may be able to pry a construction permit out of this body. Then again, as I am writing now, they may be fuddling up the rules whereby you seek a permit. Sometimes they come in the night and 'freeze' things---which means that all of a sudden your dream station is locked in the monster jaws of governmental inertia for years and

BLACKJACK

years: and nothing, I mean nothing, can ungum the process---not anger, rage, picketing, lawsuits, letters to Nicholas Johnson, political leverage, tears, desperation, and death.

Someday, someday: I have promised myself to write for you a book about the FCC, and how they lose strange applications, as they did for me: and a strange man, called John Harrington, in Complaints and Compliance. Someday. Not now.

To apply for a construction permit for a radio station, you will need the following:

A frequency;
A friendly (and hopefully honest) attorney;
A non-profit corporation;
A transmitting site;
Seven or nine good and loving people to be on the board of your corporation;
Ten copies of FCC Form 340;
Some money or other assets;
Patience. Acres of it.

It will take time, and the willingness to wait. Maybe as long as two years. But you can be doing this while you are holding down another job and eating clams on the beach and drinking yourself insensate at the same time.

A kiss is to love as a Form is to the Government, so you should get a fistful of the Form #340. They are free---one of the rare things that you and I can get for free from The Man---from the Federal Communications Commission: either through any of their 24 district ofices, or through the main octopus in Washington, D. C.--at 1919 "M" Street Northwest, zip 20554. Use one copy of this to rough out your answers. This is a good touchstone to getting yourself geared for all the steps and requirements facing a licensee of a broadcast station in the United States.

While you are waiting for your forms (since your letter will inevitably get lost somewhere in the government machinery) you should make sure that there is even a frequency available in your area. I am fairly convinced that there are none whatsoever within 30-60 miles of the following cities:

New York City
Los Angeles
Chicago
Philadelphia
Boston
San Francisco
Detroit
Cleveland
Washington D. C.
Pittsburgh
Dallas
St. Louis
Minneapolis
Seattle
Houston
Atlanta

The Kangaroo Rat of New South Wales

Indianapolis
Miami
Baltimore
Cincinnati
Portland (Oregon)

There are other cities which are marginal—even for ten watt stations. And I may be wrong on one or two of those listed above: you might be able to squeeze a useable signal into a major market even though you have to put your transmitter some distance from city center. There is another solution which has to do with trying to get some of the existing educational stations to move around, frequency Musical Chairs. But most of them, I must warn you, are so piggy that they won't even consider moving, much less give you the time of day. In this case, you have to file dreadful "Show Cause Orders" with the FCC—all of which I will explain to you later.

To figure out if there is a frequency available, or if you might have to transmit from 20 miles outside of town, or if you will have to try to move two or three 10 watt stations around—to do all this boring and complicated work, you will need an engineer who knows how to work the F 50, 10 chart, and knows FCC rules, and who will do a frequency search for you.

A frequency search don't mean turning on the radio and listening for holes. It means having on hand a copy of Section 0, 1, 2, and 3 of the FCC rules—available after a mere 6 months delay from the Government Printing Office, Washington D. C. Your engineer will have to order an official list of the existing and pending FM stations from

Tom Berry
1705 DeSales St. N.W., Rm. 500
Washington, D. C. 20036

When the FCC rules finally come, your most important information is contained in a Footnote to Part 1, Section 1.573. I have just saved you $500 in engineering fees. You can almost do a frequency search of your own by ordering Bruce Elving's excellent, concise, and complete FM STATION ATLAS from

Box 24
Adolph, Minn. 55701

for $2.50. Some engineers have to go to school for five years to learn the contour prediction method for FM found on page 79 of the ATLAS. But you are best off with the rules themselves which tell you *most* (not all) of what has to be done for an FCC application.*

There is one thing that I would suggest you avoid doing at this stage: that is, going to your local regional office of the FCC and asking for their help on your community station. For one thing, these people are notoriously disinterested in acting as information sources. They are convinced that they are overworked. They may be right: the whole FCC consists of 1100 people whose job is to oversee a million or so acres of aether. But FCC bureaucrats are a special breed: not only are they Civil Servants with jobs to protect from controversy and life,

---

*Like all religions—parts of the Bible of the FCC are not contained in the material you get from the Government Printing Office. Some is contained in their booklet on *Good Engineering Practice*, issued in the late 40s. Some charts are out of print, and some, I do believe, are locked forever and out of sight in the mind of Mr. FCC-FM, Ed Hackman.

23

but they have their own special loathing for the public. Employment at the FCC involves a deadening process, and some dreary codes made up in 1934. Worse---these bureaucrats---especially on the local level---will do anything, including making up stories, to get you out of their hair. I know of a dozen or so cases where innocents have been told, "No---there aren't any frequencies available," just so the petty official can get back to his papers.

The FCC lower echelon is a continuing paradox to those of us who have to work with it on a regular basis. One would think that they were not public servants at all: but rather, somewhat testy and very powerful nitpickers right out of Dostoevsky. There are occasional gentle and good souls who creep into the Commission on a lower level: but they are so rare that I don't even bother to ask the local branch office of the FCC for anything except forms and administration of the 3rd Class test.

Your best source for help and rumor and advice and lore is not at the FCC. Nor would it be some local broadcast station owner: those tits are all salesmen hiredhands, paid to whore their particular frequency. Nor it it the dotards in a school of broadcasting: they are paid to suck $600 or $1000 from poor you (and your desperate need to communicate) and give you little in return. And you are going to get no help from the state university School of Communication---a repository for all the troglodytes who can't make it in the commercial world.

No: your biggest help is some First Class Chief engineer. One of those who has been working at one of your local radio stations for awhile, and who loves (most of them do) to talk about the industry.

A good broadcast engineer has an instant lightsecond source for information about what's going on at the station he is working for, or the station across the way, or the one in the next city—or in many cases—some across the country. They know about equipment for sale or about to be taken out of service. They know who is running which station. They know who is going to be hired, they know who is drinking too much, and who is sleeping with whom in the front office. They know all the prices paid for radio stations in your particular area over the past twenty years. They most probably have an avid interest in and affection for more than just the technical side of broadcasting. Often, their knowledge spills out into obscure and occult aspects of FCC law.

A good Chief Engineer is independent and bizarre. He has a mind of his own, because he has the access and expertise on the *means* of transmission of sound. The Chief Engineer of a radio station has the boss by the nuts, and they both know it. The salesmen and the management and the disc jockeys hold the Chief Engineer in some awe. And they should. I have talked to several engineers who managed to wire the station they work for. And they will be *goddamned* if they will ever draw up a comprehensive wiring diagram. "That's my lifetime job security," one said. They trade on the fact that most owners can't tell a 3X2500A3 from a muskrat; they know that 98% of the jocks think that a rectifier is something for proctological examinations, and that the whole place would fall apart if they picked up and left.

In each area there are one or two or three literate, intelligent, madmen Chief Engineers (who likely as not work for several stations)---who are a goldmine of information for you. If they like you, they will not only tell you who is eating out who in the sordid world of radio---but they will help you with frequency searches, locate obscure and cheap equipment for you, tell you which station owners or managers to approach for necessary tower space. And, if they like you *a lot*---they will help you build your teakettle for nothing and maybe even contract to serve as your Chief Engineer for a small fee every month. They---like most people in radio and television outside of the moneygrubbers---are bored to death with the day-to-day workings of their job, and they share in the excitement of a new operation going on the air, even if it's just your dinky 10 watt station.

As long as we have gotten this far, I should tell you some things *not* to do. One is: don't sit around and dream of what call letters you are going to ask for. This opportunity occurs only after the issuing of an official construction permit by the FCC. Another is: don't start ordering or buying equipment. Although you must specify equipment for your application---you would be an idiot to buy any. You may end up with transmitters in the basement and antennas in the bath-tub: and no radio station at all to hook them to.

It is very important at this point that you carefully ignore any and all rumors that you hear. Well, not all: maybe 95% of them. As soon as you get your idea for a viable community outlet and start to work on it---you will be besieged with stories that flow out, through, under and around the broadcast industry.

It is as if people in radio are not content with holding down their chunk of frequency, spreading their gruel over all the countryside by electronic means. No, they also seem to need the constant flow of rumor, half-truths, misinformation, and outright absurdities in order to function. The stories you will hear will be of two types: immediately as you start on your application, you will hear of at least three other groups who are putting together *their* applications and who will be competing with you for the chosen frequency. If you check out these tales, they will probably be wrong, or greatly exaggerated.

The other story will be one of how the existing AM or FM station in town is going to make trouble for your application by filing secret material with the FCC, officially protesting your proposal.

Forget it. Even given the unlikely chance that some existing broadcaster is frightened to death of your miniscule effort, he can do little to influence the FCC either for or against you. For if the FCC is a bore and a lump to all we want to do in radio, at least AT LEAST it makes the attempt to be just, watching the wranging dogs of broadcasting and keeping them from tearing each other to bits. As long as you are honest and straight in your application, and in your pre-on-the-air contacts with the people in your community, you have nothing to fear.

You are learning from these experiences that the world of business is as gruesome as you always expected: and broadcasters take the vileness sweepstakes. The FCC has almost no enforcement machinery of its own; yet we read each week of fines and penalties levied against radio and television and common carrier operations. How do these come about?

Well, it's that radio executives, when they are not counting their millions, are sitting around in their managerial positions in the 10,000 broadcast outlets around the country, thinking how best to do in their fellow media barons. "I'm gonna tell," they say, and they do: tattling like some 5 year old kid to Mumsie in Washington. Think of the waste and tragedy of it all: all these 45 year old pesky pee-heads, in their offices, scheming on how to make trouble for that pesky pee-head down the dial who they know (rumor mill again) is indulging in Double Billing.

Because of the bitterness and childishness so endemic to the field of radio, I suggest you avoid previous controversies, previous alignments, previous battles. There are many angry feelings that have grown up like weeds in broadcasting because it is a field so rich with screw-the-world, make-a-fortune, eat-your-neighbor. It will not help you to get involved in these vicious entanglements. The temptation may be overwhelming, but hisory points in another direction; as Thoreau said:

"Some circumstantial evidence is very strong, as when you find a trout in the milk."

and want to carry it over the aether. Preserve us from such wormwood. Go your own way. Avoid past entanglements. Be careful and cool and righteous. Smile.

While this is going on, don't be getting your friends all worked up with dreams of zingy programs you are going to put out on the air. You will have days weeks months to do that while you are sitting around on your ass waiting for the FCC to act on your application. Don't get people frenzied with stories of the knockout radio station which will be going on the air in a matter of weeks. This is a genuine soul-cruncher, because applications---even the simple one that I am proposing---can get stuck in the bowels of the FCC for many months. Believe, but don't eat your heart out.

A good thing to do at this stage (while you are still waiting for your forms from the FCC) is to try and round up some money. For a 10 watt (Class "D") application, you will have to show anywhere from $1500-$10,000 in assets, depending on how much used and available equipment you can show. If, by some queerness in your head or your life you can come up with a legitimate *solid* source of $25,000—you can apply for a higher power station. If you can come up with $100,000—you can make a downpayment on an existing major market FM station. If you have assets like that, stop sitting there: call me at once: (619) 488-4991!

But assuming that you are poor like the rest of mankind, you will have to come up with at least several thousand dollars in negotiable funds or equity. This means that you will have to talk your friends and family or directors into offering you---on paper---so many dollars based on the premise that your organization will be granted a construction permit by the FCC. These can be in the form of equity in car or house (which will have to be discounted 30 or 40% for being illiquid) or in the form of stocks, bonds, or savings account. There is nothing in the FCC rules that demands you use these assets to build your station. But you have to prove that at least you *can* lay your hands on that amount of money when and if they grant you a construction permit.

Whatever you do, don't go Foundation hunting. This is the latest in the continuing series of American dreams: and like most, is a cruel and vile delusion. Foundation people spend most of their days granting monies to their friends on other foundations. They are cold and unfeeling to most of us who pound on their doors. They are elitist---and seem to spend their days plotting non-controversial projects to fund. I spent some two weeks in New York in 1966---trying to get some money for KRAB to boost its power. The most sordid experience was a half-hour with a turkey hired to say no for the Rockefeller Foundation. I remember sitting on the 49th Floor of the Time-Life Building, wondering about her dismal personality, thinking of the dozens of radio stations I could be running with the money they squander every month on their hot-shot offices. Money tends to isolate, and excess amounts of money can turn genuine philanthropy into a form of leprosy called fuck-you-ism.

There are a few good, honest, and caring funding groups in this country. The Carnegie Foundation will be honest with you if you write or visit their offices in New York. The Stern Family Fund (through The Citizen's Communication Bureau in Washington D.C.) has done more for good radio than a thousand dimwitted grants by the Ford Foundation for 'studies.' The Robert Kennedy Foundation gave $70,000 to Bilingual Broadcasting Corp. of Santa Rosa, California, for a minority owned-and-operated community station. And the madmen at Pacific Change in San Francisco have, in the past, actively funded community stations which are to be run solely by Blacks, Chicanos, American Indians, Unwilling Draftees, and other minorities.

But the demands on these few foundations are tremendous---and they can only offer some

26

seed money to deserving groups after a large and complicated winnowing process.

For gods sakes, if you know someone on the board of directors of some turd foundation, *use him*. Elsewise, your chances of getting professional funding for something as vital and important as free speech radio communication are very very slight.

More and more community group stations are coming to depend on Health, Education and Welfare grants to build or improve facilities—and Corporation for Public Broadcasting funds to operate their stations. Operating grants from CPB---as Tom Thomas will explain in detail in the Appendix to this book---are more or less automatic: if your station is run by a bona fide non-profit corporation, with an FCC "educational" license and if you meet their basic qualification in number of employees, power, time on the air, and studio facilities. Those annual grants of from $10,000-$20,000 should be pro forma.

HEW is another bag of tea. They (so far) have tended to give their largest grants to the potboiler "education" radio and television stations: the safe-and-dulls that run state or district facilities.

Jesus: look at the grant list sometimes. $350,000 to that porky Channel 11 down the street that avoids community problems, agonies, and reality like The Crud. I do believe that HEW would give more money to our type stations if they could---but they are beat upon by legislators and the monstrously powerful educational Pithecanthropus establishments to dump money into these intellectual out-houses leaving only crumbs for those of us who care for the true potential of men's crazy minds.

When you finally get the copies of the Form 340 from the FCC, you will see that you are required to have a non-profit corporation as the actual licensee of the radio station. This corporation should comply with IRS *and* FCC requirements; e.g., that it will be solely involved in not-for-profit activities, and that the directors will not over-compensate themselves. Another is that the organization will not indulge in any activity designed to affect legislation directly, or elect one single individual to public office. Finally, they require that in the event of the dissolution or winding up of the corporation, its assets (if any) will be turned over to another non-profit corporation with essentially the same goals.

You will need a lawyer to help you with this whole project. Now, you and I know that there are a great number of attorneys who are money-grubbing twinks---who work for giant firms, and pretend their honesty away. And then there are others who have chosen to be generous with small inexperienced groups such as your own who want to do something meaningful with their lives. It will be *greatly* helpful if you can find such a good lawyer to work with you---in all phases of your operation. Someone who will give you the benefit of his training, and not charge your ass for it.

For instance, most lawyers will charge $500-$1,000 to incorporate your group as a non-profit corporation, and even more to move you towards the tax-exemption status you would like to have after you go on the air. But, there are a few attorneys who will do all this for no more than the cost of filing your forms with the state. For most of our applications, we have spent no more than $50-$75 to set up the corporation.

In addition, you want to encourage this lawyer to join your board and work with the others in helping you to get the station on the air. These guys have seen all the sordid side of the money-grab world, and---if they are at all sensitive---might well appreciate your not-for-profit

BUILD YOUR OWN
EXPONENTIAL HORN

28

mentality. I know of one who claims that when he is not doing free work for our radio station, he is busy dispossessing widows and orphans from houses with unpaid mortgages.

Sometimes, you have to go to groups such as the ACLU, the American Friends Service Committee, or your friendly local radical group to see if they have an involved lawyer. It is worthwhile to be cautious in your search: you need more of a maverick rather than a politically committed madman.

When you go to incorporate, may I suggest you avoid flag-waving names. You should be modest, even a bit of a shuck, in naming your organization. "The Right-On Foundation," or "Up-the-Pigs, Inc." or "Fuck Fascists Forever" may be scenic and funny to you, but they will hardly help you to obtain a permit from the Federal Government which is understandably sensitive to these aspects of community life. When we set up KRAB in Seattle, we named the corporation The Jack Straw Memorial Foundation (*vide, The Nonne's Priest's Tale* by Chaucer). KDNA in St. Louis was owned by The Parent Corporation, and station KUSP is owned by The 'Pataphysical Broadcasting Foundation, Inc. ('Pataphysics is to metaphysics as metaphysics is to physics. That's the way that David Freedman explains it to me, but he might be crackers.)

Your corporation Foundation organization will need a board of directors. Seven or nine is usually the best size. If you have less, your meetings will turn into orgies, and with eleven or thirteen, you will have to explain everything to all the latecomers---and spend hours on the telephone just keeping them informed.

Don't look for bullshit 'community leaders' to be on your board. They will never come to your meetings, and if they do, they will probably fall asleep or dominate the whole thing with their foolish pride. What you need is a working board---not window dressing, because the FCC for one doesn't give a goddamn what hot-shot leaders you have on your corporate letterhead. All they are interested in is their citizenship, where they were born, and whether they have been accused of monopoly practices.

You should try to get people who are compatible with your desire to do good radio. A lawyer, a school-teacher, some students, some people who have learned the *aesthetic* of good living. People who are interested enough to come to your meetings regularly, who will help you in any way they can to put on a good operation.

You can get a boost to your application---and a true boost to your station---by having a strong, no-nonsense, sexually and socially and racially integrated group of intense and devoted people on your board. No matter what their age, no matter what they look like, no matter how shy or fancy: you should have good people, people that you like. You will be meeting wtih them every week while you are putting together your application, and you should meet with them at least monthly after you have your permit. We always brought a gallon of wretched red wine to the monthly board meetings of Jack Straw---which meant that they would never last for more than 2½ to 3 hours as everybody would be soused and all ready to go home by that time. Board meetings can be great if the people you are working with have *actual* power, but are sufficiently separated from the day-to-day operation of the station to have some sort of kindly overview, a 'benign eye of reason.'

I would like to suggest that you pointedly avoid putting political or social or religious fanatics on your board. Their loyalties may obfuscate their vision---and they might want to use that station for their own form of preaching. How can I explain to you that there is a specific type of person you need: slightly anarchistic, highly cerebral, cosmically gentle; such a soul is perfect for you and your organization.

You should—by all means—find out if the people you are getting on your board will cause you any trouble further down the line. Ask them if there are any problems of citizenship, or past activities, or moral turpitude (whatever in gods name *that* may be); then find out from some FCC attorney if there will be any difficulties. I do emphasize, however, that the FCC ain't too choosey about who it will let be involved in the ownership and management of broadcast properties. If you don't believe me, just let the rumor mill tell you about the personalities, ethics, and continuing business practices of the owners of radio and television stations in your area. You could people a whole new panoply of Dante's *Inferno* with these characters. If the Commision chose to look up the financial history of your friendly local broadcast magnate—all those bureaucrats would be tied up in a steamy mess for eternity, and wouldn't have the time to make trouble for you and your gentle application.

ATTORNEYS

There are a couple of FCC lawyers (those who are in private practice, but who practice principally before the Commission) who are honest, and who will not send you to penury as they help you with your application process. Try to get someone like this when it comes for you to file your Form 340. If you have no contacts in Washington, D. C., I could suggest some—or you might make contact with the Citizen's Communictions Center. FCC attorneys are like Vaseline. You can do what you have to do without them; but there are periods of extreme dryness when they can be a powerful help. Some, however, I must warn you, are more like sand.

Let's assume that you have received your ten #340 forms from the FCC. As you look them over, you will note that there is an extraordinary amount of what we politely call 'bullshit' required. You may well wonder why, (given the large proportion of apparently genuine questions about programming and the community) there are so many dewlaps and dull bulbs broadcasting over our precious air.

The answer is that you should view this extended questionnaire---as you should view the entire FCC---as nothing more than a random obstacle course, set up to discourage none but the most greedy (or the most crazy) from their goal. You see now the secret of modern bureaucracy. If quality of broadcast could be achieved through volume of paperwork, then American radio and television would best the greatest in the world, like the BBC or the CBC.

Alas, it ain't so. The FCC Form 340---like all the forms of that strange body---are exercises in unreality; tests of your patience with the bureaucratic mentality. As long as you treat these forms as a subtle type of torture, created to test your willingness to put up with damn near any indignity and foolishness—then you should do well, and get your station on the air in short order.

One extremely important thing: make your application as *straight* and *non-deviant* as possible. Put your head in the head of the government official who will be reading your Form 340, your answers to questions of financial ability, your suggested programming, your 15 exhibits.

Do you think that bureaucrat wants to read any jokes, any Existential philosophy? Do you think he wants a good boff—or maybe wants an intimate sketch of the realities of the theories

30

of Bakunin? Fat chance. That guy wants something that he can skim with his eyes, something that will not bother his mind (and more importantly, his bosses' minds—or what's left of that will not bother his mind (and more importantly, his bosses' minds—or what's left of them) in the slightest. He wants an application which conforms to all those other forms he has been Please. If not, give up. Now. You'll never have a station. Believe me, I tried. I put a collection of very clever and witty and brilliant statements in an application I filed with the FCC for a Washington DC educational station back in 1959. Would you believe that I am still waiting for them to act on that particular application?

I am now going to give you a super-secret on how to do your form 340. This secret is worth somewhere between $2000 and $4000—and I am going to give it to you for $3. Sometimes I am so generous that I can't look at myself in the mirror without crying, and wondering about my motives. Which are something else again—let me tell you.

The Big Secret on Filling Out Form 340

Every radio and television station in the United States is required to keep a public file. In this public file, they keep their applications to the FCC—along with important changes, and communications with that body.

This Public File is available to anyone, *anyone* (even you) who goes by any radio station between 9 AM and 5 PM, Monday through Friday. All you have to do is to give that station your name and address (no further identification needed) and tell them you want to see their public file. Of course, when you do this, you will be tramping on the natural paranoia of all businessmen (even educational broadcasters' paranoia) which is to deny their 'private' information to the general public. Therefore, expect—when you ask for this file—to be given the A #1 golden fuck-a-duck runaround. I mean!

They will tell you that they don't have to show you the Public File unless you state what organization you are with. That's a lie. They will state that their public file is at their lawyer's office, and he is located in Washington, D. C. You can then cite the law to them: which is that all stations in the country—both educational and not—*must* maintain a file for the public at the station itself. And you are entitled to see it. And if they continue to hide it from you, you will write the FCC and raise hell. Which you can.

I tell you all this not so you can exercise your ability to scare your friendly local broadcaster—but so you can fill out your Form 340 in the easiest and most economical fashion. And the way to do that is to go look at one which has already been filled out, and which *worked:* thus, you want to go to a station which has filled out a Form 340 (thus, it will be a local educational station), filed it with the FCC, and gotten a permit to broadcast as a result.

Pick out some turgid school or college broadcaster in your area; pop over and go through his public file. Take notes. See how *he* answered some of those silly FCC questions. See how *he* did his engineering. See how he set up his programming. See how he promised the world to the FCC—and then ended up (as you can hear) pouring out dark-grey bilge educational kill-the-mind-dead radio—a radio which is as tragic a loss to the medium of ideas as all the commercial crap.

One further point: try to pick an educational FM broadcast station which has gone on the air in the last five years. The reason for this is that the Form 340 has evolved (more complicated) and there may be some parts of those filed in 1960 which are a bit archaic, which may be misleading to you.

When you go to raid your local educational broadcaster's application for your own, don't just copy it slavishly—please. Use *some* originality. Be honest if you have planned some non-controversial programs. Emphasize your educational programs. If you are going for only 10 watts, remember there are parts of the part called V-B which you don't have to fill out. There are other differences which will show up between you and some School district...especially in the realm of funds available, and organization of your corporation.

In general you are learning with this copying act a central part of law: that is, to do something right for the government, go back to previous filings. Find the successful ones. Find how they did it. Obviously—if they got their application through the threshing machine, then it was done right. Try to do yours generally the same way.

Besides a showing of money, the most important part of the Form 340 is the showing of a definite place to put your antenna and transmitter. Your friend the engineer may be able to hook you up with a local broadcaster who will let you perch on his tower. You want to go to UHF television stations in your area first. They usually have the tallest towers—and are most willing to permit you to lease space from them, because they need the money the most. Next, you should approach your local VHF television or FM stations to share space.

In all cases, you should be willing to pay $100 or $200 a month rental for the chance to use the tower. You should explain to the station owner (or engineer) that you are not a competing application, that you are just looking for a place to put a low power station, that your antenna will not impose a large wind-load factor on his structure—in general, that you will cause him as little trouble as possible. If he sees it as an easy way to get a bit more revenue into his operation, he will probably allow you to specify his tower. And it is important to remember that the FCC actively encourages 'antenna farms'—where most of the FM and TV stations can operate their transmitting apparatuses from, causing the least amount of interference to other non-broadcast services. A broadcaster should not give you a flat no on space sharing without a good reason. If you make contact with all the antenna farms or large towers in your community—and get noes from all of them—you should try to get these refusals in writing. It can affect your attempts to get mileage or space waivers from the FCC later on. And if you get a positive answer—be sure you have *that* in writing. A short letter of acceptance can save you days of grief further on down the line if you are challenged by some poop. In any agreement you get, be sure you take care of the simple problems: who will supply the power (will you have your own meter?); will you have regular access? Will you share engineers? Who will be responsible for the security system? In all contracts (if you get that far in your negotiations) it should be specified that, if at all possible, you will not have to start paying rent until you are ready to turn on your transmitter.

If you can't find a tower, then look for a good tall building in your town. Remember that FM is line-of-sight—and that means that what you can see (outside of fog and smog) is what you can hear. If you go up to one of your tall buildings, and can see far enough to encompass your town, and a couple extra besides—then you should try to get permission from that building to use their facilities. Remember—when these managers start asking you hard questions (how much does the antenna weigh? what does it look like?) you can get acres of free help and plenty of free (and expensive) catalogues from your area Gates, Collins, or RCA sales representative. He is paid to try to sell new transmitting equipment—and might even give you a free lunch while he is trying to sell you. You can broach the subject of cheapo equipment (used) after you have eaten—for all of these companies carry second-hand equipment that they have taken in on trade from rich broadcasters. I would try to avoid buying equipment which is more than 10 years old. It sounds all antique and funky to have a Western Electric transmitter built in 1947—but I can tell you those things are murder to get running, and worse to *keep* running. The same goes for tape recorders, turntables, and the like.

But please understand that sometimes this is all you can get. And, often, you can get some equipment donated by one of your local broadcasters after you have a construction permit. Anything that's an antique, which is ruining the good will of their engineering staff, and which they can take as some sort of a write-off, they will give away. If you can reach the right person. If he likes you. If he is in a good mood.

There are some companies which specialize in used equipment: Maze Company, of Birmingham, Alabama; Broadcast Equipment Supply of Bristol, Tennessee; Boynton Studios of Morris, New York; and Guarantee Radio Supply of Laredo, Texas are the best-known. Once you have your construction permit as an educational radio station, you are automatically eligible to purchase any and all equipment, furniture, shoes and bib overalls you may need from your state surplus distribution agency. Federal surplus supplies are divided among the fifty states, and are administered by those states: HEW—through the Broadcast Facilities Improvement Division—has a list of all names and addresses. Ben Dawson said he found 150 feet of brand new Heliax 3-1/8" cable at the Seattle office for $50. However, some of the state agencies can be junkers: the California State Agency for Surplus Property in San Lorenzo was offering nothing more exciting that 3-foot-tall landing strobe lights for $10 the last time I was there. (Of which I bought three).

Back to transmitter locations: if worse comes to worse, you can try to find your own hill outside of town where you can build your own tower and transmitter site, with remote lines to your studio. This can be El Paino #1 however. If you take a virgin hill, which overlooks everybody and his brother, you have to have all sorts of permits: tower approval from the Federal Aviation Authority (Form 7460-1); approval from your local city or county for use variance on the property; probably an ecological impact study. Then you have to have permission from the owner of the property and adjacent property owners to bring in roads and power and telephone lines. Then you have to be sure that you set your transmitter up in a bunker so that curious idlers won't take the whole things home with them one day: bunkers, concrete-&-broken glass fences, land mines. You think I am kidding—but out in the country people can render your remote transmitting apparatus nothing but broken shards and a few pieces of beaverboard. Pardon my paranoia—but some of us saw the bombing of KPFT—Pacifica's station in Houston—as a scarey message from the future. As our lives and equipment get to be more complicated—it becomes easier for one individual to fuck us up. The disjointing process of a single individual with a single gun would have been impossible before the invention of the jet. And the disruption possible through one fool willing to embark on his own brand of vigilante law—as in Houston—becomes more possible as we invent machinery to survive away from daily engineer care&loving (which is the case with the present state-of-the-art which is remote pickup and transmitter operation.)

Speaking of mountains—there is another person in your community who is a goldmine of lore on remote transmitter and receiver complexes. He is a man who works (probably) for or with the city or county government. His department is called something like land-mobile-radio, or 'communications.' He is paid to know the availability of every mountain top or building top location in your area. He may be your local Motorola Mobile sales representative. He may work in some obscure part of the county government. Everyone recognizes him as the authority on point-to-point and point-to-mobile-unit communications. He knows every high point within 50 miles. He has personally travelled up to and through some of the most impassible heights in the country. He is a quiet expert. Most of the city or county government doesn't even know who he is, or how important he is for their communications machinery. Find out who he is. Go talk with him. He probably likes his job. He probably loves talking with strangers—like you—and impressing you with his knowledge. Let him. He might even find a place for your transmitter. Rent free. He just might have one of *them* onions up his sleeve.

Transmitter location. As soon as you have that taken care of, you will have to go to work on 'the engineering' portion of your application. You will need maps. Find out the address of the nearest office of the Office of Geological Survey (Department of the Interior) in the white pages of the telephone book under UNITED STATES GOVERNMENT. There you can buy—for 75 cents each—the 7½ maps that your engineer will need for the plotting of his radials (if you are going for more than 10 watts). They also have the large Sectional Aeronautical charts—for your contours, and the landing charts that you will need for Section V-G of your application. I have always wanted to buy a warehouse so that I could purchase all the 7½ maps—highly detailed, beautiful large scale maps—glue them together, stick them on the ceiling, lie down on the floor, light up a toker, and have the omphaloskepsis of the world. What a dream.

In all these, you will have to determine your latitude and longitude. This is done by finding the exact location of your site on the 7½ map, and by means of rulers and such, figuring out the degrees, minutes, and seconds from the margins. I think you can work it out for yourself without me giving you a dumb lesson in Topology, eh?

When your Form 340 with its 16 or 18 exhibits is completed, you will want to look it over to make sure that the following steps have been taken:

1. The Board of Directors of the applicant corporation will have to meet—note time and place—to resolve together to file the application.
2. All the questions in the form that do not specifically apply should be blocked out with the words DOES NOT APPLY or DNA (after the famous non-existent spiral di-oxy ribonucleic acid.)
3. Your corporation should have specific power to own and operate a non-commercial broadcast station embodied in its articles of incorporation.
4. Any and all persons or organizations that have pledges to your group, to help finance the building that stations should show balance sheets as part of your application. These balance sheets should show *plainly* assets and all CURRENT AND LONG TERM LIABILITIES.
5. You can cut corners financially—but you have to show how you are going to do it. Do you contemplate paid staff? If so, how many and how much? Or will yours be an all volunteer organization? If you plan used equipment, prepare filings to show that it is available as cheaply as you list in the Section III part of the form. I know one station (KPBX, Spokane) that got a permit for a ten watt station showing net construction expenses of $1500—but they had documented proof that the most expensive equipment (transmitter, antenna, control board) was available for free or loan.

The Fork-Tailed Indian Shrike

6. The engineer who signs your section V-B can be an official registered professional engineer, or even a 1st class engineer—but anyone who knows what they are doing can sign, and check off the space Technical Director. I've even signed as one myself, and the stuff I know about engineering could fit into a midge's cupola.
7. Be sure that the date on Section 1, Page 2 is *after* all the other dates of the rest of the form exhibits. Many applications get returned without even a file number to correct this deficiency.

You must send an original and 2 complete copies to Secretary, Federal Communications Commission, Washington DC 20554 Keep another copy for yourself for your public file. If there is some serious deficiency in your application, you will hear from the FCC within 6-8 weeks. Otherwise, they will merely send you a card giving you the official file number of your application. If you don't hear from them six months after you have filed (seriously!) you might write and ask what has happened to it. The last time I heard, low power FM educational applications with no questions as to financing, ownership, or interference were being acted upon within 12 weeks.

But god knows, don't get your hopes up. The mysterious paper-pooper in Washington marches to a far different drummer than you or I ever dreamed of. I have seen permits granted 5 weeks after application. I know of an application for an AM clear-channel duplicate frequency that has been in the hands of the FCC since 1947. So, the best you can guess—if you must guess—is that you will be hearing from them sometime between 5 weeks and 27 years after you file.

As I look over the things I have written to help you with this project, I think I may have been giving you too much of a scare about the FCC. Remember this about them:

1. They don't care about you;
2. They have heard your story before;
3. They don't care about you;
4. Influence means little to them;
5. You mean more work for them;
6. They don't care about you, *at all*.

However, *however;* if you submit a clean, respectable application, and pray fervently to The Great Aether God— you may well get a permit to go on the air. They aren't out to help you until you have a construction permit, but they do recognize that they have a job to do. And if you are right, and honorable, and persistent, then they'll get tired of saying *can't-have-it can't-have-it can't-have-it* like some miserable child and will grant you permission to do the thing that you wanted to do all along: that is, to broadcast, to transmit the sounds of generations across the halls of our ages. For even they can get tired of saying NO, and one sweet day you have found the secret to getting your chance to play your fingers down the timbrels of the aether, to make some dramatic, unearthly music; to create a new community of feeling, thinking individuals, and to create words and ideas and musics which have never been made available to that community before: all those feelings of love and bitterness and rage and beauty. You, with this giant palette called a frequency. And it just waiting there until you and your friends come along with songs and poems and statements and moans and howls and words and feelings to transmit to all the hungry ears, all around you. They, and you, could ask no less.

Part Two

In the Kwa language of Yoruba, there are two words for radio: "Ghohun-gbohun" (snatcher of voices), and "A-s'oro ma gb'esi" (that which speaks without pausing for reply.)

After arrival of a construction permit, the most important thing for you to do is to build the mother.

In the appendix to this book there is an extended essay by Ben Dawson on what equipment to buy for your station, how to start the construction of studio and transmitting facilities, what to look for in tape recorders and the like. This material begins on page 256.

WHAT TO DO WHEN THE CONSTRUCTION PERMIT ARRIVES

When your construction permit arrives, 1) Tie on a Big One; 2) Take an evening out to raise Homage to The Great Aether God, viz:

*Make a burning cup of gold,
consisting of 14 coils of
Double-ought wire, one 6SJ7,
and eight-pound choke, and a
1927 Carbon Microphone. Set
them all in the middle of a
Van de Graaf generator sparkling,
sparkling—and intone the following:*

O Great Aether, vast sea of mhos;
   We bespeak your ever modulating waves
   Burning, burning black against
      The Bright Skies of Rectified Universes.

O Grave Aether, we sing homage
   To your every watt and wart—
   To the twin goddesses Nora Blatch, Sybil True.
   We gesticulate your Dolbears and mini-Hertz,
      Mega-Hurts and Milla-Harts.

O Grey Aether, we raise sparks of micro-micro-micro
   Farads in eternal crossings of Bars
   Of Wheatstone Bridges: Reddi Kill-O-Watts
      Come transmitting tonite, on little cat-whisker feet.

O Great Grey Grackling-cracking Aether:
   We degauss you forever in our coils
   Of thought and sleep, forever wireless
   God of our aetherized patients, lying
   On Tables of sines, and countersines,
   And cosines of sparks, and gaps;
      Aethered gaps of all sparks. O God!

Naming the new Baby

During a recent meeting of The 'Pataphysical Broadcasting Foundation, Inc., of Santa Cruz, holder of a permit for a new FM station in that city, the following discussion of possible call letters took place:

David F.: I prefer KAZU. We could have Jack play some Shostakovitch symphony on the kazoo for background to our I.D.s.
Me: How about KVÄS? You know, the European drink...
David F.: It's already taken by another station...
Me: Not with an umlaut over the 'A'...
Mary: I am damned if I am going to be involved with a radio station with silly call letters like KZOO. How could I ever go out on the streets and ask for 'Contributions for KLIT?'
Me: That sounds like a good idea...getting you and Cese and Jane out on the streets for money. That should bring in some $5 a week. How about KCHU?
Len: I don't get it.
Me: Kah-choooo! 'The wet spot on your dial.'
David C.: And we could get sneezes from all the famous people in the world. Like: The following sneeze was brought to you by Jean-Paul Sartre for KCHU.
Marcia: Why don't we take a vote?

[*Ballots are passed out, inscribed, and returned.*]

Marcia: We have two votes for KAZU, one for KCHU, one for KPFA, and one marked 'no.'
Jane: If that's my vote for KAZU, we have to change it. I meant to say 'KZOO.'
Me: How about KTIT. You know: KTIT needs your support.
David C.: Or KBRA.
Mary: Can't we be serious? KCBS. That's a nice one, a serious one.
David C.: They've been going in San Francisco for 20 years.
Mary: Or KSOL. We have a lot of Soul. And it means, what does 'sol' mean in Spanish, doesn't it mean...
Marcia: Why don't we just write down all the choices. That always helps; to see what we are talking about.

[*She takes a large white cardboard, and a marking pen, and writes down 'KUSP.'*]

Marcia: All right. We have KUSP. What else is there?

[*No one says anything for at least a minute.*]

David F.: Well, I guess that's it. Now we can talk about something important, like money...

[*From KTAO Guide #82*]

SEX & BROADCASTING

41

Part Three

*"Not for anger or despair
but for peace and a kind of home."*

—Suicide note of Lewis Hill
Founder of **KPFA**, Berkeley
1957

42

The following material was included as an operating manual for Radio Station KTAO written a week before I was to leave the country for a dismal 6-months trip to Spain. Throughout, for KTAO, you should read K-MEE or K-DOG or K-FOO or whatever it is that you have decided to call your station.

Dear Friends

A radio station should not be just a hole in the Universe for making money, or feeding an ego, or running the world. A radio station should be a live place for live people to sing and dance and talk: to talk their talk and walk their walk and know that they (and the rest of us) are not finally and irrevocably dead.

A visit to your typical American radio or TV station is a visit to a morgue: alll the good and joy and fun that can be COMMUNICATION has turned into a corpse-run for money.

We have tried to do KTAO differently. We have tried to find humans, who think and feel like humans; we have tried to give these humans a small speck of the air to let us all know of their *liveness*. We have tried to revive the body of American transmission.

It has happened. Compare the door of KTAO to the door of, say, Channel 11, San Jose. There, the steel-eyes and the Zombies chase you from their sight; they don't want you to know it is *your* air they are milking of all them bucks.

Compare that with us, once a month, at least, going on the air and asking people to come down and help us run a radio station. That simple device: we need volunteers, so we ask for them on the air. In that act there is the pristine beauty of what radio *could have been* if the force of history had been a little more considerate in 1934.

It's a Drain. A Large Drain. Having all these people, with all their needs filling up our rooms 18 hours a day: and sometimes you find yourself thinking "if another goddamn people comes in the door with their needs like dogs peeing on the walls and chewing bones underfoot: I'm going to go crackers, simply crackers." They don't stop, of course, cause you don't want them to; and you don't go crackers—cause you don't want *that* either. But sometimes it is a Drain.

When one works with so many peoples coming in and out all the time, one gets addicted to Long Notes at the watering hole (the control room) for everyone to read. Since I was leaving, I figured I needed an Extra Long Note to tell everyone about Operating Spaceship Tomentose for the time I was gone, and then I thought I might as well put it out in *The Radio Times* so that everyone could have a copy. And—as well—there would be a few left over for those 3456 people who write me each month asking how to operate a community radio station in Dingbat, Kan. This is for them, too.

It must have been Sartre who said, at the commencement of such a Cyclopaedic journey 5000 miles to the east: "Let us ignore the Kirkegaardian precept of EITHER/OR, and remember that the real stuff the glue that sticks our days together is more EITHER-AND-OR."

As long as we remember that, we should be safe, if not sane.

43

THE EPICENTER RADIO UNIVERSE

In any and all universes, there are Epicenters: those slots of energy that make the globes turn and sigh in the wind. In a broadcast station, those centers are
— The Schedule of On-the-Air People
— The Checkbook
— The Mail

There are others. You will think of others (I have just thought of seven others). There are the control rooms and the tape recorders and the microphones and the people who give orders and the people who take orders. There is the transmitter somewhere alone sitting at some hot peak and making munching noises to itself as it sends the transmissions of generations up to the line to squirt out to the edge of the visible horizon. There's that Epicenter: but it is of a nature and order beyond this dissertation—and must be taught to you at some Tube&Amp School.

For our purposes, the heart and soul of community radio are the schedules (both for on-the-air and the production room), the checkbook, and the incoming mail. A knowing manager will try to get his or her sticky fingers into all three.

ON-THE-AIR SCHEDJANGLES
VOLUNTEERS

Without volunteers, a community radio station *could* function: but it would be expensive, devilishly expensive; and it would be far less varied and interesting.

Volunteers are one of the rich antennae you have in the community. If there are four dozen of them, then you have four dozen more sources of news and informations and ideas and thoughts and speeches and juxtapositions of music. The tedium of most American educational radio and television stations is that they refuse to ask the community to come to them: they feel that they, the bureaucratic educators, with their G-9 salary schedules, are sufficient to run that 100 kilowatt radio or television station. And they are wrong.

At first community radio stations had volunteers because they were necessary. There was no other way to keep the programming stoked and burning and *there*. But the early stations always pretended that once they got financially strong enough, they would dispense with the volunteers and have only paid staff.

Fortunately, it has never happened in the 25 community stations I know about. Our country has changed so much in so few years: one of the great changes is that most everyone can survive without working—if they choose to do so. Food stamps and welfare mean that there are dozens of middle class drop outs who have the time and inclination to keep community radio alive. We are being subsidized by a kindly welfare government: and nothing has been more important for free communication.

In the course of a week some 15 people will apply to you to go on the air. You should take the best: you have a great opportunity for quality control at this stage, and you should exercise it.

A volunteer is a mini-Program Director. He has four or five hours a week—and what he does with that time affects the sound and orientation of the station.

The people who come to volunteer at the station will be graduates from the Columbia School of Broadcasting who want to be Boss Jocks, and listeners who love the station and want to be part of it.

The Boss Jocks should be discouraged. Because they don't really care for community radio, or the Oud music of Upper Nubia. Their time at the Columbia School of Broadcasting has been a brainwash time: where they are turned to plastic to conform with the plastic voice and the plastic vision of American Radio. I am not saying that they cannot be changed: they can, if you have the time and the patience to do so. With 20 or 40 hours, you can (perhaps) erase that terrible blight imposed by The Commercial Sound on some young person—and give them the feel for real radio.

But you are better off not to try. It will be a struggle and a fight. They will want to talk over records: even if it's Shakuhachi music—they will want to fade them down at the hour and do an ID and exude their own miserable personalities on the frequency. You can train them out of it—but it will be a pain.

You are better off with people who have never been on the air before: people who have no slickness, who have never run a control board, never been heard ten miles away. You are better off with people who love the radio station—who have heard it, and listen to it regularly, and heard your plea over the air for volunteers, and answer because of that dark worm inside all our minds which says COMMUNICATE!

Some of the younger people who come to you may be marginal: they may want to be djs, but they may not have their brains fixed yet. Those will make your best and most loyal volunteers. Once they get into the idea of the radio station that does not compete and does not sell—they will find themselves hooked. For after all, this type of radio is an Ideal: and they, the young, are masters of the ideal.

The best way to interview a prospective volunteer is to ask "What kind of music do you like?" If they say rock, or jazz—you will have a tough job on your hands: because to get someone from Alice Cooper to the Music of Chad is difficult. And unless they have that sympathetic chord, that chord which will respond to voices and musics far from their own, it will be a battle. The best way to tell, the only sure way—is The Eyes. What? People who will be right for your station will have nice eyes. Do you see?

Once you decide to give someone a chance on the air, you should stick them in an afternoon slot. In community radio, there are preferred times: just like at the big stations. Everyone wants to be on the air in the late afternoon and early and middle evening. Everyone wants to be on the air on the weekends. Part of that is because many of your volunteers will choose to hold down Real Jobs; thus, they are available only after 5 and Saturdays and Sundays.

But there is a demand for these times because of The Press of Listeners. Volunteers, all of us, know that the time of greatest listenership for FM radio is evenings and weekends. Like all of us, they want to be heard; they know—instinctively—that there are more people out there at those times.

The On-The-Air Volunteers Schedule is vital to the orderly operations of a station such as ours. It should be typed up once or twice a month, in duplicate, and posted where everyone can see it. It should list the times on the air, and the name of the volunteer to fill that time, and his or her home telephone number. That is the orderly wheel of KTAO.

Your best and most faithful volunteers will get the best times. And those times should be protected: you should not wrench someone up from their scheduled time without advance consultation with all the people involved. After all, many volunteers come long distances and forego other pleasures to give their time to your station, and they should be protected and loved. Arbitrary dealing with friends of good radio can and will make that radio less good, and more political (in the internal battle sense).

You will have a waiting list of volunteers outside those you have on the air(right now, we have 46 persons on the air, and a waiting list of 18). Try to put the newer and untried people on during the early or mid-afternoon. This gives them a chance to learn the board and the act of being a radio person without scaring too many of your listeners away. When a plum of a time comes up in the evening, I try to put the best and most faithful on-the-air persons into that time. It's quite a juggle.

Since KTAO is the type of station it is, and the on-the-air people are the type of people *they* are, I try to fit the musical interest of the volunteer to a certain time. I try to put the jazz people on late at night: since we allocate 18–20 hours a week for jazz, I try to put that where people seem to like it (and when I am asleep). Classical music people I try to get on in the morning or afternoons, ethnic music people in the afternoons and evenings. Those people who have a single taste in music should be scheduled at that time which is *right* for that type of music. It's like the eyes.

One of the great duties of KTAO and its sister stations is to give women and minorities a chance to be heard on radio. The broadcast industry is a hotbed of tokenism: and I think we can help break down this tawdry state of affairs by getting as many women, blacks, chicanoes, American Indians, and Orientals on the air. Within the standards laid out above—I will move people around, step on a few feet, and defy the desires of some of the audience to get these voices heard. Almost half of our on the air people are some sort of minority [I include in that the 12, 13, and 14 year olds who do programs directed towards *their* minority] and I think we would be remiss to have less.

I must emphasize, however, that pure rhetorical minority programming can be damaging to the excitement of our form of radio: and for that reason, I tend to shun volunteers who come to us missing that single ingredient that seems to mark our sound: that is, humor.

The humorless revolutionary is blind to the capabilities of radio. He hurts his cause and us by getting on the air and blithering. He does not care for the art of radio—and is thus anathma to KTAO. But how do you tell: how do you tell if some volunteer will be lively and alive on the air, or deadly and mordant and ideologically tedious. I think it has something to do with the eyes—and the 20 or 30 minutes that you talk with that prospective volunteer before you say Yes. For once you say Yes—that person will be with you a long time; and to say No two months down the line is very very difficult. Impossible, really.

Once a volunteer gets on the air, he or she should be supported as fully as possible. You should be listening to the station all the time *all the time* so that you can criticize, and praise, and suggest, and compliment. When someone does good—you should tell them: a great speech or a nice juxtaposition of music or even an interesting, well-done Minimarathon.

When you criticize, your criticism should be done in person *if at all possible* and not over the telephone, and NOT SECOND HAND. You should, yourself, present logical criticism, and offer positive supportive suggestions for change and improvement. It is a chance for you and the on-the-air person to grow—and it should be seen in that light; not as a time to tear down and destroy.

The most common criticisms will be on musical programming (even your best volunteers will try to smuggle crappy music onto the air) or the style or approach of announcing or reading or interviewing. In all these comments, I try to reinforce the purposes of KTAO:

- We are not here to compete with existing radio stations, but to supplement their presentations;
- On the air people should never never talk down to the audience: but speak to them conversationally, as if they were in the room [they are.]
- Interviews should be good drama in themselves—and not just an opportunity for some hype group to get free air time [thus the typical KTAO aggressive interview is an exercise in Artaud Theatre]
- We can and should be serious, or silly, or funny, or waggish on the air. We should never be cold, or crude, or patronizing. We are friends in peoples' homes, not colonialists.
- We are here to present whole parts of men's minds. Thus, entire musical works, and complete speeches should be the rule. Five minute news break-downs and fade-out movements should be avoided at all costs. We are not the Readers' Digest of the air.

Volunteers must, if you will pardon the expression, be loved. Because they are always underfoot, it is easy to snort and snap at them. But they are the blood of the station, and its life: if you care for them, they will care for the air. And it is a garden that needs tending, constantly and well.

PLUMBING DARK CHECKBOOKS

Now who would think that a dumb checkbook could be an important part of a community radio station. But it is: it should be cared for, and filled in honestly, and kept up to date.

Your operation is bizarre, and because it's bizarre, it is subject to more questioning and examination than the regular money-glub business. ITT and Ling-Temco-Voight and PennCentral have always been able to do unusual and creative things with their balance sheets: they can buy other corporations with Funny Money called Inflated Stock, and add their assets and hide their liabilities, and jack up their earnings per share without a query from the IRS or the SEC or the NYSE: but that's big business and you're small pimple business and therefore you will be subject to far more scrutiny. And thus you must be careful and honest. At least, until you get to the size of Bernard Cornfeld's enterprises.

A check is a reward to people or institutions for work they have done for you. The Internal Revenue Service and the State Withholding Gouge Department (in California it holds the ironical title of Human Resources Development) demand that you deduct a usurious percentage of money from each and every paycheck. We live in a country where the lower class and the middle class are taxed exuberantly to pay for the sins and pleasures of the Upper Class and the land and air exploiters. And your checkbook must reflect these cruel withholdings.

Your accountant will tell you how much to deduct from each employee's salary. We try every legal way to reduce these taxes—but the system is weighted solidly against those who lose their tax before they see their salary, who don't even have the chance that the rich enjoy of playing with the money as it passes through their hands before it goes to the feds. There is little we can do to change this: but remember that *bona fide* non-profit corporations don't have to deduct social security if all the employees so choose. Since social security is one of the especially burdensome taxes whose benefits are almost all mythological—everyone is better off to save up what they would lose to Social Security and invest it in 90-day government notes.

Every amount of money spent should reflect in the checkbook. Checks should be written from the right hand side of the register with date and explicit purpose of checks which are not self-explanatory. All income should be written on the left-hand page, with the specific name of subscribers and contribution received so that you have a double check against your file system.

It is always wise if not always possible to keep $100—$200 in the account at all times to protect you against poor mathematics. Because you are small turnip business, the banks and other institutions are prejudiced against you. Wells Fargo Bank will happily loan out your deposits to the creeps at Equity Funding, or allow Lockheed thousands of dollars of overdrafts without charge—but will always stick you $4 for a bounced check. [If you wish to read how large depositors are protected by the large banks, read *The Great Salad Oil Swindle* or *Do You Sincerely Want to Be Rich?* Those $4 charges go to pay off other bank irregularities—and there is no way that you can win against these doobers except through minor check kites which are expensive if you lose.]

Because we never have enough money—you have to be selective in the bills that you pay. Landlords and utilities require instant payment: else they will cut you off and put you in the street. Mortgages can often be delayed for months—until actual judgement proceedings—but these can be expensive because of the late charge penalties which amount to 20 or 30% fees. Late charges on *non-contract* bills (regular bills from stores and suppliers) are illegal under California law and do not have to be paid. People who work for you usually need their checks desperately on pay day, but will often take two week's worth temporarily if you are strapped. The KTAO bills are paid in the following order:
1) Studio rent;
2) Mortgage on the transmitter site;
3) Salaries to Cese, Doug, John and David;
4) Line charges to the transmitter;
5) Telephone bill;
6) PG&E bill;
7) Stores & suppliers;
8) Cream & fruit-juice fund (Grocery store);
9) Printing bills;
10) Legal fees;
11) Equipment, repair & maintenance;
12) Typing supplies, stationery;
13) Bill for pens I ordered from Trenton when I was drunk.

In some cases, the larger the company, the longer you can delay payment. IBM can be kept up to 2 months, and we have two different billings for the different telephone lines so that one can be taken out without affecting the other one. Sometimes you can delay paying a bill for up to six months by bitching about it, or calling it 'a computer error'—a method I use with great success on some large manufacturing companies like CCA and Gates. "Collection procedures" simply do not exist for bills under the $200-$100 range—you have to play it by ear. In general, the further you are away (in physical distance) from the creditor, the safer you are: cross-country legal actions are rare for relatively small bills.

DER U.S. MAIL

About 60 percent of the mail that arrives everyday is nothing but dogshit. Press releases from people who desire free airtime for their money making (or powermaking) activities—but who don't have the time or inclination to listen and see whether KTAO *wants* their mailer.

Thus we get handouts from Tunney and tapes from Clark Bradley and a battery of daily moneywaste releases from the publicity department of the Internal Revenue Service [notifications of tax apprehensions always increase around tax time] and mountains of useless information about useless activities of the West Valley Christian Youth Fellowship and the Cooperative Moneymake Art Gallery of Saratoga.

These people value sending their 'releases' out into the void, and hoping that they will strike target for $50 of free airtime at KLIV or KXRX or KOME or even us. We do our best to disabuse them, which means that we throw away half of our mail each day.

There are groups which are genuinely trying to change the world which we try to help: social and political groups such as the ACLU and the YAF and the Peace Center and the Birch Society and the Urban Coalition deserve to have their announcements read on the air. But the flower clubs and the State Attorney General's office and the U.S. Army Recruiting Service are trying to extract free time for their nothing programs and should be disposed of. Instantly.

The mail also brings subscriptions, letters from listeners, cards from friends, notifications from lawyers, and checks from home. The letters from listeners should be read and posted—for to write, and write to faces unknown at some radio station indicates an interest and devotion that deserves our support. Subscriptions should be carefully noted and carefully carded: when someone goes to the trouble, the extra trouble to support us, we should be careful and loving with their check and correspondence, and not lose it in the production room or underfoot.

After awhile, you will be able to see from the cover which letters are important—letters which were really meant for KTAO, and not just sprinkled out into the void without reason or care. Sometimes, however, the junk mail can bring indirect rewards: when The Messenger for Christ offers us a free subscription to their weekly tape service for inspirational messages, *take it!* Because it means at least 4800 feet a month of useable, eraseable tape to add to our all-too-scanty collection. The same goes for sample tapes from hype-promo program services: they will give you a half-hour tape as an example, and usually the tape is superior quality to be erased and put in our blank tape bin.

Free discs which are hidden commercials for some large company seeking illegal airtime often make excellent humor records, to be played late at night by some of the KTAO crazies. Recruitment records may provide the same sort of boffs, as do the 15 minute programs *God's News Behind the News* before erasing.

Bizarre services abound if you go through all the mail. Last year we got 3 cases of Pimple

Cream simply by asking for it: we were all pimple-free for months afterwards, and it didn't cost us a cent. Other samples abound.

Unsuspecting Chambers of Commerce and Speculator Promoters will often ask us to free cocktail parties or luncheons. I find it often helps to pick out the hungriest if not the scroungiest volunteers and offer them these free-booze-and-food handouts so that they can add a little wit to an otherwise dull party, and spread the reality of KTAO beyond these doors. Billy Iberti—when he was doing the bluegrass program—latched onto a free meal and unlimited champagne condominium opening party, and had to be carted home in a basket, much to the disgust of the hapless promoters who, once again, didn't take the time or the trouble to see where they were sending their generous offer of freebies.

RECORDS

KTAO has 7000 records in its collection. Most of them are rare and original. For instance, we have the entire UNESCO Bärenreiter collection; we have all the Arhoolie, Biograph, and Historical record collections; we have every record issued on the Telefunken label.

Most discs for most radio station are given by manufacturers or distributors to promote sales. But since Folkways or The International Record Company are not 'promotional conscious,' we have to purchase discs for our collection.

Columbia or Monitor or RCA or Atlantic cost $1 each. Folkways are $2, and Bärenreiter asks $3.60 per record. Even at those lower prices, it's obvious our collection is an expensive one. It is also a tasteful and rare compendium of all that is best in American and International recording art. Through Peters International, for instance, we have bought some 300 Odeon imports from India, 100 Regal discs from Spain, and the complete Topic catalogue from England.

To maintain this collection, we have to nag the on-the-air people constantly, as much as we have to nag them to go into the collection and play records that they may never have heard of.

The Record Nag is a constant, and has been a constant theme, the Muzzein's moan under North African life; it is sung as follows:

Records should never be
   Left out of their jackets;
Records should always be cleaned
   Before playing on the air.
     (CHORUS): Records should always
        be clean, I mean!

There should always be a paper
   Jacket to protect the records;
The records should never ever
   Be fingerprinted, ever.
     (CHORUS): Our records will always
        be clean, I mean!

Records should always be cleaned
   With your breath on the Preener;
Records should never be set on
   The counter, should always be clean.
     (CHORUS): Our records are always
        clean, Geraldine!

And when we go to record heaven,
   With glistening grooves, and
Print-free surfaces, the divine
   Matrix will press down, and say
     (CHORUS): Your records were always
        clean, you bean!

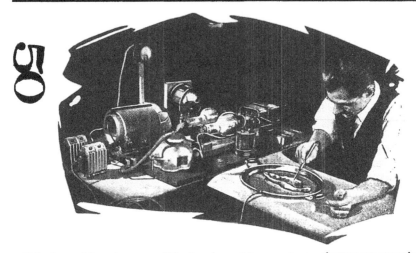

We've learned from too many difficult and agonizing experiences that we can *never* lend out records: they always get left, or lost. Good Will disappears when someone has that rare 3-record album from Burundi, or the entire *Roman de Fauvel*, and you ask that they bring it back, and they keep forgetting. You understand. Besides, you never know the kind of equipment that people have at home, and what they'll be playing the precious records on: we use Shure M-91s with a tracking weight of 1 gram. I've seen home hifi's made out of them 13 pound magnets they used on 78s back in 1935, with old golfing spikes for needles. *Never loan out records.*

The KTAO records are filed by color-code, a system devised by Fritz Reuterman at KDNA in 1969. We take colored Magik tape, and give each record a major classification, a minor classification, and a sub-classification. This means that one avoids an *exact* and time consuming position for each record, but a more general spot within a sub-classification of 30–50 records.

The classical records have three major classification (uppermost tape on the spine): Yellow (Medieval through Baroque); Green (Bach through late Classical); and White (Romantic and modern.) The next tape down usually represents instrumentation: Pink is chamber music (of course); White is mixed.

The ethnic records are arranged by country or region: thus Black-Blue is India and Black-Yellow is China, Japan, and the Far East Pacific. The third tape sub-divides these regions or countries into further regions, or specific types of music. Thus Black-Blue-Yellow is the Punjabi region of India.

Blues and folk music are divided by type of artists; jazz is alphabetical according to the first artist on the first band on the first side. Spoken word has its own major classification, as does classic jazz.

The color code system is a great invention for those of us who have no librarian blood in us: we can immediately tell when a record is misfiled, and replacing records after airplay is a quick and simple process. Sometimes you can tell when whole sections are lifted: as when our Django Reinhart section disappeared one night.

The snitching of records (and equipment) is one of the most difficult problems facing a community station. Because it is a real and nagging problem: people working for nothing sometimes think they should get paid—somehow—and solve the problem by spiriting away precious discs when no-one is looking.

It tends to poison the atmosphere when records disappear, and you try to watch or accuse people. It just don't work. I think the rip-off rate is directly proportional to the unity and contentment of the staff and volunteers. Sometimes I say (not too publically) that rip-off records are the price that one pays for a huge volunteer unpaid staff: to run a station with a paid staff of 12 costs some $80,000 a year—and the cost of lost records will never equal that. Or (again, to avoid poisoning the atmosphere) I say that the records that get taken are too popular for our stations: that their disappearance is a symbol of their commercial success, and that since we are running a community, non-compete broadcast outlet, we should welcome their absence. These are sour grapes, and just point to the unwillingness most of us have to believe that our friends would ever do something so grim and unequivocal.

I have seen different stations handle the problem in different ways. WYSO locks up all their records all the time: four staff members have keys. KPFA locks up all records and tapes when the staff goes home at five pm. KDNA locked *everything* in the building, but they were in the midst of a ghetto environment. KRAB, for some reason, locks the front office papers up, but leaves the tapes and records unlocked. Since the beginning, we have tried to avoid any lock-up at KTAO: the first thing we did when we came to town was to dispose of the locks on the file cabinets and get rid of the metal lock-up desks. This means that anything, even our miserable secrets, are available to everyone at all times. We try to avoid leaving cash out on the desk—such temptation is cruel to those who have none—but the lockup (perhaps because of the flower-garden paradise called Los Gatos) seems unnecessary. We remember the words of L'ao T'se:

> Lock love lock rose-petal
> The sun swims through fog:
>   Plop.

CPA'S & ATTORNEICZ

No matter how alternative we are, we have to slide through the traps and weeds of this society: for that reason, we have one paid accountant and two paid attorneys.

Accountants are necessary because the IRS and the HRD are not flower children, and do not accept innocence as an excuse. Worse, they have the power of the padlock and attached accounts to enforce their doubtful rules. That's why we have an accountant.

The forms from the IRS and the HRD are cryptic and full of traps. The annual Corporate IRS Small Business Form is full of more traps. The rules for withholding are too mysterious for the rest of us to figure out. That's why we have an accountant.

The traps of the FCC and the FTC and corporate law and equal time and licensing and Form 301 require a Washington D.C. attorney. That's why we have an attorney.

Lawyers and accountants are good for something else besides filling out forms and occasional emergencies: they are great repositories of File Cabinet space. Everytime anything of any importance happens to KTAO (strange letters, newspaper articles, magazine articles) I send them to Washington where they get stored just as if they were in a government vault. A

**AGENTS! DEALERS!
BIG MONEY IN RADIO**
Demonstrate Metrodynes at home. Big profits. Demonstrating set on 30 days' free trial. Write for money making proposition.
METRO ELECTRIC CO.
2161 N. California Ave.   Dept. 48,   Chicago, Ill.

complete history of community radio is stored in those commodious metal hulks at the back room of Haley Bader & Potts.

The best rule for you in reference to attorneys is: never sign anything without specific legal approval. (Sing):

*Never Sign Anything
Until Again
Haley Bader & Potts
Sez Dott's All Right.*

TELEPHERNZ

"How strange life is, and changeful, and the crystal is in the steel at the point of fracture, and the toad bears a jewel in its forehead, and the meaning of the moment passes like the breeze that scarcely ruffles the willow."
—ALL THE KING'S MEN

You cannot trust the telephone. The telephone is your enemy. People will pick up the telephone to call you and say the most doubtful things. And you risk your honesty if you listen to them.

You can go on the air and do dingbat for thirty hours straight: and someone will call you to tell you that it is great. And then, when I complain to you the next day, you will say, "But I got a *whole bunch* of telephone calls."

And at that moment, you have fallen into the trap of the telephone. There will always be a soothing voice on the other end of the line to tell you that you are doing just fine. And you're not.

When you do great programming, you will know it. And you have no use nor need for the disembodied 'Hey, Man, Great!'

It will confuse you and mislead you—the telephone will. The bells will toll for thee as you are doing that spectacular reading from Richard Wright: and it will be some petulant teen age pimple asking to hear Joni Mitchell, or asking for a ride to Tofu, Calif. Ignore the telephone's impatient ring, for it is a misleading, misguided, misanthropic ring. It tells you nothing.

When they call you, the six or ten or twenty-nine: when they call you to tell you how wonderful you are on the air—you have failed. If you were to lead—in the proper way of radio to lead—they would never call you to praise you. They would be too thunderstruck mad blind knocked down dead to call.

It is only those who have not heard you that will call you.

PROGRAV GUITCH

The subscribers to the station need something concrete to show that they have contributed to the operation.

53

The usual form is several sheets of paper stuck together and covered with detailed listings of the programs—for a month or two weeks.

KPFA, KPFK, and WBAI all started out with a 24 page brochure, called *The Folio*. KPFT was the first to break away to newsprint—since all the KPFT originators were newspaper people—and then to the folded *"tab"* size of 8, 12, 16, or 20 pages.

The original *tab* newsprint program guide was put out by Tom Thomas at KDNA in 1970. Gradually, all the other stations followed suit so that now the only community station with a small stapled booklet is KBOO. KUSP in Santa Cruz puts out a poster.

All these newsprint guides are a concession to the central fact of printing: *viz*, if you staple, you need a people to staple. This doubles the cost. The tab size newsprint fold-16 pager needs no staples, is not handled during production except at the end of the line where 200 or 300 are tied together by end of the line-men.

The value of the tab newspaper is that there is—for very little extra cost—a great deal of space for ideas and pictures. Layout can and should be done at the studio using the techniques made possible by zip-a-line and rub-on letters and half-tone pictures. The web press revolutionized printing by getting rid of the funnel: usually a crotchety old union man paid $10/hr to sit before the linotype and distill your words and complain about the dirties. The revolution in four-letter word print came about through the web press technology and the elimination of Uncle Straight with his hot lead at the end of a chain.

The Pacifica stations favor extensive descriptions of programs—and, indeed, the guides ("Folios") have little essay material. KRAB does the same, as does KBOO—but with less detail as befits their size. The KDNA-imitative stations (KOPN, WYSO, WRFG, KTAO) use less program listing and more verbiage about the nature of the universe, the agony of radio, or the transmission of generations. KUSP sets graphics above everything else.

Planned program block listing used to be significant when these stations had no competition for their sterling, brilliant, controversial programs. However, with the dilution of the intellectual monopoly especially noted in areas such as the Bay Area, the Los Angeles Basin, New York City, Puget Sound—the competing voices of 70 stations makes exact listing of programs a luxury.

And it is a luxury. To find a person who is willing to sift through data concerning a month three months hence and lock in some 350 programs with specific dates and times is to find someone who has a secretarial telephoning list calendar mentality: and that type of talent is expensive. Creativity is always cheap: dull-bulb bureaucrats extract high wages for their dull-bulb lives and pursuits.

I have always seen the KTAO Reginald A Fessenden Memorial *Radio Times* as a literary gift to those courteous enough to support the station, and a source of information to the many people interested in community radio in other parts of the country.

No matter how you do it, it is important that you use the U.S. mails as cheaply as possible to distribute your material. If you are a non-profit corporation with tax-exempt status, you can get from your local post office permission to mail at non-profit organization rates—which means that each program guide will go out for 1.7 cents each instead of 4 or 5 cents or whatever it is that regular bulk mailers pay. This may not sound like much to you—but if you ever get 1000 subscribers, and mail out twice a month, you will be saving at least $200 with your non-profit imprint. There is more work involved: you have to sort and bundle the material—the post office nearest you will tell you exactly how—but you have a special government subsidized mailing privilege with this non-profit status; since so many bullshit religious orders make use of it, you might as well, too.

You should search carefully for the offset printer nearest you to print up your guide. Find one with a "web" press. You will have to deliver to him camera ready copy: but after your first two or three program guides, your mistakes (which at first will be so blatantly obvious to you when you get back the printed guides) will go away and you will be a print person as well as an electronic one. Most of the community stations that put out a guide make the same mistake of using the program guide as a place to *further* demand monies of subscribers. There is always a paragraph or a page demanding contributions.

This is a waste of space, and an insult. Those who are getting the guide have already given you money. They don't need more exhortation—but rather, less. In addition, printed demands for money never work. At least, I have never seen those that do.

SOME CUM INN

Since we ask them to, people often come in the door of KTAO. Which leaves us with the problem: what do we do after we have shaken their hands, thanked them for visiting, and shown them the two rooms that make up KTAO.

I try to give them presents: like the pencils that say
Time flies like an arrow
Fruit flies like bananas
We also have yardsticks that say
NIHILISM MEANS NOTHING TO THE
DANCING PEASANTS
on one side, and
NIHILISM MEANS "NOTHING" TO THE
DANCING PEASANTS
on the other.

There are some visitors who do not mean you well at all—for these stations with their doors wide open are a watering-hole for all the dissident, disenfranchised crazies that the State of California keeps dumping on the streets. You remember the guy with the flower in his hair and the '45 in his briefcase: that same briefcase left open in the studio to the fearing jape eyes of me and Bill Wade. Or how about that Mike character who would dump our coffee in the toilet and bars of Lux soap into the coffee maker (have you ever tasted coffee made in Lux soap?) Or the people from Chile just last week who I thought were going to bop me because I told them that their tape of music of that country was "too commercial."

The doors of these stations have to stay open: that is the strength of community radio. But there are times when the open door policy can actually be dangerous. It's one thing in the flower-power-paradise of Los Gatos, but at KDNA an open door could mean a lost VTVM or a female volunteer dragged out on the stairs and raped.

I have said before and I will say again that the strength of KBOO and KTAO and KUSP is not only their dictator directors ("benign autocracy") but the fact that they are physically small. This means that all work is carried on in the same proximate area; telephone calls can be overheard; visitors have to address themselves to the crowd of hangers-on; there is little room to plot and scheme. KRAB and WBAI and KPFA are so huge that countless plots and counterplots may add spice to the daily activities of those stations but certainly detract from the programming.

These stations have to be built on trust and one either a strong central authority or a strong collective (as in the last days of KDNA). Politics and infighting may be exhilirating—and a symbol of the importance of the station to the outside world, but it hurts the overall creativity of the operation.

MONIRAIESES

SATURDAY, NOVEMBER 9, 1963. Starting at 12:01 AM, KRAB (Seattle) will conduct its first gigantic marathon in order to raise $1000 for the station. Most of the regular programming will be suspended, and for 42 hours we will play music and parts of the most unusual lectures, discussions, concerts, commentaries that were heard over the station for the past year. Pledges will be solicited at disarmingly frequent intervals, and we will have a squad of motorcyclists (to be identified by hearts & "KRAB" tattooed on their arms) ready to go out and pick up the pledges. The money is needed to buy essential tubes, tapes, and to pay a bad assortment of bills that have collected in our TO BE PAID file. The Marathon will extend until 6 PM on Sunday.

—KRAB Program Guide #23, Nov. 2-15, 1963

Now you know what the good Maj. Edw. Armstrong felt when he stumbled on the principal of feed-back circuitry: when the earphones damn near blew off his head. Without knowing exactly what we were doing we invented a new technique of fund-raising. It worked, and it is now being used and misused by an alarming variety of non-commericial, educational, and community stations to wrest money from an unwitting public. This is what we wrote for the next program guide (KRAB #24):

"We suppose that by any contemporary standards the marathon was a rousing success. $1,079 pledged in 42 hours, (over $200 called in in the last hour), about a thousand dollars collected on these pledges.

"It was an interesting experience in many ways. The chance to be accessible to the listener every hour of the day and night; the chance to meet many of the people who had heard us for some time and had never been by the station;...the ability to test our own ability to go on and on for hours on end...As one listener said: "It's too bad there can't be a Marathon every weekend." As another listener said: "I hear you've made your goal: can I have my money back?"

"And yet, there was one thing distressing about the whole Marathon. It was the fact that that 42 hours was as close to commercial radio as KRAB has ever come. Every half hour, religiously; sometimes for as long as five minutes, there was a heavy appeal for funds: every

half hour, we explained the purpose of the station, the crying need for funds, the necessity of some money in order to survive. Some pleas, expecially in the last hour or so, were so impassioned that they made us want to weep with the self-pity of us. We learned the degrading lesson of advertising: tell people how good, and noble, and pure you are, for long enough...and soon you will come to believe it yourself...

"Until the time of the Marathon, we avoided selling ourselves: as we have said so often, KRAB was established to traffic in ideas, not in commerce. We have always limited ourselves to three plugs a day—three quiet explications of the idea that KRAB, and of our need for funds, and the address for the mailing of checks. And, in all that time of moderation, eleven months of quiet appeals, we have only been able to muster slightly over $4,000 in subscriptions—about one a day. In other words, in 42 hours of standard advertising technique, we obtained almost a quarter of what restraint and hesitancy had brought in over the preceeding year.

"We find it depressing to think what contemporary techniques of advertising have done to Americans: even in our own listeners, the advertising klaxons have instilled an automatic blab-off: any appeal for money opens the circuits unless it is repeated again, and again, and again. People are dying from an over-profusion of words.

"The Marathon has affected us in many ways; for one thing, we never knew our own power: the dry voice of dissent lost somewhere up there on the end of the dial does have some listeners. We know that now; further, we know the strength of outrageously repeated appeals; finally, we are tempted to change our whole way of raising money: to go along, quietly, on what we have now—saying nothing about money for a month or two or three; then, when the bank is looking worried and our account is looking vacant, to crash into another Marathon; then, once again, drift along again, until the till runs dry. On and on, drifting between poverty and prosperity, between the non-commercial and the commercial. It would, after all, be a perfect symbol of the schizophrenic nature of listener-supported radio."

—Program listing #24
KRAB, Seattle, Washington
November 21–December 4, 1963.

O PROPHECY! WHERE IS THEY DING-A-LING-A-LING?
WATCHMAN OF THE NIGHT, WHERE IS THY STING?

The jewel was in the forehead of the toad. The liquid gold vomited from the bowels of the asp.

If we did invent the Marathon (and there is some dispute: evidently WBAI in New York did one, their first, around the same time) it is not too heavy to say that we were present at the Alamogordo of non-commercial radio.

The technique has been refined. KRAB still does them—but with begs for money every 5 minutes rather than every 30. The motorcycle crew has been lost somewhere down the line, giving rise to the uncomfortable fact that some 25% of the people who pledge do not send in their money.

The good programs are still reviewed: but the whining demanding voices are somewhat less subtle and far less nice now. The typical marathon on KPFK will wheedle and cajole endlessly—for a whole goddamn month.

The sad fact is that the Marathon is no longer a stop-gap emergency money-raise for the community stations. Rather, it has come to be a method whereby these stations can budget ahead of time an extra $20,000 or $80,000—knowing that a month's beggery will raise some sum like that.

And the Marathon is no longer a spontaneous outpouring of love by an appreciative listenership: they used to bring us bagels and pies during our non-stop in Seattle; now it is careful calculating tool of the bureaucracy of these stations to perpetuate themselves (the bureaucrats, not the stations).

"What have you done to my baby?" Lee De Forest used to ask of the commercial broadcasters: "You have sent him jitter-bugging in the streets, demanding money."

We can ask the same questions of contemporary community stations. If they are not—like KQED-FM—selling goddamn airtime for $45 an hour, they are doing the Money Demand trick for weeks on end, and *not shutting up the rest of the time.* I think we oughta send those motorcyclists with their tattoo hearts out to trash some of them non-stop demands.

SUBSCRIPTION

The great brilliant insight of Lew Hill (*Gran Pere* of community radio) was that if you ask for money from the faithful listeners, they will support you. This had only been tried erratically before 1949—and that by religious shuck stations in the Middle West.

Hill and Partridge and Chiarito and Meece had to learn this the hard way: KPFA started as a commercial station. They planned to sell ads during the day to support the tremendous programming at night.

Soon enough they learned that commercial programming infects the non-commercial programming, and they dropped the sell-ads idea at the beginning of their on-the-air time. We had to learn the whole sorry lesson over again here at KTAO last summer and fall. We sold the first 7 hours of our broadcast day to a Jesus shuck commercial group, and it cut the heart out of the station.

So we are back on the dependence on goodwill of the listeners. But, as a reaction to the disgraceful humbuggery of the standard Marathon, we have evolved the Minimarathon (basically an invention of KDNA, St. Louis).

Minimarathons are never announced ahead of time. The method is one of dark surprise: after a good speech or a fine set of music, the on-the-air peep stops whatever it is that he is doing and announces that he will not continue with the programming until he has one or two or three pledges.

The secret of the Minimarathon is 1) a definite and oft repeated goal for you (and the listeners) to work towards; 2) a constant repetition of the telephone number; 3) good humored—never threatening nor oppressive—plea for calls; 4) an effective follow-up technique.

Ideally there should be no more than two mini-marathons a day. The rest of the time, the listeners should be left alone, except for i.d.'s.

The theme should be that since people are listening to your voice, the programming is obviously important to them. That this programming is not free, but rather—costs some $3000 a month. That in return for their subscription, the listener gets something tangible: *viz*, a bimonthly program guide.

Minimarathons should be run until the announced goal is met. The telephone number should be repeated at least once a minute. And the resultant pledges should be rewarded by several hours of *shutting up*.

The follow-up is as important as the pledges themselves: because people often forget, and the longer you wait to remind them, the greater the chance is that they will send us nothing.

The first (and most important) tool is the name and the address of the pledgee. We stamp white 3x5 cards with the following:

KTAO

NAME_____DATE_____

ADDRESS_____

_____ZIP_____

CITY_____

AMOUNT $_____

MASTER CHARGE ACCT. NO._____

_____SIGNATURE

The cards, when filled out, go into the PLEDGE pigeon hole of the rolltop. Cese goes in there every two or three days, and sends to each pledgee a current program guide staple with a return envelope that looks like this:

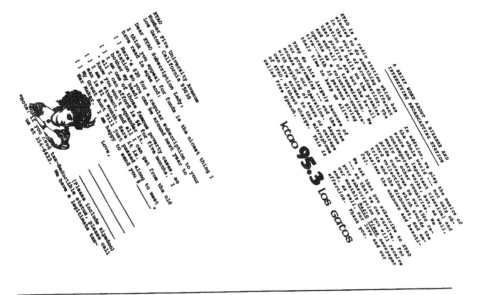

with a handwritten note thanking them for their pledge.

The 3x5 card gets marked "Reminded" with the date. We stick them in the *First Reminder Sent* part of the desk. If the subscription comes in, we find the card and discard it.

If the pledge does not come back within ten days or two weeks, we send out a second reminder, which is the following stamped on a piece of stationery, or a 3x5 card:

> FROM
> THE KTAO SUBSCRIPTION LADY
> Five University Avenue
> Los Gatos, 95030
>
> DEAR FRIENDS:
> Recently, you pledged $_____ to KTAO, and we were and are very thankful for that. However, despite at least one very earnest reminder, we have yet to receive your pledge. Please remember that KTAO depends on its subscribers for at least 50% of its operating budget — and if you fudge, we feel miserable (and supremely let down). Please do honor your pledge.
> THE SUBSCRIPTION LADY

We fill in the pledged amount, and send out another return envelope. We put the cards in the *Second Reminder Sent* slot of the desk.

We haven't yet figured out what to do next if no money comes back. Doug suggests a firebombing. Cese suggests a goon squad. I have always felt that we could make up a rubber stamp that says:

and send that out with *another* return envelope.

Indeed, the problem of what to do with unmet pledges is a difficult one. After all, what is a pledge but a spur-of-the-moment promise made by a stranger to a stranger over the telephone in the heat of passionate interest in some program or another. I tend to get mightily pissed at those who break their promises—but only because the check book register tells me that we take in $1500 a month, spend $3000 a month, and each deficit carries us that much closer to an eventual liquidation of the debt of KTAO by you-know-what.

When the subcription checks come in, the donors should be thanked at once, with a handwritten note, and the newest program guide. They are carded by name, address, zip (essential for 3rd class mail—which is the cheapest way to send out the guide) and amount given. I also put comments on the 3x5 card before filing it: if they send a note, or a present (one subscriber gave us a *Tomentosa* cactus, which I thanked them for, noted it, and was able to tell them a year later that their cactus had died) I always put that on the card too.

All the new cards go into the space marked "New Subscriptions" so we can make metal address plates from them.

The subscriptions come to die once a year. For that is the other important invention of Lew Hill: that support of a community station should be like a magazine subscription: something to be renewed each year. Something that had to be confirmed annually.

I go through the current subscribers file once a month and pull out those cards of people who are expiring *two months* in the future. I get Cese or someone to write them a handwritten note, with a return envelope, telling them that they are about to expire, and inviting their resubscription.

I wait for a month. If they still haven't resubscribed, I take them off the mailing list, and send them a program guide with the following stamp inside:

> THIS PROGRAM GUIDE IS THE LAST ONE YOU WILL EVER, EVER RECEIVE UNLESS YOU RENEW YOUR SUBSCRIPTION. PLEASE DO: WE NEED YOU AND YOUR LOVING SUPPORT. LOVE AND KISSES,
> THE KTAO SUBSCRIPTION LADY.
> 5 UNIVERSITY AVENUE
> LOS GATOS, CALIFORNIA 95030

It used to be back in the old days we had funeral notices with black borders which said

IT IS WITH GREAT REGRET
THAT WE TELL YOU
OF THE DEMISE
OF YOUR SUBSCRIPTION
TO RADIO STATION KTAO.

And in the corner:

At home:
The Subcription Lady
Five University Avenue
Los Gatos, California

I am not sure what the results of resubscription are for the other community stations in the country. Ours are extraordinarily bad: I think we have a renewal rate of about 20%.

Part of our problem, I guess, is the disappointment that people feel with the guide. I mean, they *really* want a guide, that will tell them what the programs coming up are going to be. They find *The Radio Times* to be somewhat of a disappointment because we see it as

61

something else again, I'll tell you!

FOOTNOTE TO MONEYRAISE
(LETTER RECEIVED BY KTAO
February 10, 1973)

BANKRUPTCY SALE
BUY STOCK NOW

SOUND RIDICULOUS? THAT'S YOUR SALE PITCH. WHO THE HELL WANTS TO SEND MONEY TO SOMEONE GOING OUT OF BUSINESS?

KTAO, THE ONE-WAY TALK PROGRAM. YOU TALK, WE LISTEN TO YOUR GREAT VERBIAGE. SHIT. K-BOR.

KTAO— LISTEN TO WHAT WE WANT TO HEAR, WHAT YOU TOO SHOULD LIKE. SCREW IT.

KTAO— THE 2 PLUS YEAR SWAN SONG, THE I'M-GONNA-TALK-UNTIL-YOU-SEND-IN-MONEY

S E X & B R O A D C A S T I N G

62

THAT SENDS ME, MY FINGERS, TO THE DIAL.

K-TAO R.I.P.: DEAD OF SELF-INFLICTED WOUNDS.

ROCK GARDENING
IN THE UKRAINE

KTAO and the other stations like it are community stations. Like the others, KTAO is essentially self-contained: and is the anathema to the community it lives in.

You might as well be Rock Gardening in the Ukraine. The people who live and work and play in your community don't give a damn about you and your music of Tasmania and grouchy interviews with candidates for public office and 73 hour festival of music of the bunraku theatre and gagaku.

Americans have been brainwashed for the last 40 years to expect entertainment from their radios. They want radio as a background, or something to shop at the Alpha Beta with, or to eat their frozen fish sticks by. They are not even believing that a radio station can care for the society, and the environment enough to do the type of grumble-and-grump that makes up a radio station like KTAO.

With all its faults, the Federal system of licensing broadcast outlets is the best. Because it is free of all the dibbly politics that would hurt a station like KRAB or WBAI or KDNA or KTAO...if KTAO had to go to the Santa Clara County Board of Supervisors for a permit, we would be playing Al Hirt and five minute news features and Forward in Faith. Robust broadcasting couldn't exist on the local level.

This is a bit contradictory to other things I have said about the FCC—but the difference between you as an 'Applicant' and as a 'Licensee' are enormous. For one thing, once you have your permit to construct—the FCC is all for you. They will expedite many things (application to move transmitter site, if you have lost your original one) in order to make your going on the air as quick as possible. And, as long as you don't make up stories to them, or fudge obviously where you should be straight, they will not bother you except with occasional FCC inspections which are required (and should not be a source of blind terror to you). For the FCC is primarily and patently an organization of establishment and conservation: e.g., the gummint. They were set up not to harrass existing broadcasters, but to protect them. When you have your

MUSIC FROM THE AIR PRODUCED BY WAVE OF HAND

construction permit, or your license, you are an 'existing broadcaster'—and all sorts of goodies will flow your way. You are part of the machine now, and unless you are stupendously incompetent, you should be able to maintain your license against all the fools who will (I tell you they will) threaten you.

The only radio stations in the U.S. which have lost their licenses are those which have nakedly and officiously defied the F.C.C...either by staging grossly illegal promotional contests, by ignoring a simple plea for renewal forms, or by not attending to the easy strictures of the Fairness Doctrine. As a matter of little-known fact, the most trouble visited on existing stations has fallen on the heads of the militantly conservative—the Carl McIntyres who, some of us believe, should not have been handled so militantly. After all, they are experimenting, as we all ought, with controversial radio: their one failing is that they refuse to believe—religious militants are such dolts—that there are other points of view that deserve airing.

This is a very important point—and I will repeat it here. If you're like 93% of the community broadcast people that I have met, you have a full-to-overflowing cup of Paranoia. I don't know where you got it, or why you must cling to it—but you have it, and seem determined to carry it with you to the grave.

Because of this, the Big "P" as we like to call it, you will get some feelings of unreasoned fright from time to time—like three times a day. After some controversial program or another one of those unnecessary "fucks" that went out over the air (in full-color stereo); or maybe after some especially bitter controversy among your staff, where one of the volunteers has gone off down the street, shaking his fist at the front door, and vowing vengeance on all it contains: it is at times like this that you will figure that your station is doomed for sure.

Please understand that these conflicts are intrinsic to radio—hell, intrinsic to any endeavor where people give great hunks of themselves. These conflicts always erupt over the question of money and don't-scare-the-sponsors (in commercial radio) and internal politics and don't-insult-the-subscribers (in non-commercial community radio).

I will discuss the *agon* of these conflicts later on. But I must stress at this point that one of the traditional duties of the FCC has been to protect the existing broadcasters—even mouldy old you, to protect you from losing your license.*

If you and your group are licensed by the government to broadcast with a certain power and a certain frequency with a certain organization—even god and his grandmother cannot pull that license away unless you are a down-home blind staggering idiot.

I know: for years I though (I carry your monkey paranoia on my back too) that the FCC was going to steal the KRAB license from us because we were doing such controversial, outstanding wild-man programming.

*Footnote: During a particular bitter squabble at KPFA—in 1954—Lew Hill resigned. Several others remained behind, doing some great programming, programming more like what we expect from community stations nowadays. Lew Hill—who really wanted KPFA to be the BBC Third Programme for the United states—got so pissed off that he filed a complaint with the FCC. He asked that they put KPFA off the air.

The FCC dutifully sent an inspector to Berkeley, who went through the records of the station and wrote up a report which said that, as far as he could see, the station staff had done nothing illegal, and there was no reason for the FCC to revoke the KPFA license.

Talk about the FCC being an instrument for conserving existing broadcasters! Even the founder of KPFA couldn't eat his own baby; or get anyone else to eat it, for that matter.

I was wrong: except for one or two genuine turnips in the Broadcast Bureau, most of the staff was lukewarm—neither for nor against us, wanting most of all to get our files off their desks so that they could turn to something more simple, like that renewal in Foo Foo, Mo. or that payola case in Bakersfield.

You are not an object of inner political workings and destruct mechanism at the FCC. You just think so because your head is screwed on so funny, and your fears follow you so noisily down dark alleys.

The only counsel I can give you with regard to these beasts that trail you so closely is—try not to let them become so grand and so real that they create the very situations that you fear.

BEING BOSS

The only way you as part of the radio station can influence the station is by listening to the station all the time. Monday and Thursday, 10:30 AM and 3:30 AM; Sunday afternoon at 4:44, Wednesday evening at 7:53. That voice of KTAO should accompany you at every step of the way. That is your promise to the universe of radio: to listen and praise and criticize.

The volunteers need your ideas and your hearing. When they do masterful programs, you should be listening so you can praise at those times of magnificence, and condemn at times of ghastliness. You should never stop listening: even at 1:53 in the morning when you wake up because of some dark dream nudging against the towers of your mind: you should reach across the specs and water at bedside and push the button so for a minute, or two, at least—the voice of KTAO comes flowing across the marble of the aether, dripping across the floor into your ears. The midnight people need to be heard as well.

You listen for the errors. You call about the most serious: you learn to train your ears against the equipment so that you can call when the slightest thing goes wrong: when the voice is distant or there is a scratching against the music or the transmitter-receiver gets locked into KPOO at the wrong time. You have to be an aural doctor, to distinguish the moment there is a failure of equipment and operation, or those bad times when substitute replacement due in at 10:30 at night doesn't come and you have to call Julie who's been on the air for 4 hours already and tell her you will heave yourself out of the warm and the cozy under the Hudson's Bay Red-Black-Stripe and take the baby and nurse it for an hour or two until the alternate substitute replacement can be found for midnight until three.

You are the it, and the maintenance of the voice of KTAO. Fortunately—running a radio station is so different from any other job, real or imagined—because it can be carried everywhere with you. The voice of your responsibility can be jolting along in the truck with you, eating oxtail stew with you at the round table, setting in bed and gabbling at ten in the morning, swimming in the icy waters of Lexington Reservoir: your responsibility can and does move everywhere with you: and you should move with equal felicity with *it*.

DEALING WITH CITIZENS GROUPS

Citizen groups are the natural outgrowth of 40 years of media screwing, bruising and abusing of the minorities by the owners of the nation's radio and television stations. If they didn't exist, we would have to create them to be our conscience, our guide, and our terror-masters.

For citizens groups are truly terrifying: or wait until you are in the same hot and sticky room with 40 pissed-off blacks who want to know why you, honky you, haven't gotten any of their black brothers on your staff, why you do no third world programs at all, why you are drinking that bottle of pig coors beer. I mean: I have seen the strongest of network broadcasters wet his pants when Marcus Garvey Wilscher comes in and *doesn't miss a trick.* He knows his power and his rights at license renewal time: and he is goddam well going to get jobs, money, and programs for him and other minority groups.

You may think that as a non-commercial broadcaster you are naturally exempt. After all—you are a good guy for all the minorities, right? Wrong!

Because you are one of the weakest broadcasters in terms of financing, and bureaucratic papershuffling bullshit ability, and your doors always open, and you have that burning liberal conscience: right?

I can't tell you how to deal with this. You and I will be faced with groups and individuals angry with 300 years of social injustice and 40 years of media injustices. They wants their rights, and they want them NOW!

The best advice I can give you is be willing to meet with minority groups—try to be a source of information to them, to show them how to use your air, to show them how to get other stations in your area to open up to them. Don't say no to them: that is an excuse for rampage—they've been hearing no from so many people for so many years that you will be doing a disservice to you, and to your station, and to them by being negative.

Recently, a minority group came to KPOO in San Francisco, asking for a program. The station manager made the singular error of saying no to them: no air time, no nothing. When the group then came to us board of directors, we thought it over for a while and then said, shit, why don't you just take the station. The staff was outraged, the black group amazed: and only us, old hoary greybeard us—knew that we were unleashing on this particular pissed-off bunch of blacks, and chicanoes, and American Indians the pot of burning gold which would ultimately, and finally (as all knowledge and truth must) end up burning the hands and eyes of the beholder.

Because someday, someday—a minority group, maybe a group of whores, or dwarfs, or an entire leprosarium, will come to KPOO, and ask for some airtime. And let the new managers say no to them: let the new staff once-minority-now-majority say no to anyone, anyone at all—and watch the petitions to deny fly, watch the telegrams and long complaints to the FCC pour forth.

Or to put it another way: the best way to handle the disenfranchised minorities, is to co-opt them: give them airtime, and power, and a position on the board. They will either blossom into life and humanity with their new belonging (and belongings)—or will sicken and die of the tedium of FCC forms, and nagging *their* volunteers to fill out logs and the usual bore-and-anguish of being one's own boss.

67

Part Four

"If parallel lines do not meet
it is not because meet they cannot,
but because they have other things to do."

—*Nikolai Gogol*
by Vladimir Nabokov

68

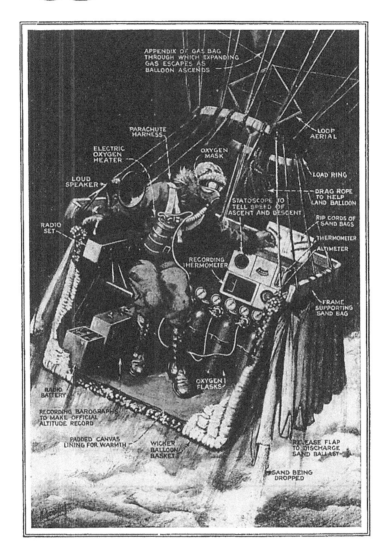

Prográmminck

69

THE MOST VITAL PROGRAMMING

You read in the newspaper that a local realtor has been accused of discrimination: Call up the realtor. Interview him on the air. Call up the person discriminated against. Interview him or her on the air. Make your questions biting and tough. Try to remain unbiased. Let them talk as long as it is interesting. You have done a public service, community, in-depth discussion of an issue of some importance.

A hot-dog from Washington is coming to town. Find out who is making the arrangements. Ask if you can interview that hot-dog. If there is a rally, send out two volunteers with a remote unit (cost: $100-$200 for the unit). If you don't have that, send them out with two sony cassette recorders. Interview the people who are there, why are they there, what do they hope to accomplish. Try to interview the policemen, the secret service, the dogs, the on-lookers. Take the listeners there with you.

There is a continuing scandal going on in your town or city. Having to do with unresponsive bureaucrats, or lack of action on some important issue such as housing, adoptions, welfare, civil rights. Talk one of your volunteers in doing a in-depth study of it. He or she goes out with a recorder. Or calls up people. Interviews them. Interviews people on the street. Comes back with 10 or 20 hours of raw tape. Edits it down, artistically, to 30 or 45 minutes. Interposes some music which has bearing on the subject. This is called documentary reportage. This is when you are doing your duty by your license—the duty that all the other broadcasters forgot 40 years ago.

Your program day is dull. It seems that all you are doing is playing records, or old tapes from your 'archives.' Take one of your microphones. Stick it out on the street. Put a sign on it: saying "You are on the radio." Put the microphone (bird calls, cars, jets, trees soughing in the wind) on the air. Let it set for an hour or two. See what the kids do (they always discover a new technological device first).

Ralph Nader, or Ralph Bunche or Ralph Duggins or Ralph Revolutionary or Ralph Birch is coming to town to give a speech. Find out who is sponsoring the speech. Get their permission to tape the speech. If it is great, play it on the air. If it is dull, erase it.

70

The state highway department, against all reason and logic, is going to build an 8-lane freeway right through your neighborhood. No one seems able to stop them. The local newspaper reports a group newly organized. Call them up, interview them—preferably live (live interviews are more alive than telephone interviews). Call up the highway department, the AAA, the state senate committee responsible for highway affairs: get their side of it. Keep it on the air. See what happens. Interview the grannie whose house is going to be demolished by the bulldozers. Let her speak her piece.

A local minority person is shot by a frightened or heavy-handed policeman. There is a rally. Go record it. If it is good, play it on the air. Call the police department—try to get their side of the story. Call the policeman, try to get him down for an interview. Call the police chief. Try to get him down for an interview.

There is a hearing in Washington that no-one is covering. It looks interesting: you have read about it on page 106 of the *New York Times.* Call up the staff counsel of the committee in Washington. Ask him to let you interview him on the air. Ask him hard questions. He appreciates your interest, and the chance for publicity.

You have read that Dean Martin or Dean Burch or Dean College or Dean John in Washington has said something interesting or outrageous. Call him up: ask him if he will let you interview him on the air. He might appreciate your interest, and the chance for publicity.

There is a member of the Birch Society, or the SWP, or the local Conservation group—who has been denied a voice in the community for all of his 50 years. Ask him if he would like to do a 15 or 20 minute commentary for you. Set it up for an exact time. Let him tape it ahead of time if it is more convenient. Your first request should be for a one-time commentary (no interview: just him speaking). If he is great, or even just good, and wants the time—sign him up for 3 months. If he is still good, ask him to stay on. Neither you nor I can predict how someone will do on the air in a commentary job: that is why your opening offer should be what it is to all your program participants: a one-time on the air spot. If they deserve it, ask them back. Again and again.

Miss Golden Egg's press agent calls up. Want you to interview her. You know it is your regular hype (he has never heard of your station before). Go ahead—give her an interview time. Don't let her read her hype statements over the air; instead, ask her why American eggs

SEX & BROADCASTING

taste so terrible in comparison to the European variety. Ask her how many chemicals they shoot into the chickens—and what that does to the eggs (and to us). Ask what they do with all that chicken shit—whether it is polluting our water supplies. Ask what it does to the flavor of our eggs to have all them chickens shut up in coops like Howard Johnson Motel rooms shrunk down to chicken size; ask if that is what is fucking up the taste of our eggs, and chickens.

An important jazz or rock musician is coming to town. Interview him if you can: but don't ask the usual dipshit questions that are asked on the other radio and tv stations. Get him to reveal some parts of himself that he doesn't say to those commercial boobs. Give him enough time to be a human. Realize that you cannot talk about music: you have to play it. He is separate from his instrument and his artistry. Talk about that.

The television stations are all doing convention coverage. Get two of your funniest volunteer on-the-air people. Set them up in the studio with a television set. Have them talk out over the air what they are seeing: invite your listeners to turn down their television sound, turn you up. Blow their minds.

Read something great over the air: something like *Winesburg, Ohio.* Or the poetry of ee cummings. Or *Black Boy.* Or the memoirs of Salvador Dali. Or the poetry of Marvell. Or the diaries of Lee de Forest. Or the novels of Curzio Malaparte. Or *The Tale of Gengi.* Or the journals of Columbus. Or Jean Genet, Knut Hamsen, *Lazarillo de los Tormes*, Lytton Strachey, Nathaniel West, Robert Benchley, S J Perelman, Thomas Mann, you name it. Don't read more than 5 or 8 minutes at a stretch. Intersperse it with a cut or two of music which is (or seems) appropriate. This make great half-hour or 45 minute radio.

There are 450 little mags (poetry, prose, essay) in this country—thousands in the world. Get their addresses (from The Committee for Small Magazine Editors and Publishers—COSMEP— Box 703, San Francisco, Calif. 94101). Send out a bulk mailer, asking for free subscriptions. Say you will read them on the air if they are good. When you find an especially good or interesting mag, read it on the air. Call the poets involved. Ask if they are ever in your area, to come by and do a reading for you. If they can't—because they live in Muckleshoot, Washington, ask them to record a tape reading for you. Play it on the air if it is good.

A local representative of the Flower Garden Bucolic Society comes by with a PSA (Public Service Announcement). She has never heard of your station. Ask her if she will let you interview her on the air. If she consents, when you have her firmly chained down in front of the microphone (most people do not realize that they can get up and walk out of a live on-the-air interview) ask her about the starving children in India. Ask why, with 5000 children, in the world, starving to death daily—she is bothering herself, her head, and you with the Flower Garden Bucolic Society.

A media salesman, a would-be Texas dj, an interesting kid, an old man comes into your station. They want to be on the air. You don't have an airshift for them—but you can and should interview them on the air. You have to be sensitive: if they are interesting (separate yourself from you-in-the-room-with-that-presence; think of you-the-voice-in-someone's-home) ...if they are interesting—let them talk on, guide them slightly. If they are stupid or dumb, cut them off after 8 or 10 minutes.

Every person who signs up to run for political office is asking to be heard: their candidacy is a cry for help: LISTEN TO ME, they are begging. Give them all equal time (you have to give them all an equal amount of time by FCC rules) on the air in interview. Get your toughest people to interview them for 20 minutes. Don't let them get away with any statement, any promise. The biggies and the smallies will both come to you. Roast them all. Take what most stations make into dull programming—and make it into a vital exercise in sharp, deep, inquiring, live radio journalism. The listeners (and some of the candidates) will bless you for it.

EDUCATING THE LISTENERS

As a broadcaster, you have to put your best out every day. What a drainpipe! You will have 40 or 50 or 60 volunteers, and limited facilities, and terrible finances—and still, you are going to want your programming to be as best as the BBC or CBC or whatever it is the models we have to work with.

Which is another of the problems facing broadcasters in this country: we have no real models. American educational broadcasters provide no leadership at all: they have no nuts. The commercial television stations have show us how to fit important events into 2 and 3 minute news clips, but that doesn't help you with your more leisurely use of time. Radio as it existed in this country 25 years ago, when it *was* the heart-soul of the Aether god, before those terrible dj minds came along to obliterate it—was the best example of creative use of the sound media. But it is gone now—except for a few scratchy transcriptions. And you have to create the new art of it.

It is a weakness—but it can be your strength. We are manufacturing the artistic side of broadcasters, that subtle act of transmogrification from the ear to the hand-and-mouth of listeners. The quality of the listeners that come in the door of your radio station to be broadcasters, that subtle act of transmogrification from ear to the hand-and-mouth of broadcasting: that is your responsibility.

Your listener-*cum*-volunteers must be trained. I don't mean the Ron Bailie School of Broadcast type of train—where the creative parts of their personality are shaved away, and they are stuck in the grease-bottle of commercial deejayery.

No, I mean that you can train people on the techniques of tape recording and being themselves on the air and the massive mixing of ethnic, folk, blues, classical and jazz recordings by showing them how, and letting them go. We found at KTAO that we could train people to be a control room person in 5 or 10 minutes: all that stuff that Columbia School of Broadcasting charges $1000 for—we teach them in less than an hour.

But your audience has to be trained as audience, as well as volunteers. If you do as KDNA did—you make the telephone a true part of the day-to-day. Jeremy built a special control board which could be operated by a child, and which had the capability of handling four on-the-air telephone callers (plus moderator) at the same time.

People would be invited to call up. One or two or three would. The moderator—often—would begin to fade himself out of the picture, stop exerting himself (as the voice of control, or of reason) on the listeners who were becoming over the telephone program participants. I hear many fascinating telephone programs over that station, where 4 strangers in their four strange homes would be conversing with each other for hours, ranging up and down the most fascinating variety of subjects, fighting, yelling, laughing. The four of them with their telephone world, which also happened to control a 63,000 watt transmitter, and the ears of some 10,000 or 15,000 people. The listeners were the station.

I have said that when people come to help you, you must determine early-on if they can shuck the bad parts of themselves and become radio people. It is your job to give them the chance and the need to become a good radio people. You should always be careful in pitting the world out there (you have so much high art of that world represented in your record collection) against the much more limited artistry of the people coming to you. After all, those tapes you play from New Zealand Broadcasting Corporation are real art, bought for millions of dollars worth of producers, and equipment.

You have always to protect your volunteers and staff from envious comparisons with this high-power stuff represented by the dead tapes and records in your collection. You have to hold these records and tapes up as ideals, or standards—but all local programming must be given the benefit: if it is a bit long-winded, or boring, or ill-produced—it has just that much more validity because it is *local* and *live* and *love*.

Local, live programming! Used to be the bell-wether on FCC renewal forms. Local and live. The best that you can offer your community. They deserve it: their own voice. If you are broadcasting some flute and guitar with poetry reading live in your front studio, or if you are interviewing some codger on the history of your community—and it doesn't have the smooth-and-silky of that tape you heard from the British Broadcasting Corporation last year: remember that you are only trying to compensate for 40 years of American broadcasting hurt by governmental inattention, and commercial exploitation—an exploitation as deep as the exploitation of the land, air and water. But the fact that the art of American broadcasting has been given such a shoddy shaft for these 40 years means that we have to move all that much faster to catch up. We have so much to do, and so few resources to do it with: which is where the change comes from. You are it.

The tension dialectic: between the best of the world's resources in your library—and the down-home community access: give them every chance to the access. It and you will be powerfully rewarded in the process. You have a great lack of community models for your local programming. Which is exactly why your debates and panel discussions and music-talk mixes and out-on-the-street remote broadcasts and shit-kicking interviews will be that much more interesting: you will be the first in your community to experiment with the limits of the community and its ears. It and you deserve no less.

DOING NEWS AND PUBLIC AFFAIRS

Doing news is not my strong point. If I have my druthers—all community stations would indulge in obscure 16th century music, and music of the Ituri Rain Forest, and live 25 minute interviews with candidates for public ofice, and city or county council meetings.

But I do know enough about news to know when it is being done badly: and it is done *very* badly on PBS and most of the educational stations in this country. Once again—they feel for some blasted reason that they have to do the 5 or 15 minute wrap-ups—invented by commercial broadcasters for commercial reasons—the international/national/local news (10 minutes) sports (3 minutes) weather (2 minutes), despite the fact that sometimes there is 15 minutes worth of news—sometimes 50 minutes worth. There is no reason why you have to lock yourself into a time sequence invented to sell cars and soap.

If you can afford to hire a wire service ($200-$300/month) do so—but don't take the "radio station" service: pick up the newspaper "B" wire, preferably UPI: it won't have those lame summaries that make all radio stations sound like porridge at the hour. If Reuters is available in your area—take it: it has a far finer international service, and it is unlikely that any of the broadcasters in your area buy it.

For local news, you should hire someone fulltime. Someone who won't steal rewrite articles from the local daily—but who will go out on his or her own and do investigative reporting. The great waste of the 7000 am stations in our country—many serving small, newspaperless communities—is that they restrict local coverage to the Junior Hi-Y annual Golden Ball Fete and crop reports: and ignore the often lively political and social controversies of small town life. The best programming in this regard is done by the local and regional services of the British Broadcasting Corporation which cut away from the network four or five times a day to let people know what's going on around the corner. Most American broadcasters don't have the nuts to do this sort of public service—and if they did—they wouldn't spend the necessary $10,000-$15,000 a year for a good newsman to organize the material.

For national and international news, it is good to subscribe to some serious journals and newspapers. You can have one volunteer who weeds through these every day for interesting articles to read—and for $200 a year, you'll be far ahead of the other radio and television stations in your area. I would suggest for background and feature articles:

>The New York Times
>The Washington Post
>The Wall Street Journal
>Harpers
>Esquire
>Rolling Stone

75

The Los Angeles Times
Whole Earth Review
The New Yorker
The Congressional Quarterly
New York Magazine
Washington Monthly
Saturday Night Magazine [Canada]
The New Statesman [England]
The Manchester Guardian Weekly [England]
Le Monde [France]
The London Observer

One of the best sources for international news is available to you. for free, right now, with 15 minute reports, in-depth investigations of international news events, and special commentary features. It is available as a package, and you need spend no money (besides basic receiving equipment) and need no special permission to re-broadcast it. It is the stupendous British Broadcasting Corporation International Service on Short Wave. They are beaming about 5 one-million watt transmitters at you this very moment. With a good Heathkit SW receiver, you can pick up and retransmit at least one of their news programs every day. The weekly *Letter from America* by Alistair Cooke is a great piece of spoken drama journalism, and if you write them in London, they will send you the schedule of the programs they beam your way.

You are well advised to retransmit some of the other short wave services. Radio New Zealand, Swiss Radio, Deutsche Welle (Germany), and the French Broadcasting System all have good commentary programs. Radio Tirane (the most powerful short wave system in the world with 8 gigabillion watts) is good for occasional laughs over their tape loop broadcast which daily cooks up Russia ("Imperialist Revisionists") and the United States ("Imperialist Capitalists") with equal—if monotonous—vituperation.

In your news, you should follow the same proscriptions as in the rest of your programming: do something *different* and *better* than your area broadcasters. A cheap and highly newsworthy program is available to you for free, on a weekly basis. That is rebroadcast of your local city or town or county council. For three or four hours, you can stick your ear, the ears of your audience, into the steamy mess of democracy in action. Some of it will be boring and soporific: other sessions will be dramatic, the unusual shout-match of people competing for money, power, glory. Usually you can talk the council into paying the 15 k.c. line charges from their chambers to your studios—they are always delighted to know that their words are being heard beyond those particular four walls.

You should check out all the other radio station area news coverage. Let your imagination run free and think of the thousand ways that you can do programming of a distinctive and higher intellectual nature. Schedule the news at a different time: say at 8:30 in the evening or 1:30 in the afternoon. Give people the option of your coverage at a more convenient time than the other ten stations that all do the News at Noon or Six or Eleven.

Not only will you earn the respect of your listeners—you will enjoy their subscriptions if you provide a true and honest distinctive service to the community.

Facts and Fads for Radio Fans

Above, Bureau of Standards' Measuring Apparatus for Checking Broadcasting Stations

Bernays Johnson, at the New York Radio Show, Lights a Lamp by Means of Energy Transmitted through the Head of a Young Woman Who Does Not Seem to Be Disturbed by the Experiment; Above, Right, Radio Receiver on One of a Number of Autos at Paris Contest; the Sets Were Tested with Cars Running

Left, Captain Autrey, of the U. S. Signal Corps, with One of the Latest Types of Radio-Receiving and Transmitting Sets Designed for Use in Trenches; These Two-Way Short-Wave Sets Are Portable and Mark an Advanced St⁓ in Radio Communicatio⁓ Applied to Field Use ⁓ War Conditio⁓

OBSCENITY ON THE AIR

Some of us puritans think that you don't have to have a fuck-a-thon every day or so to test our relationship with the FCC. That's the way that some of us (grey hairs, fear) avoid entangling problems with the government.

That is not true for all the community stations. Some of them love to be real. To be real, means that you do not use a tape-delay for your telephone call-in programs. If poetry has a full panoply of them four and five letter words, out they go! (Out on the air, that is: not into the trash can.)

I am allergic to most hair-pie poetry and hairy expletives undeleted. Not really because I am scared of the FCC—but rather because I always like to imagine my granny listening to the station, and liking what comes over it. We come into her living room. If I lay a big fart in her living room—an electronic fart—I feel a bit abashed. She never did like those childhood indiscretions anyway.

I feel that those things can be said without resort to the heavy words that scare some people so much. And yet, on the other hand, I think that the fucks and shits and motherfuckers are indeed the language of the people—their real language, pure and unstrained. When I am interviewing some poor bastard that has been a sailor all his life—I am damned if I am going to try to shut him up from talking of his own feelings in his own language.

And I am damned if I think that the government can or will even try to tell me what the language of certain citizens is or should be.

For historical perspective—you should know that up to 3 years ago, the FCC regularly sent out "20-day letters" to broadcasters who were heard to emit such bleeps—and who were unfortunate enough to have a bad-ass listener who wanted protection from the furies within himself.

A '20-day letter' said "It has been alleged that...we have made no judgement one way of the other on this matter...we are making inquiry so that we can hear your side of the story...you have 20 days to respond."

In the dark days before the '70s, before people realized they could fuck, and only thought they could have intercourse, a 20 day letter could be a prelude to a hearing. For years, the Pacifica stations and the Jack Straw stations were plagued by these particular nits. Finally, the FCC, in 1969—refused to give KRAB a full 3-year renewal on the basis of five complaints about this type of broadcasting.

We decided to fight it—and in 1970 the FCC held a hearing in Seattle, Washington to determine if KRAB should have a 1 year or a 3 year license. The case brought up the silly complaints (a bluegrass program, where Dave Wertz told the one about marrying a donkey to a jar of crunchy-style peanut butter: "You get a piece of ass that really sticks to the roof of your mouth...") and the real and tough ones—including a speech by James Bevel on being Black and on the castrating parts of University life. As well, there was the (mumble) pressing case of a 35 hour autobiography of Paul Sawyer which we put on the air and chicken me took off. Sometime if you want to read an artistic account of the events of that winter, read the chapter on KRAB in Michael Arlen's brilliant *Living Room War*. Or if you want the whole Hearing in the theatre of your head, read it in the Reference Room of the FCC, Docket #18943. The final dicision of Hearing Examiner Ernest Nash, handed down March 22, 1971, said that when a station is daring enough to do true community programming, then the FCC should be daring enough to tolerate some vulgar language—and I have included much of his history-making decision in the Appendix to this book. Anyway, KRAB got a three-year license renewal, the 20-day letters tapered off, and with a couple of exceptions, the FCC seems to have forgotten its pretended role as the Mother Superior and Soap-In-The-Mouth for non-commercial radio stations.

I have to tell you: some KTAO crazy once though we should petition the FCC to set up a list of the naughty words that it wanted us to keep off the air, and a list of fines to be imposed if they were uttered. What we could do then is tote up the whores ($25), the shits ($50), and the fucks or motherfuckers ($100). At the end of each month, we would just mail them a check for $500 or $600, regular penitence for our dirty mouths.

One of the real questions of "obscenity" is whether there are whole minority groups for whom these words are not foul, or evil, or wrong—but rather, their own true vocabulary. Or to say it another way—should we penalize a ghetto person for speaking his life-language—even though it might bother puritan us.

I've gotten to the point now where I tell on-the-air people that they can poop out one of those words if they have to: but if it is used as an obvious attempt to gross-out the mothers out there, then it's N.G. As I say—I am always thinking about my granny, worrying about her ears. But maybe you don't even have a granny to worry about.

TO GET THE SMARTS

Back in the old days, the days before Tim Leary was the Great White Farther of us all, he used to teach psychology and counsel at Harvard.

Students would come to him bright-eyed and ask for advice. They were uncomfortable—they couldn't figure out what they wanted to do with their lives.

He would say, "Well, what do you want—what do you really want?" And they would say: "I want to be a lawyer." Or "I want to do something for mankind." Or "Psychology...I want to play the psychology game."

And he would counsel them—he would say that if they really wanted to know about the mind, and psychology, and people's heads—that they should go into a mental institution. To find a job mopping floors in some booby hatch—or to get themselves committed for a year or so. "That way," he said, "you will learn more about psychology than you will read in a thousand textbooks written by professors who have never seen the inside of a state funny farm."

That is the secret of doing what I am telling you here. I have given and will give you enough materials here to do the job well enough, to slop together some radio station operation. But I can only show you so much in print. If you want to be an artist of the media, the Rembrandt of Radio, so to speak—then you should send yourself to graduate school. And I ain't talking about the Wharton School of Business.

You want to know about the reality of radio. Work for some junker daytime AM during the summer, selling time and playing crappy rock. Or take a year or two out and go to the Big Pimple, the pustule from which flows all the ideas and rules and practice and traditions that I am telling you about. Go the Washington, and become an FCC legal-technical nut.

Get yourself a crappy room on Q Street. Start hanging around the FCC. Go to the Reference Room once or twice a week; order out files at random, hearing cases, petitions for rulemaking. Pore over them. Immerse yourself in that paper world of pleadings and filings and responses.

Get a list of the practicing professional broadcast engineers...the engineers who are paid $2500 to prepare the technical parts of the forms we are talking about in this book. Go around to their offices: find the one or two nice ones (there are one or two who aren't absolute pimps to the media whores). Ask them to hire you on at $2 an hour, to be a journeyman apprentice engineer. Work in the office on a part-time basis. Help with the shit-work that goes into those applications. Learn about frequency searches, and drawing radials, and computing distance to the 3.16 mv/m contour.

Learn the Ten Gold Rules for putting together the engineering portion of an application. You may never have to do one after you leave his office—but whenever you see one, you will know what has gone into that simple, apple-shaped curive and those ragged 2-10 mile graphs. You will know that for the population figures given in response to a single one-line question on the Form 301, there are some 8-10 hours of dumb clerical computations done with Standard Statistical figures from the Census Bureau. Learn it, so that you can learn what a shitty dull job it is; so that you can walk away from it in 4-6 months and do real radio.

Try to find a lawyer who is a member of the Federal Communications Bar who will let you work in his office—back in the library where you won't scare the clients. Maybe rustling material out of Pike & Fischer. Likely as not (they are even more uptight than the engineers) you won't be able to find even one who will help you in your informal apprenticeship. To hell with them. Go over to Ila at Al Kramer's office, or ask Sue at Citizens Communications Center, ask them to let you do some free research project for them.

If they like you, they will set you up with a legal fact-finding job that will test your ability to marshal facts, that will work you to death 7 days a week, 12 hours a day, and on Labor Day too.

And, at the end of your apprentice year, you will know as much about FCC history, tradition, and practice as those boobs who have been practicing their fake law for the last 30 years in the hearing rooms at 1919 "M" St. NW.

Because a month with Kramer's bunch or the CCC or with Sam Buffone or Bob Stein can teach you so much about the bad and the good of American Independent Regulatory Commission procedure.

You will have, at the end of your self-education project, at the end of those 12 pound filings made with the FCC after 36 straight hours of work and research and reading and typing: you will have more information and ideas and thoughts about communication than the radio and television station owners and managers who, after all, came to their egregious position through sales and the run-of-the-mill thievery which is the tradition of American broadcasting.

You can and should learn what they never seem to have the will or perception to learn, and what the communications schools will never teach you. Your FCC and engineering and legal and reference Smarts can lead you to a far more rewarding sense of what is true communication in the U.S. With those Smarts, you can do more, more feliciciously, more diligently, more brilliantly—than the bubs who have been denying us free communication all these long and heavy years.

"Bill Carola"
Landaus
Seranaders

"Up to now, most community free-form broadcasters had to depend upon small, ill-built, often low-power FM radio stations for the transmission of their ideas.

"This effectively limits the listenership to the rich and the electronically elite—those who can afford a Scott tuner, or who are willing to set up an AR-15 with Yagi antenna. The poor and the minority have been left out of the whole community broadcast experiment.

"Now, we are about to do something different. UHF television—for so long the graveyard of the fortunes of innumerable Jewish dentists—is coming into its own, much as AM came into its own in the mid-20s. The all-channel receiver bill insures that by 1980 UHF television will be as commonly available as VHF.

"There have been some few attempts to create a television station which would not be some awful hawker's box, or some pasty educational outlet laboring under the heavy hand of the school or college bureaucracy. Viewer-sponsored television of Los Angeles is one,   Ch. 20 in San Francisco is trying, albeit poorly, to do for UHF what KMPX did for FM.

"But the application of Double Helix Corporation for a UHF tv station in St. Louis, Missouri will be a first experiment in fat-free controversial low-budget visual transmission: through a thousand Porta-Paks, we will try to bring the hot new inventive 525-linear experience into thousands of homes of the culturally and socially deprived."

—Jeremy D. Lansman, Founder of KDNA, announcing the application for KDNA-TV at the first annual Non-Institutional Media Council Meeting, held in San Francisco at the Cathedral Apartments, 12 February 1973.

SEX & BROADCASTING

Part Five

SEX & BROADCASTING

82

Some Other Media Games

83

SHORT WAVE

One of the most exciting finds for me for transmission and broadcasting was in July of 1973—when the FCC announced that once again it was going to open the door for applications from private groups or individuals for short wave or international broadcasting.

This was announced obscurely enough with the release of *Report and Order,* Docket #19530. If you are interested in what I am telling you here, it is essential that you write the FCC (% Dockets Division) and get a copy of this. Ask for it by name. It is free.

It will tell you that each day, 100 frequency-hours are open for non-government broadcasting on international bands. And if you want to transmit to the world, you are free to do so: after filing application on FCC Form 309 and getting a permit.

The U.S. is one of the few countries of the world to open short wave bands to private groups or individuals. It is a great pity that the four stations on the air right now from this country are all fundamentalist religious crumbs or jesus E-Z listenin' operations.

Short wave is important to many people in many parts of the world—because their own local service is non-existent. And short wave is organic broadcasting. When you make you application, you will have to make provision for the seasonal changes in short wave transmission (all international broadcasters do frequency musical chairs four times a year). In addition, you will vary your transmission according to the 11 year sun-spot cycle: 1975 will be low, 1981 will be high—and all this will affect when and where you can get a frequency.

The FCC serves as an adjudicator for the private U.S. services that want to get on the air. It is not licensing in the truest sense—but if you are granted the right to transmit, Lloyd Smith of the Safety and Special Services Division will take your name and request to London with him to deliberations with the other 144 international broadcasters who subscribe to the International Conference of High Wave Transmitters or whatever it is that decides who goes on what frequency.

As with your other applications to the FCC, you will have to show adequate financing to get your station on the air, and your engineering will have to be in keeping with current short wave standards. There is one man who is Mr. Short Wave as far as frequency allocations and applications go: his name is Stan Leinwoll, and he operates out of 22 Sterling Rd., RD #1, Princeton, N.J., 08540.

One of the cheap aspects of short wave broadcasting is that you do not have to get "type approved" or "type accepted" equipment. These devices of licensing were first set up by the FCC to insure quality transmitting by American broadcasters; soon enough, the 'type accepted' concept was taken over by the manufacturers so that the rest of us cannot build cheapo antennas and transmitters for our AM and FM and TV stations: we have to pay the jack-up jack-off of 400% to RCA or Gates because they have been licensed.

Short wave has no such restrictions: so you can build your set up out of scratch. There is one requirement which will be expensive to you: you will have to broadcast with at least a 50 kw (output) transmitter—which means that you will have to feed in at least 150 kw from your local power company. We figure that power from this source might cost as much as $5000-$6000 a year. Which is why we are looking to our own generators, or solar or wind power to operate the short wave station.

Before you go out today to get the stuff ready to transmit, you should listen to short wave. Realize that your transmitting into the particular void which is International Broadcasting will set you miles apart from other American radio people. It might be lonely out there.

I shouldn't forget to tell you about land requirements: you are going to need some 10 acres or so of flatland, away from airports, military installations, and high population areas, in order to have sufficient antenna space. As with some educational FMs and most AMs in this country, you will utilize direction antennas to beam your signal towards a certain "target area" (there are some 78 in the FCC rules). This means that your transmitted power will be souped up in that certain direction, with almost no signal going in the other directions.

UHF TELEVISION

Most of the major markets in the country (and almost all of the minor ones) have an extra, spare, lax, unused UHF television channel that you could apply for. There are a few problems involved—you have to have a transmitter site, and a great deal of money in assets to make application.

However, we are convinced that no one has really tried to make application to the FCC for a UHF television station with assets of $50,000-$75,000—and we are planning to make such a test case.

The poverty of such a station will be its virtue—and this is the way we have visualized it:

"You remember when I told you about the ideal television station. It would be alive: no canned movies, no 4-hour cold news clips. A live station, with live people, and live events.

"The whole thing would take place in a warehouse room. The camera would have been going all night: the picture transmitted would be from the camera placed in the corner of the studio-warehouse, looking out over the empty desks. Then at 6 or so, you'd see a door in the corner of the room open, and one of the news people, or secretaries, or some on-the-air person would come in: hang up his coat, start up the coffee machine, clean up his desk from the trash of the night before.

"Some people would come in: walk around, arrange their work for the day to come. And all the time the one wide open eye the camera over in the corner would be transmitting the various acts of a television station coming to life, the staff and the soul of it coming it, stirring up the broth of a video day activities.

"The camera wouldn't move—be moved—until the first program was ready: about 7:30 or so, it would be time for the first news. So the camera, heretofore motionless, would be trollied around to the news desk, and the man there would look up and start reading out some things that had happened during the night. Then later on the camera would move over to the kitchen, or to the bandstand, or to the office: anywhere around the television warehouse where there was some activity, some life.

"At those times when there was no news, or live programming, or cook show, or someone to talk or sing or play, the camera would be left motionless: focussed on some desk, or on the technicians setting up for the next program, or the manager haggling with someone who wanted to get on the air, or the person at the front desk who took the telephone calls, or the back alley where they were bringing in some musical instruments for the late afternoon program. All people who visited the warehouse-studio-television station would be faced with themselves, being transmitted all over the community, being shown as it, the event of communication, the isness of television, were being carried to the world out there.

"And the program-non-programs would be going on all day, all evening, far into the night: the television station would make no concessions to recording, the dead art of magnetic tape: all would be live and current: and at those times when there would be 'remote programs'—in the same way the distant cameras would carry the material live: the setting up of some meeting, the meeting, complete with intermissions, and the final desultory sweepings of the end of the operation.

"And the end of the television operation would come much as the beginning. The camera would be wheeled back to the corner of the huge room: one would see various member of the janitorial staff sweeping up, emptying the trash cans, dusting the desks, locking the door. And finally, at 2 or so, the last of the clean-up crew would go out the door, shutting it, locking it behind him: and you, the omniscient viewer, would see, across the empty room, the door clang shut, hear the lock click: and then be faced with silence, and the empty television operation, waiting there, motionless, for the next day and the next complete television act to come into being out of this silent, white, desolate room."

—P P McFeelie:
Prospectus for
KTAO-TV

KTAO-TV

Channel 48 is available in San Jose. What this means is that any group or individual who has financial capability can apply for and probably receive government permission to broadcast over that frequency.

There are tremendous numbers of pitfalls. Television is not just twice as expensive and twice as complicated as radio: since expense and complexity quadruple with the number of units necessary to transmit picture and sound, the possibility of trouble and poverty is enormous.

But the potential is as great. There are a hundred viewers for television for every one pair or ears for radio. UHF television might well be at the point of FM radio in 1957: where everyone knew of its possibilities, but so few were willing to gamble because of the many many failure stories from 'The Dark Days' back in 1950 or 1951.

Telvision does not have to be so preposterously expensive. Early FM failed because everyone wanted to run it like network radio from 1945. UHF television should bear little relationship to the immensely successful VHF television. The average Ch. 3 or 8 station in a medium-sized market will encompass an employee staffing of 75-150. The annual budget is close to $500,000—and salaries are a great part of that budget.

Channel 20 in San Francisco is the first experiment with small staff television: but it is still commercially oriented, with all the expenses and restrictions that that means. No one, with the exception of the early educational television stations, has experimented with a paid staff of 6 or 8 or 10, overseeing a volunteer staff of several hundreds. And no one has had the courage or insight to play with 8-hour-a-day (at the start) all live television. Or rather, those who have gotten into UHF television have ignored the message of listener-supported community radio—that is, there are great numbers of talented, thoughtful, free, willing volunteers who will give time and effort to the *act* of communication. All these volunteers are available to those who make use of the medium to *develop* the medium.

For if television managers would come out of the money pits for a moment (even to go to the reception desk to see those trying to get in the door) they would see that there are literally thousands who can give freely and willingly to communications. A real community station can manifest this talent by making *specific* requests on the air.

Thus, if there is a need for a moderator, or for someone experienced in world affairs, or for someone who can explain some domestic issue—the television staff has merely to go on the camera and ask for help. And it will be forthcoming at once.

It's that myth of expertise. Television staff and managers in the commercial and most especially the 'educational' stations see the public at large as a nuisance: the great beast out there who takes up their time with telephone calls and visits which distract from the great business of making money, or 'educating.'

The true community station should and would open its doors to the people: so that they can learn the visual experience first hand. As the community station should ask monies of the general public for support, so it must ask the willing hands and minds of the general public to add to the variety of its programming.

And since there will be only live programming, there will be an army of willing helpers to keep the camera busy and active—moving from here to there, seeing the events in the room as the events in the world. The television station should be a microcosm of the tormented living excited upending outmoving world out there: not some bucolic 'pure' and 'isolated' studio which filters out the madness of the outside world and thus protects and stultifies its viewers.

KTAO-TV may or may not come to fruition. It is an idea and a pleasure for us to play with over the next few months. But even if it doesn't happen, someone *should* and *must* experiment with honest television: where the camera and the world become one, and the action of the streets is moved into the living rooms of the audience who, in part, must become participants in the action channelled to them.

In true community listener-supported television—as in community radio—the audience must and will have that funny sinking feeling of being part of the entertainment-events-news-excitement-life. They will no longer be afraid to enter into the activities of the camera: they will be invited, expected, demanded to join in the events which pass across their screens. And if they choose to join the army of volunteers, they will find their lives changed, and the community growing closer and warmer around them.

By their own personal part as a unit on the screen going out some 100 or 200 miles to some strange livingroom or bedroom or kitchen, they will find themselves parting with some of the isolation imposed on them by a sociey of specialists and role-players. With their participation in the giant studio activities, they will find their own expertise is no better nor worse than those who talk to us of news or weather or sports from the commercial outlets.

The television 'personalities' will become all of us. We will with our own television community (which is the community of all the searching and feeling us) divest ourselves of the roles demanded by a society which can only visualize the activities of mankind in terms of cells, not unlike the octagonal waxworks of the bee. Man is more variegated than that: but heretofore, a severely structured society has not discovered the electronic-interface man.

The investigation leading to KTAO-TV will take us to the single point: given the refusal to mount a 100-man-paid-staff-commercial outlet, but rather an 8-man-under-paid-staff-community-outlet-with-volunteers: can it be done on a small financial base? Can we put together such an ideal group for $50,000 or $70,000? And can we maintain such a cheapo station without bankrupting the whole operation?

Television, American Television, the Great Eye of Plenty, still has a few holes and snips and nits of freedom left. We should grow to experiment with one of these holes. If we do not, given the nature of fear and timidity and greed that marks American communications, perhaps the chance will be lost to us and this area forever.

NB: The first listener-supported community oriented television station in the U.S. has gone on the air in Los Angeles—KVST-TV, Channel 68. You might write them at 1136 N. Highland Ave., Los Angeles 90038 and ask them what it's like.

FINDING A CHANNEL

*Who is to imprint the first foot on the snow
where no others have gone before?*

All the television channels are pre-assigned: that means that there is a table of assignments for the whole country. This appears in Section "A" of the Broadcasting Magazine *Yearbook*. The table does not address itself to which station are already licensed, those which are held as construction permits, and those which have applications pending for them. You will have to figure this out yourself by looking back at the existing stations list (also in Section "A")—or by writing the FCC direct in Washington D.C. and asking if there are any pendings on your hoped-for channel.

You do not have to apply for the city listed: you can apply for any town or community within 15 miles of city center of the assigned frequency. If you wish to go further afield than that, it means a petition for rulemaking...which gets complicated and might cost you a few hundred dollars.

There is not much more on television in this book because we haven't tried it yet. I have a feeling that once one community TV station gets started—there will be dozens of others to follow. But, as the Sasquatch say:

Who is to imprint the first foot on the snow
where no others have gone before?

CABLE RADIO AND CABLE TELEVISION

I have never been able to figure out why people have gotten so excited over cable television and cable radio. Because what is missing in these is the central pleasure of regular over-the-air through-the-aether transmission: that is, the availability of unknown random groups and individuals, in unknown random places. Or, to put it another way, it is the effect of the letter we received in fall of 1964 from central Utah reporting dx reception of KRABs programming from Seattle, from 1500 miles away.

I contend that for the money you would get together to do cable television—you can apply for (and get) a non-commercial FM community station serving some 1,000,000 people. Right now, there is a frequency just sitting there, waiting to be tapped, in St. Petersburg, or Memphis, or Nashville, or Birmingham, or New Orleans: all desperate markets, surrounded with thousands of frozen minds—which need the type of defrostation that you could provide.

For you to spend another 500 spasms of energy putting together programs and equipment for cable radio—you could have a permanent FCC license. And if we are able to prove in our application that we can construct and operate a community UHF television station for much less than $200,000 (asking $150,000 of that from Health, Education and Welfare)—then your excuse for your going into community cable television is lost too.

However—if you are addicted to cable, need those wires for your security, there are a shit-load of information sources available to you. The attached list was provided by Barry Verdi—who actually is hired by a piggo Teleprompter-style group to open their public access channel to...who else...the public. I have read Shamberg's book: it is great; the rest you will have to try on your own.

Still, and I have to say: if you want to do cable, you mought as well be Rock Gardening in the Ukraine.

INFORMATION ON CABLE TELEVISION AND CABLE RADIO

THE SPAGHETTI CITY VIDEO MANUAL by *Videofreex*
    Praeger Pub., Inc.,
    111 4th Avenue, New York City 10003

MAKING THE MEDIA REVOLUTION by *Peter Weiner*
    MacMillan Publishing Co.
    866-3rd Avenue, New York City 10022

GUERILLA TELEVISION by *Michael Shamberg*
    Raindance Corp. (Publ: Gordon & Breich)

    Also available—with all the old copies
    of RADICAL SOFTWARE—from
    Suite 1304, 440 Park Avenue S.,
    New York City 10016

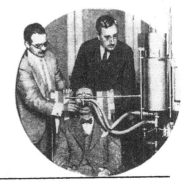

89

PETERSEN'S GUIDE TO VIDEO TAPE RECORDING
  by *Charles Bensinger & Editors of Photographic Magazine*
  Petersen Publishing Company
  New York

VIDEO TOOLS *by C T Lui* — all issues published by
  C.T.L. Electronics
  86 W. Broadway, New York City 10007

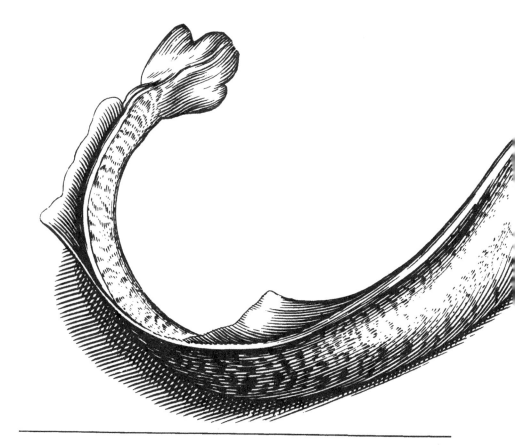

SEX & BROADCASTING

90

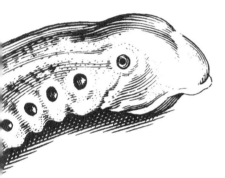

Mental Telecommunications

Some people do not need an RCA 50,000 watt model 2XC44-B transmitter in order to send pictures.

SEX & BROADCASTING

92

Communications with Space

Once a year Jeremy used to aim his highly directional rhomboid antenna up in the direction of Ceres Minor and transmit Indian music. Just for a few minutes. Just to let them know. Just in case they were listening.

You might do the same thing from time to time. So they can be sure.

MAU-MAUING YOUR EXISTING LOCAL SCHLOCK STATION

This book is primarily about how to start your own station—not on disturbing the crooks who are already operating in your area. However, if you have no resources, no resources at all, and if you want, sincerely, to better your local radio or television, I will tell you how to do it.

The following article was printed as part of the Bay Area License Renewal Challenge Buckshot campaign launched by the most excellent *San Francisco Bay Guardian* on the 27th of September 1971. You should write them and ask for the whole issue, if they still have any: their address is

    1070 Bryant Street
    San Francisco, California 94103

Over the years, I have worked with and talked with a variety of community groups in the United States who are out to improve local broadcasting. They always make three terrible, stupid, classically stupid mistakes.

1) They always try to go after the strongest stations in their area—I don't mean strongest IN TERMS OF MONEY OR RESOURCES, I mean strongest in terms of visibility and public service commitments;
2) They always ignore the most vulnerable radio and television stations in their area—namely, the shit elevator music stations (also called functional stations, or background stations, or E-Z listenin') or the turd religious babble stations; and,
3) When they go to visit their local broadcaster to ask for some concessions, they ignore the ground rules for decency and patience that will win countless victories.

I want to emphasize those first two points—no, I want to emblazon them on your naked ass: if you want to win, really win, don't start chasing after your local rocker that fired one of your best friends; ignore the case against the highly visible newspaper-television combine who does enough controversial public service to *guarantee* renewal.

Any station you can hear on your radio or see on your television is fair game: and the ones in your market *you never think about* are the ones that are most probably truly vulnerable.

Mood music background stations do their damndest to stick any and all talk programming on Sunday morning or late Sunday night. That makes them vulnerable.

Religious shit stations ignore the Fairness Doctrine and Equal Employment Opportunities regularly and savagely. That makes them vulnerable.

Your local country-&-western station assumes their audience is so dumb they don't have to do prime time public service. That makes them vulnerable.

Your local top-pop or hits-of-the-fifties stations loathe controversy because they don't want to threaten their $100,000/month income. That makes them vulnerable.

Listen to your radio for a day. I mean *listen to the stations you have never listened to*. Try to realize that all of them, no matter what they play, no matter the image they are trying to convey, have one singular thing in common: they are all licensed by the Federal Communications Commission, and have to abide by its rules for public affairs programming, and minority employment, and fairness doctrine, and 'public-interest-convenience-and-necessity.' That makes some of the damnedest stations vulnerable to your visit and your ministrations.

In this article, I am concentrating on methods of improving your local broadcasters' product. I slide over the whole process of going after the license of a broadcaster—what's called a "competing" application, in which you actually file a Form 301 or a Form 340* with the FCC a month before their license comes up for renewal.

Your competing application—one to take over the offender's frequency—is a mess. If you decide to embark on such a voyage, you should look forward to 5-7 years in hearing and a few private detectives following you around, along with a few other life destroyers. I don't recommend it.

FOOTNOTE

*The form 301 is the one used to file for a commercial AM, FM, or TV station. Form 340 is the one used for a non-commercial AM, FM, or TV station. Form 301 is far more complicated—and requires a hefty filing fee. Form 340 is simpler—both in things like community interviews, and in engineering. If you are going to challenge an existing radio or television station, you must file one of these forms 2 months before the expiration of their station license. If you decide to go after a commercial station—asking to replace it with a non-commercial one—you must file on the Form 340. This means that you can never do commercials on that station—nor sell it for a profit. You wouldn't want to do that anyway, would you?

96969696

There are some lawyers in Washington D.C. that specialize in competing applications.** They are viewed by the rest of the FCBA (Federal Communications Bar Association) as *goitre* cases, and even if you get a good lawyer, even a great one—your chances are hazy at best. Phil Jacklin and his COM crazies in San Jose have made just such applications in the Fall of 1974—and you might call them up and ask them what they think of the process.

I used to be all in favor of competing applications. Until I had one or two filed against me. I then saw, of reality, that the competitive process was a knife that cut 12 different ways come Sunday. Or maybe you have heard that the Washington *Post* which runs three of the better commercial television stations in the country found just this sort of spook knocking on their back door. For no other reason than the fact that they felt not so long ago, that a man named Nixon was up to some Hanky & Panky, and needed a closer examination. That is why there are three of these applications pending against the Washington *Post* television stations. If the competitors win, we all lose.

**Note: sometimes, the commercial people and the Washington DC lawyers call competitive applications "Strike" applications. When they do this, they are sucking you into the system through lingual (not oracular) means. A "Strike" application is a legally technical term that was used in the old days to describe applications that were filed with the sole purpose of impeding a legitimate application—*viz*, if you were a broadcaster in Dewlap, Ga., and someone made application for a station in Nosetree, Ga. just down the line—then you got your cousin Sim to go in there and make application for the same frequency—you were filing a "Strike" application which was and is illegal under the rules of the FCC. A competing application is perfectly legal, and was, in fact, envisioned by the original Communications Act of 1934 which established the FCC.

HOW TO TERRORIZE YOUR LOCAL
BROADCASTER FOR FUN & PROFIT
(from the San Francisco Bay *Guardian*)

I don't know if you have ever tried to get access to a radio or television station. I mean just get in the door. Those boogers will use everything possible to keep you on the other side of the desk from their precious frequency. Their frequency. I keep saying that. I keep forgetting that the frequencies, the air, the aether—belongs to all of us. Or should.

What broadcasters usually put in your way is a bleach-blonde secretary with a frozen smile and a retarded mind. She is programmed to keep you out. The idea is that if you got in there and started trying to do something creative with the air, they might lose some bucks. Might lost their 30% or 50% or 70% annual profit margin. Because that is what broadcaster in big cities manage to pull out of the air. Your air.

A bleach-blonde secretary, with silver fingernails, and an infinite number of ways of saying no. You could be looking for a job, or asking that they do some public service for you, or maybe you have some fine creative idea for stirring up the air, for getting some ideas from here to there, across the sky. You may want to talk, or argue, or sing, or convey some idea (your idea) of the nature of the universe, the purblind foolishness of man.

You may have some such grand idea to propound, and you know they are not going to let you in. They are going to create a wall called mystery. The mystique of the broadcaster. The wall that says, "Only I know how to program this frequency. I paid a great deal of money to get it. And if you can't buy it, get out." That's the way their heads work—those tits: or do I have to explain to you that more than 75% of broadcast management get to that spot through being hotshot salesmen. That tells you where their minds are.

They create the aura of magic. All that equipment. All those engineering feats. All those weird things you have to do with your voice before you can get on their air. Like having to go to some awful broadcast school like Columbia School of Broadcasting or Bill Wade or Ron Bailie—those awful rip-off schools where they take your thousand dollars, cram some honey-voice in your throat and ship you off to Montana to work at some AM 250 watt daytimer. And you think: "This isn't what I meant when I said I wanted to broadcast..."

...to Transmit sound ideas. To get across the barrier of broadcast mystery which, I must tell you, is a shuck and delusion. For the magic of radio and television is not locked in the hands of those twinks who run KFRC or KEST or K101 or KPIX or KNBR or KGO-TV or that miserable KNTV in San Jose.

Those people are sucking millions of dollars from a precious resource, the broadcast frequency—and you are being denied the chance to communicate. And all they have is the bucks to buy and hold on to that frequency. I'll try to give you some ideas about how to get in those sacred doors. I will try to teach you a few tricks which are the crumbs left over from the singular fact that there just aren't enough air spaces to go around—and the government has to regulate who gets on the air and who doesn't.

Up to a few years ago, the only people who got to broadcast were those who had enough of a gold mine to buy a frequency. And, because the Federal Communications Commission (like most independent regulatory commissions) tended to be run by those it was supposed to be regulating, there was little redress for the disenfranchised public.

Like, the commissioners would be having lunch with the president of ABC, and then go back to the office to vote on some permit for ABC. That was back in the Eisenhower years, and those were drab years for broadcasting in this country—far drabber than now.

Some court decisions, and some good guy commissioners have made it possible for us to get a toe in the broadcaster's door—and although they will bitch and moan and lie to you, you should do everything to get in: if you find their programming to be deplorable; if you have some sort of an alternative to offer them.

The public record ploy is perhaps the best way for you to terrorize management of a radio or television stations, and, if you know what to look for, it can be the best way (perhaps) to get some action to improve that station's programming.

My friend Bill Wade wandered into that screaming-meemie junkheap called KFRC the other day and said to the secretary that he wanted to see the public file. "What?" she said. "I want to see your public file," he said. "I don't know if we have one," she said. "What is it?" "You had better have one," he said.

All stations are required by the Federal Communications Commission to have such a record (of their dealings with the FCC) available at the station itself during working hours (9-5,M-F) for any member of the public to inspect. And if they lie, as so many broadcasters do ("It's in Washington," or "It's locked up in the president's office" or "It's at the transmitter") all you have to do is tell them that if they don't produce it for you, you are going to raise hell with the Complaints & Compliance Division of the FCC (William B. Ray, Chief, FCC, Washington D.C. 20554). Usually, they get the message and let you look at it.

When Wade finally got past all the bureaucracy at KFRC, he ended up in an office with two or three feet of documents, and the manager of the station, nervously popping in every now and again to try to figure out who in the hell this donab was (the last time anyone had asked to see their public records was back in 1968).

Wade kept mum, giving them only his name and address and none of the other information they asked for when he made his request (there is no provision in the FCC code that says you have to reveal why you want to see the public record. Broadcasters' fear and trembling make them try to find out.)

What you should look for in the public record is *PROMISE vs. PERFORMANCE.*

Since this is license renewal time for all California broadcast stations, if you want to point out to the FCC any gross discrepancies in a broadcaster's promise to do public affairs, or performing arts, you might get the license renewal of that station held up for a few months. And I want to tell you—broadcasters hate and despise anything that tinkers with their (what they believe) God-given right to get their license renewed rubber-stamp like. For instances, Wade noticed that the license application for KFRC promised the Commission that the station would be doing 3 hours 15 minutes a week of public affairs programming. Since the manager had heard his laughter down the hall, and had looked in to see what was so funny, Wade asked when that 195 minutes of public affairs was actually being broadcast over that station, what with all the shriek and holler and those $50-a-minute commercials they run 10 or 15 times an hour.

Well you know all that fancy public affairs programming goes on at some ungodly hour of the morning, like 3 am, or maybe Sunday morning when everyone is safely in bed or in church and the bucks won't be affected. And bless my soul: the station manager couldn't account for more than an hour or so—and was super-irritated that Wade had the knowledge of where to look. In the public file for the trouble-making stuff.

If you are willing to go to the trouble, there are some other ways you can get broadcasters' goats. And sometimes, you can get the license renewal held up, but best of all, you can sometimes get them to give you or a group of yours some time on the air to really talk about public affairs.

For instance, minority employment. That is a real doozer right now. The FCC has required all large radio and television stations (more than 5 employees) to state how many minority employees they have, and to explain how they are striving to train and hire black, chicano, and other minority groups. These questionnaires are part of the public file, and you can always ask that station management to describe to you what they are doing in this area.

Concomitant with this—although not specifically laid out—is the question of minority programming. If a station is lily-white, and aims none of its programming to any minority group, or makes no contact with minority group leaders to develop programming aimed at the poor and the dispossessed—one can complain and ask for time for this.

A very shrewd group of chicanos has wrested five half-hour prime time programs from that enormously wealthy (and enormously piggy) Channel 11 down in San Jose. If you are associated with any such group, or have a general interest in seeing such programming on any local station, this is the time to raise hell.

As in all of these rhubarbs I am giving you, remember that your most potent device is reason, and persistence. For instance, if you look at the public file and find that a station has promised the FCC back in 1968 to do an hour program a week aimed toward blacks—and if the station is doing nothing of the sort—then you can write to them and ask for time for such a program.

Wait a few days, and for God's sakes, keep a copy of your letter. If there is no response, call up the station manager. Tell him your idea. Ask his cooperation. If he acts like you are insulting him for asking such a thing, or if he is evasive, put your facts together and write to the FCC in Washington, D.C. and explain exactly what has happened. Send a copy to the station in question. Be persistent. Remember whose air is being prostituted.

One of the best ways to get air time is through the Fairness Doctrine. This states that, if a broadcaster puts a program on the air on any controversial matter—then he must provide time for opposing points of view. If you hear a program on most any subject, and if you think that the material presented is controversial and bears refuting, write a letter to the station and ask for time to reply under the Fairness Doctrine.

Once again, wait a few days. If nothing, call. If you are getting the brush off—write up a letter (once again, reasoned and logical) to Mr. Ray at the FCC, with a copy to the station. Be persistent. They'll figure out any excuse not to give you time. They want to sell it out from under us—remember.

There are some other ways of getting your voice heard. One of them (a good one, if you happen to have the money) is to buy a broadcast station—or, if you live out in the wilderness—start on of your own.

And, if you are really foolhardy, there is the famous "competing" application...in which you file against an existing broadcaster for his FCC license. If you plan to do this, get prepared for years of bureaucracy and hearings and appeals.

It's really dull stuff, but you might end up with a million dollar piece of broadcast property if you can prove to the FCC that the existing licensee is a money-hungry twerp, and that you are going to do a better job of broadcasting.

I don't recommend it: it would cost you about $20,000-$50,000 to go for an existing AM license, and another $50,000 to get the thing on the air. But what is even more unfortunate is the creeping kafkaism which gets in your blood after you go through such a hearing. I went through a five year FCC hearing and found myself starting to agree with the rightness of the essentially mad situation which is called comparative hearing. It took me months to recover.

If there is any chance you can get air time for yourself or for your group or for any worthy controversial program by simply asking to have it, without going through the FCC—then do it. But if you can get such airtime for free, especially from a local AM or TV station, then you are far more charming and witty than I am.

Because I know that broadcasters are essentially greedy and insensitive boors. They are in the business because of the enormous profits to be made by those who run radio juke boxes or television 9th run movie houses. They know the great public demand is for acres of goo—and they have little concept of leadership and less of journalism.

Leadership—in the tradition of the English or Canadian broadcasters where the aether is seen as a giant pallet—a monster canvas made of clear air on which those who care can draw great sound and visual works of art. Drama and public affairs and interview and that great wandering microphone and camera—peering in to all the holes of all the lives around us and pulling outsome great truths and some great beauty: some great knowledge about where we are, or should be, or will be—as we wheel down the beaten silver tracks of our days.

Broadcasters. They are such greedy magpies. Not realizing that they are hurting themselves and us by *not* presenting the wide-ranging intelligent in-depth experiments in sound and light. Fooling themselves into thinking that their cash flow and their ugly monster LeSabres and Country Squire Wagons are so important—not knowing that they are building up a negative charge in the Eternal Universal Bank of the Aether God...a cruel but just god with blank visage and the ability to powder their eternities to so much drip drip and dust and eternal ephemeral boredom for all the boredom they have given us.

BUYING AN EXISTING STATION

People from other countries find it inconceivable that Americans can buy and sell radio stations just like they were so much raw land. But it's true: it is one of the most carefully protected rights in the FCC book of rights for monster station money owners.

There are two ways that this buy and sell aspect of frequencies can affect you: one is that the transfer process (which usually takes at least a year) is that time where stations are the most incredibly vulnerable. The soft underbelly of the whole system is contained in FCC Form 314 or Form 315 and a variety of good guys and bad guys have made life miserable for station owners by getting involved in this delicate stage of the capitalist process.

If I felt a little more confident of the world out there—I would write here, in so many words, the true and fool-proof method of fucking up any transfer while it was before the FCC; I could show you how *you*, acting alone, could block, single-handedly, a deal which is worth a million, or five million, or twenty-five million dollars.

But because I have been skewered on this particular shaft, I don't think I'll tell you. Once, I had to sell one of my stations. Everything was going along swimmingly—and there was a whole bunch of money that was going to be set free to pay my own bills and to establish ten community stations to pay for the one just sold.

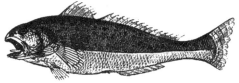

And some dick, by the name of Mr. Creep, wrote a letter—a series of letters to the FCC—about our application. And delayed the whole box of tea for six hungry months. While we chewed our nails, ran out of money, had to put the station off the air, and one of our volunteers got raped on the doorsteps of the station because of the deteriorating neighborhood.

One dinky little creep, with a wretched little hemorrhoid of revenge: he cost us thousands of dollars and untold misery. The aether god is going to get him and roast him in transmitter hell—I can tell you.

I can also tell you how to do the same thing: to make the richest and most powerful broadcasters loathe and hate you—revile you and gall you. I can teach you that terrible lesson: but I don't think I will. At least, not until I know you better.

The other thing you should know about buying or selling a radio station is that *you* can do so: if you have the assets and the interest. When you read in *Broadcasting Magazine* that some local AM rocker sold for $75,000, or some major market FM sold for $750,000—don't be discouraged. Few of these sold for that amount cash: most are transferred by the ancient and time-honored American Way called E-Z Squeezy...twenty or thirty percent down, the balance to be paid over 5 or 10 years. As is usual in these cases, the listeners end up paying for the station: only they don't get it. It might as well be you.

To find a station to buy—there are some buzzards around called Radio-TV brokers. They are all listed in Section "E" of the Broadcasting *Yearbook*. Even if you don't buy through them (and you are well-advised to avoid their 5% commissions) you can get a shit-load of mind-boggling information and gossip and knowledge of the hi-jinx of stations and broadcasters in your area. All you have to do is call on one of the salesmen and tell him you are interested in buying a station. Tell him what you can come up with as far as downpayment. If you don't have it, make it up. If you are convincing enough—you should get at least one free meal out of him, and have the chance to look into the dark pit which is radio-TV ownership and management: a murky area which most broadcasters try like hell to keep you from seeing into.

If you can come up with $10,000-$50,000—you have the assets to do some serious looking for an AM or FM station. Most businessmen buy into radio or television because of the guaranteed six-jillion percent return on invested dollar. You probably cannot afford the stations that have that guaranteed return, so you should look to a 'special situation'—e.g., a broadcaster in trouble, or one who has gone into bankruptcy, with the license being sold at public auction. When I was hot to buy a certain expensive station in a certain market—a friend advised me to hold off. "If you want to get into that market," he said: "Wait, and listen. During the next year or two, there will be at least one radio station out of 65 on the air that will be in trouble. That will be the cheap one to buy." He was right (but I didn't have the sense

enough to wait—and lost my ass in the process.)

All applications to buy a radio station must go through the FCC, on Form 314 or 315 which are just as complicated and dull as all the other forms I have been describing. It used to be that if you had a few thousand dollars, you could pick up a failing FM station in one of the top 50 markets. Those days are gone forever—as you can see right now by checking out the *For the Record* section of *Broadcasting Magazine*.

Most of us who are radio nuts have damn near memorized the going prices for going stations in different size markets. We read—hell—we *devour* the Buy-Sell section of *Broadcasting* every week. We know what that daytimer in Truth-Or-Consequences, New Mexico sold for last year. We know about that suburban-FM steal (just outside of the Bay Area) from last January. We wonder about that asshole religious group that paid so much for an 1000 W. AM in Dallas six months ago—for we have seen the downward trend of AM daytimer prices.

We may lack the hard knowledge of the contract: what percentage down; length of payout; interest on unpaid balance; whether it was a 'bare-bones' (license only) sale, or 'equipment-rich' sale or high income property sale or a highly-leveraged situation. That's where the sensitive spider-web radio gossip networld comes into play: those of us who know or have a lawyer in Washington and a friend or two who works in different markets and the ever-valuable First Class Info-Flow Engineer—we know most of the intimate contract data before it is placed in the Public Reference Room 239 in Washington for confirmation.

You could do worse to involve yourself in watching station buy-and-sale, notices of bankruptcy, and hurry-up sales for $0 down. You have nothing to lose from answering those broker ads each week that appear next to the *Changing Hands* section around page 40 of *Broadcasting* weekly. You probably can get to where you can try to figure out the call letters of the station listed for sale merely as "Chicago Metro Area. AM/FM combination. Profitable Operation—suburban location. Asking (large print) $135,000 (small print) down... Hamilton Landis & Associates." After awhile certain patterns will begin to come out in the price, location, power, gross, and potentiality of the stations listed, and those sold. And you can start to fantasize bargains of your own—like we did when a friend in Fresno found a "grandfathered" 100 kilowatt FM for $106,000. Or when Family Radio's brilliant Carl Auel found that AM daytimer outside of Boston for less than a quarter million. Or when my friend Len dug up that Portland suburban AM for $150,000.

It's a rich and fascinating game—as fascinating as the others I have described in this book. And someday it may come to be a necessary adjunct to your non-profit community station. For some of us' have found that the buy-and-sell capital gains routes the only easy way to finance another dozen minority-community radio projects. It beats trying to hang out with those boogers at the Harvey Coleslaw Foundation who could care less about you and that piercing need to transmit the sound-&-wonder of mankind.

RELIGIOUS RADIO

When the first edition of this book came out, I got a spate of letters from religious broadcasters asking how they could get their own stations on the educational part of the FM band.

I was very unwilling to send them the book—because most of these people are truly evil: they use their religion as a blanket to draw around themselves to keep from doing good radio. They absolutely refuse to allow dissenting voices on their thick effluvium covered air. They make thousands of dollars off their "good-will" offerings and shield themselves in non-profit religious organizations to keep from paying taxes on all their swag.

The FCC has not been able to staunch the torrent of religious shit (even if they wanted to) because anyone who is legally, technically, and financially qualified can own a radio or tv station in this country.

However, with a few pregnant exceptions, most religious dildoes have been kept off the 'educational' band—because of the dubious methodology of their 'schools' and financing. Let us pray that the FCC can keep these Jesus nits in their place—but they won't be able to do it without some help from us.

When I speak badly of the religious people—I don't necessarily question the broadcasts of the right wing religious freaks: most of these people make good controversial radio—and I think we need more of that. Life Line and the Wooooooorld Tomorrow and Unshackled make great programming, and should not be condemned with the other porridge of Jesus crap.

Rather—I am talking about the e-z god mush goo which is put out by many of the major market religious broadcasters. You should never be fooled: the commercial religious stations are most appallingly profitable broadcasters—second only to the 50 kw network o&o operations. Gorp pays handsomely, making some of us wonder how these crackerboxes plan to make it through the eye of the needle they are always ranting about; in other words, no religious broadcaster has been a true mendicant.

I blush when I tell you that one time—in order to pay the bills of KTAO—I rented the first seven hours of the broadcast day to such a christer. Great God Fessenden—forgive me! This is what I wrote when they finally left the air:

Actually, we thought it might be somewhat of a lark. We knew there were listeners out there: but we had failed to give them any sense of responsibility. The most we ever got in contributions during any one month was $2000 in January of 1972. The rest of the time—dribs and drabs. "We'll show the malefactors," we thought.

Well, we did. For nine months, KTAO was commercial in the mornings. I mean commercial. 12-15 ads an hour. High-pressure echo stuff. And Jesus every minute.

The tradition of American Commercial Jesus broadcasting is always a puzzle to foreigners. They can't understand the rich mixture of Calvinistic morality in U.S. Gawd-worship. They are puzzled by success in money terms being equated with success in divine terms. They can't believe the simple concept of *If God loves you, he will make you rich*. That's alien: it's pure America success art: and who can fight it.

Yesterday I wrote an article about *My Nine Months with Jesus* and it was a bitter and vindictive piece of writing. I tore it up. Because, of course, I gave as much as I got, and we suffered together. Having ten or a dozen time-warp 50s Jesus Christers in KTAO from 5 AM to 1 PM, seven days a week: having them and their bibles and curseless ways underfoot 49 hours a week, 196 hours a month was as tough on them and our listeners as it was on us. There was a bit of Mau-Mau each day each way—and I imagine they had some sufferance

too.

They just *appear* to us not to suffer so much because at least six times a day, when things are tough tiddy, they get down on their knees and they pray for us. I mean they pray for me and Doug and Cese. To help us. They are out there right now praying for our conversion. Can you imagine the weight of all those prayer-waves beaming in on us right now as we are sitting here, glorying in the now-free KTAO?

Big Jim was the one who started it all. He left KEGL. He was a time broker, and was casting around for a place for the Lord's Work & Word. He heard from the radio grapevine (transmission time somewhat superior to our own signal) that KTAO was having problems, and he offered to buy the first seven hours of the broadcast day for $1000 a month for the first three months, and $2000 or $10 an hour thereafter.

For a Christer, Big Jim is actually not a bad sort. I guess he has the true belief: and when he passes on to The Upstairs, I would hope that there is some time-sales-brokerage for him to get involved with, because he is then as now going to sell the hell out of paradise—and god himself will find his ways and words being sold to the far corners of the divine pastures.

When Jim first started in on KTAO on the first of April, Paul and Fritz were running things. Paul and Fritz are both Fallen Catholics: and there is nothing worse for the pentacostal fundamentalist Jim to come across—especially in the realm of power.

Like all of the gospel males, Jim was overweight, and had ulcers. They all had ulcers. Since their interpretation of the bible did not permit them cursing, adultery, drink, and all the other sensual pleasures that keep the rest of us going, they took out their aggressions on their sales. And, as I have said, they sold the hell out of god, finding 'Christian' businessmen in the most unimaginable places in Santa Clara County. My favorty ad was for some ENCO station in San Jose, whose motto was *Let Jesus fill your heart, we fill tanks.*

They always had listeners. Saturday morning was request time, and the telephones never stopped ringing. Saturday morning was always the low point in the week for me, because KTAO would be packed with Christers. The adults—at least most of them—I could handle: whenever they asked if I had been saved, I said 'Of course' and they would drop it, knowing from the light in my eye and the whiskey on my breath that I had not. The kids though: nothing is more tedious than a rose-cheek blue-eyed ten-year-old girl asking if you 'had been washed in the blood of the lamb.' Over and over again. I could have strangled her when she opened an especially fine edition of *The Reginald A Fessenden Radio Times* (the "Fun & Excrement Pain" issue) and offered to burn them all because the cover picture struck her as being a trifle irreligious, being a picture of a swooning maiden on a cross.

There was little interaction between the 'morning' people and the afternoon and evening people. Although we co-existed in the same two rooms (especially at the changing of the guards, between 11:30 AM and 1 PM)—we rarely exchanged any warmth. Christers are really not very funny people—and they loathe any merriment at the expense of their particularly harsh, big, and no doubt ulcerous god. I always wanted to get one or two of them in some mind-fuck session, for ten or fifteen hours, and explore the depth of their belief in divinity; but they were as chary of the opportunity as I. Proving, of course, that we are reluctant to test our prejudices and gods after we pass through puberty. Why should we, even if we are killing ourselves?

I guess I was disappointed by the strict narrowness of the morning programs on KTAO. Where the rest of us would gladly, from time to time, put in some religious music (a Bach cantata, the superb harmonies of Bahamian cult music, black religious music like Rev. Kelsey)—the morning people would never deviate from their sterile, monotheistic, drab, 'gospel' music. White Southern Baptists have fried away their god-given aesthetic sense with too much sow-belly and grits: and their music shows a pale deathly washed-out boorishness. If their word is the Ultimate Word—then the rest of us are doomed to an Eternity of turgid rhetoric and music as vital as yesterday's boiled spaghetti.

The men have ulcers and fat; the women have that stricken look of born sinners: after all, for the fundamentalist, a woman is the root of sin—and the Christers' women looked as if they were struggling to atone for being the diddy of all our troubles. We had a particular cross to bear during the nine months that God-&-KTAO were One: that was a wraith by the name of Judi.

Judi came to Jim (and to poor us) out of Synanon. She had dropped that and her Judaism for the particularly bleak form of Pentecostalism espoused by the KTAO Christers. She prepared the logs, did all the continuity, handled the telephone and was thus underfoot a great deal of the time. A pretty thing, if you favor the schoolmarm types...but she was hell if you crossed her puritanism.

One day Doug and Jane were doing just that. Fondling in the front room: hugging and a-kissin' before the outraged eyes of Judi the Just. She had a tantrum, said some naughty words (which I gleefully reported to Jim: we all became tattletales and children under the bleak eye of Fundamental God).

I of course was pissed off at Judi trying to regulate the love-making activities of the KTAO regulars; she was pissed off that any of us should impinge with *our* morality on the sanctity of the morning watch, and especially (as it happened) in front of a sponsor. When she blew up at Doug, I told her that she should go back to Synanon, she stuck out her tongue at me and told me to fuck off. I reported *that* to Big Jim, and he came out and tried to be the Prince of Peace: "Judi," he said. "I don't want you to ever use that sort of language. *Ever.* I was in the Navy. Heard it all the time. Even saw leprosy. And there is no excuse for a woman (or a man even) talking like that." I stuck out my tongue at her again. God we were children. All the KTAO regulars had to stay out of the station until 11:30 after that, and when they didn't, Judi wrote me a note:

> Lorenzo—
> I said 11:30. I have kept my end. Keep yours and
> keep them out."

I posted this one on the board, and you can imagine the graffiti that sprung out of that one:

> "Can we arrive early if we promise to pray & go to
> Sunday school?"
>   and
> "This morn' my children—let's have a 'silent' collection
> —no coins. Just folding money."
>   and the inevitable
> "Jesus shaves."

That was and is and always will be the low point in the history of KTAO. The station has always been a gathering place for some of the best people in the disenfranchised drop-out community of San Jose-Los Gatos-Santa Cruz: and it made us miserable that some Jesus-smitten biddy, with her sexual hang-ups should try to regulate our morals and our visitors and our access to the station: I mean is it worth $70 a day? I vowed to end this goddamned marriage.

Like I say, we can dwell on the badness, but the high comedy is more worth remembering. There was Lois. Dear Lord: a regular grandmother type with orange-blonde hair who squatted in the control room like a toad, and jumped like one anytime we came in. Of course we had to giggle when she was around, and she had to always come out and tell us not to be so loud and then she turned up the heat like my third grade teacher who kept the heat all the way up to 90 or so and all of us would get dropsy from the heat and all and fall to sleep and she would shake us awake because we weren't supposed to sleep in school: Lois was like that and I imagined that if we didn't stop giggling she'd come out of the control room some day and shake us all until our teeth rattled.

There was Rev. Donaldson who used to do testimony on the air

> *Pray for Ronald Bobbitt*
> *that the Devil will not*
> *cause him to have an*
> *asma attack on Sunday*
> *so he can go to Church and*
> *Serve the Lord*

—Salvation
Caroln Lovern wants prayer for Gorri Tally her sis-in-law in L.A. on account of witnessing her sis in law will not have her in her house.

—Michelle Campbell
healed when you prayed for that little girl who was having trouble sleeping. Michelle had this problem also and she put her bible under her pillow and now she sleeps well

—Mother called in and asked prayer for her son Rich who is in jail He was once saved & filled with the Holy Ghost

—A friend called and asked prayer for David—who is a backslider & Thomas—who needs salvation and Lucy —who needs deliverance from evil oppression

—Esther
(Woman with nervous condition that you prayed for)
Testamony:
You prayed for her last week for a dental problem God healed her and took the pain away

—Ona—Arthritis of spine also her husband—Harry bad alergys, and sinuses

—Pray for a lady who has just had back surgery & her nerves are giving her a bad time. Also her husband needs a job.

—Pray for the KTAO Banquet—to be a success & to glorify the name of Our Lord Jesus Christ. on Sat nite Nov 25th

(I typed up some of the cards left over from this program and you can see the questions and the spelling that the good Reverend had to field: Jesus, we assume, will do dental work if one prays hard enough.)

They're all over at KFMR now. Jim and Lois and Charlie Henry and Joel: playing Jimmy Swaggart and "I Should have Been Crucified" and Bill Gaither "Build My Mansion (Next Door To Jesus)" and Wendy Bagwell and the Sunliters and "He's More Than Just a Swear-Word" and The Singing Rambos. All those Christly songs that wake you up at 3 AM the words going around and around in your mind until you think you are going crackers: "Get all excited/Go tell everybody/Jesus Christ is Born." That simplistic didactic pedantic fundamentalist Pentecostal harmony: The true words of the Born Again Christians who, god knows, given their singularity of purpose and humorless drive will take over the world and condemn all us sensualists and jokers and passionpots to the sewer-pits of hell and brimstone that we so richly deserve.

The thing that ultimately drove them away was the one thing that is so typical KTAO that we didn't even think of it at the time: the steam-driven 1910 wood-burning transmitter. Just after Judi had given out with her pronouncement, the great Aether God leaned down out over the steel-grey sky over Mount Umunhum and smote the KTAO transmitter a fearful blow so that the cooling fan wilted and died in a blaze of black smoke.

It took us three days to get a new one and Jim was watching those dollars leak down the drain and he got so impatient that he went over to KFMR and they said they'd love to have him and the word got out and I knew that in due time the contract marriage bond between Jesus and KTAO would soon be properly and perpetually dissolved, and the Oak Ridge Boys and "I'll Keep Holding on to Jesus" and The Downings and "Jesus Is a Great Big Man" would be off somewhere else, out of our hair, where they never goddamned belonged.

If you want to hear what you are missing you can tune down to 104.9 sometime and hear the happy christian sound as they have come to call it. I for one found that my nine months with Christ confirmed everything I ever knew and heard about Jesus-radio. It was said for us by one of our long-suffering subscribers who wrote last September and I was going to put it on the bulletin board so that Jim and Judi and Lois and Dan and the whole gang could get the word of The Great Aether God who, I believe, is a little hotter and far more divine than Jimmy Swaggart & The Happy Goodmans—but I knew that we all embrace our delusions and prejudices to our withered dugs as long as we are able, as long as they work; so I kept the letter in the file back of the desk so that when the day came that Jesus got run over to another spot on the dial I could pull it out and let us all see it: because we know that we can't change them and their dark grey ulcerous god and they can't change us and our lewd loving ways but at least, by Christ, *now* we have one last sweet sensuous laugh:

Date: 14 September 1972
Attn: To Whom It May Consern
Subj: Your religious program is imbecilic
It is also time wasting recommunicating anything to the untouched. You won't reach those not in your psychotic/delusional/hurt level—because your approach is aimed not at those out of the flock, but those already there. Dig? Your trip is like a political speech, only the faithful show up. For the others, your message and means of transmittal of the same is vomit inducing.
Take your fantasy trips at home.
And step up to functional reality. [God is in you, you shit head. Not out there].

With love,

Richard
(From KTAO Guide #111)

ILLEGAL TRANSMITTING

You understand—I would be the last person in the world to counsel you to go into illegal transmitting. I think that there are enough mouse-holes of legality for us to fit into—to make our stations a permanent part of the aether.
Still:
We have all heard of those 10 or 50 or 500 watt am transmitters, set up in some kid's apartment, in which he transmits to himself and his friends and neighbors. My friend Fred Oyster claims that there is a part of the FCC rules dealing with 'closed-circuit' (college dormitory) broadcasting in which you can outfit a fairly high-powered am station *legally* without going through the FCC. Write him at Box 191, Middletown, California to find out what he means by that one.
There are stories—some aprocryphya no doubt—of stations like K-FAG or KIEF or W-LSD or KPOT that come out of trucks. Vans outfitted with turntables and tape recorders and microphones. And a high-powered FM transmitter that can be driven up to some hillside late at night. To broadcast to whole valleys. Like the Bay Area. Or the Los Angeles Basin. I know nothing about these—but I know a few have been caught. They don't go to jail.
Then there are stories of the AM vans. The ones that can be parked 125 miles from a population area. The signals—through skywave—go up. And then mysteriously down: down as in the earliest mysterious waves of sound that bounced, like cosmic red rubber balls—bouncing down, apparently from heaven, from aether heaven, into the surprised ears of those tuning across their old superhetrodynes. They say that you can do the same— you feed your

signal from your van into the dark skies, with perfect angularity toward the city of your choice. Send them your message of truth—at some dry dead crackling place on the am band. Your voice of reason comes in. I know nothing about this of course. But *if* you do it : please do something worthwhile. Don't use it to broadcast Grand Funk, for god's sakes. I often thought that if they took all my stations away from me, sent me to broadcasters' eternal limbo—forbid me the sacred microphone ever again—that I would have one of my wise engineer friends set me up with a VW camper. Fully equipped. I would become the ghost of community radio. Drifting around the country with my message which would peer into this or that city a finger of reason, fading in and out, the ghost writer in the sky. What would I do? What would I tell the surprised band tuners from 200 miles away? I guess I would play, and replay, and play again—the radio drama by Sam'l Beckett. *All that Fall.* The BBC did it in 1957. The Copyright expired in 1962. I know a friend who has a tape copy. Faded and worn—but perhaps what radio is all about.

I would play that. And some Water Music gamelan from Java. And that talk by James Bevel where he says "Can they take away your balls without it really hurting. NAW. That's what happens to the school administrators. They take away their balls. And you *know* it hurts." I would play that one, entire.

And maybe that dreamy reading of poetry by Robert Creeley. Or perhaps that silly *Three Men in a Boat* on the Argo record. I would read some myself—some of those letters in the now out-of-print book *The WEAF Experiment.* So people could know something, anything about early radio.

O, it would be great stuff—my pirate-drift ghost broadcast. I would have to move fast, in my beat-up van, to get away from the sage-riders of the FCC, in their black Ford truck, painted all black, even the windshields—the Black Maria, with the half-wave dipoles and the whip antennas and the rings sticking up all over, like the horns of some enraged bull, chasing me around the ring of our days. I would have to move fast...

...if I had to do that. Fortunately, they license weirdos like me, so we don't have to do it illegal. Although I can get a feel for the thrill of it. On that dark butte. Alone—me and the crinkling stars, blatting out electromagnetic waves, cut out of the same steely cloth. Me and the stars: all of us blaring forth our messages. Carved in black aether-gold.

Because of the microtechnology available nowadays, one of my friends has said that you could send up the perfect weather balloon. Broadcasting your message. All across the U.S.

It goes up from the far east end of Long Island. Launched from some deserted beach near the Hamptons. The balloon carries with it a tape cart machine. With an hour or hour-and-a-half endlessly repeating message. Over and over again the tape goes, activating the tiny transmitter (FM—line-of-sight, microwatts), pushing the micropower out into the fan shape antenna that springs from the base of the balloon when it is safely above the sea.

The winds take your message of love to the east—for a few hundred miles. And, then, when it gets into the area of the stratosphere, the prevailings start carrying it, your white balloon, and your message transmitter back over the entire United States. Like all those satellites—broadcasting beeps and wheeps of reason—at, maybe—87.9 mHz, or maybe 108.0—or if you want to blast them all, around 92. Or even Ch. 4 sound. Moving back, slowly, painfully slowly, crawling cat's feet in the cirrus across the entire face of the U.S. With a message, your message, which could only be wiped out if they sent up one of those SAM missiles. Or the entire U.S. Air Force. The might of military, sent up to destroy your silly message, and tiny balloon. What a fine flapdoodle.

Or maybe you would consider the singular weakness in the U.S. Intercontinental microwave relay system. In which there stands—on a thousand mountains chained across the country—a dish receiving unit, and a dish transmitting unit. The dish receiving unit is getting in weak signals, from 120 miles away. If you were to beam a signal at it, on the same frequency, from only—say—300 yards away—then your message would be picked up by that microwave receiver: you would blot out that mini-signal with your own. And maybe your signal would be more pertinent...a word or two, to bloom forth on network television programs in 50,000,000 homes. A word of reason, or interest, or amusement. A short poem, perhaps. A simple statement. You are just doing 21st century leafletting. You understand I know nothing at all about the technology of this.

Supermarket shit music systems are usually pulled in from some FM subcarrier. The FM signal with its bilge-full of bilge music are weaker than a beamed transmitter-antenna unit—often just a watt or two—aimed at the subcarrier receiver. Your signal could easily over-ride the music. "The beers in the cooler are laced with 59 varieties of chemicals as preservatives and retardants. Do you know what those chemicals are doing to your body? Neither does the government," you could say, in a stentorian voice. God, would the packages drop, the manager runs all over: where in the hell is it coming from? Or: "That Roach-kill. That they tell you to spray all over your kitchen cabinets. Do you know what those chemicals are doing to your sensitive body. Or the even more sensitive bodies of your children?" They would be pulling down receivers, tearing up the wiring. You would drive them bonkers for awhile, especially with that tape loop (left behind, hidden behind the canned string beans) about the legal manoeuvres that Shell has gone through to hide the dangerous effects of their Pest Strip. You would blow their ever-lovin' minds.

Fools always say, "Why don't we go out to sea—out past the 12 mile limit. Broadcast with 500,000 kilowatts." That's foolish advice: it costs millions—you could buy up an existing AM clear channel for that. Really: you could buy up WLAC for the cost of outfitting such a ship. And you wouldn't have the Coast Guard to contend with. Ask Carl McIntyre. If you are absolutely denied access to the media in this country, but still have money—you can go to Mexico, lease an existing AM or FM station there, and beam your stuff into the Southwest. But this is tricky: you can never own all of one of these stations (you have to have 50% Mexican national ownership) and there is a good chance that the Mexican Government is far less tolerant of free speech than our own beloved FCC.

We know little about foreign broadcast privileges. We know even less about foreign short-wave opportunities. The experts on this are the poops at World Wide Radio, Inc. Given their fundamentalist drip-dry programming, and the sucrous attitudes of their staff—I haven't gotten up the balls to ask them which countries we can go into, yet. You might try it—though. They can be reached through Billy Graham who is a media man himself. Or do you remember when he said that Jesus has a tape recorder going all the time. Recording all our words. And our thoughts and evil deeds, too. Eternal playback, I mean!

111

SEX & BROADCASTING

112

Part Six

> *Please order your minions of Satan to leave my station alone stop you cannot expect the almighty to abide by your wave length nonsense stop when I offer my prayers to him I must fit into his wave reception stop open this station at once*
>
> —Aimee Semple McPherson, telegram to the Dept. of Commerce after they padlocked her radio station for frequency drift (1925)

SEX & BROADCASTING

114

Some Essays on Radio, Music, Programming, Et Al

115

EXTERNAL TRANSMISSION

I got a funny feeling about the power of radio once: a thousand years ago in another country. I was on a journey, one of those Cyclopaedic journeys kids dream up in order to torture their parents, or themselves, or their American-ness. I was looking for something: and I am damned if I can remember what it was.

It started in Frankfurt, where I got a new shiny Blaupunkt radio installed in my car. It was a beautiful radio, with lots of black-shine buttons, and a single red eye that lit up to let me know that the world was crowding my dial. There was AM for static, FM for white noise, and the Long Wave band for the cryptic messages from the BBC.

I went into a dark corner Frankfurt bar to celebrate my new radio and my new car. I remember (vaguely) drinking an inordinate amount of icy clear schnapps intermixed with draughts of warm dark beer. I remember a foggy mind and a foggy night through dark windows. I remember an old man with a potato face huddled in the corner of the bar. I remember him mumbling, and I thought he was talking to me. I turned my head, smiled, said something in lousy German. Hands across the sea.

"Juden," he said. He said a great number of other things, other words, but *"Juden"* he said most of all. He didn't smile. *"Juden,"* he said. I knew it was time to leave Germany, even though it was 1961 and there should be nothing to fear.

"There's nothing to fear," I told myself, and I drove south and west. Running from Frankfurt to Neunkirchen, from there to the burning valley of the Saar, the hot black no man's land that still—Common Market or no—gashes between France and Germany.

"There is nothing to fear," I said, as I raced south from Saarbrucken, down through Macon and Lyon and Avignon—down towards the sun and the light. It was a time of great, aggravating isolation. My sole riding companion showed a single red eye, and spoke to me in many languages and many musics.

I began to warm, to feel warmer and safer, when, in Dijon, I heard the voice of Radio Diffusion Francaise. The voice was feminine and husky—and spoke to me at seven in the morning, when the mustard-colored sun grew up over the streets of Dijon. "Beethoven," she said: and I heard a trio by Beethoven. "Rameau," she said: and the harpsichord music played around my ears. "Handel," she whispered: and a cantata of Handel sped me towards Nimes.

After the Pyrenees, I passed into the broadcasting garbage heap of Spain. "Buy!" screamed the radio in garish (and incredibly low fidelity) Spanish. "Now," screamed the accents of Barcelona, Alicante, Granada. "Cheap, buy now!" screamed the voice as I raced towards the pillars of Hercules, ran to the strange territory of North Africa.

I want to tell you something funny about radio, about one of the evenings of my life which is frozen in my head. Like Wordsworth, I close my eyes and see it all painted on the inside of my eyelid.

Five of us were gathered in Casablanca for the trip across the Atlantic, back to the bang-and-whimper we call America. Our ship, or purported ship, was operated by the Yugoslav government as a cheap freighter, with just enough room for twelve passengers. It was mid-winter, so there were but five of us. The liner had sunk somewhere in the maw of the Mediterranean. "It will be tarde two weeks," said the ticket agent.

The five of us took the car into the mysteries of North Africa. I have no end of stories to tell you about Meknes and Fez, about Marrakech and Tangier. Stories about French colonial culture beginning to decay (but not quite), and Arabic culture beginning to reaffirm itself (but not quite), and the dark twisty streets of the *medinas*. I have a great deal to tell you...

...but not yet. What I really want to tell you about is one afternoon. When I got the feeling for communication, and the culture gap, and history, and the gape of time, and all that sort of stuff. We are heading west from Fez. The five of us—suspended at eighty miles an hour on that straight and true asphalt highway. The French *colons* had left behind a great number of dams and power stations and highways. Highways only beginning to fall apart under the forces of self-determination.

The sun is setting. I have the sun visor down. The sun—fat and red—stains the lower part of my face, just below the eyes. On both sides of the road, the culture of Morocco barely moves. Shepherds in their djellabahs stand at the side of the road. One of them may be holding a rope to a burro. One of them may be standing beside a hunched and veiled wife. She has a giant load on her back, a load of sticks and leaves. No one moves. The sun stains them and the land crimson, and they don't move.

It is Ramadan. The ninth month of the Islamic calendar. Fasting, abstinence, appeals to the Whomever it is who broods over Islam. From the radio pours the voice of a man chanting: a single voice moving slowly, randomly, through a strange musical scale, singing to some strange eye watching the crimson land, the crimson peasants, a crimson car racing along a black strip of magic over an ancient land.

The voice was to continue for nine days. But what I remember most is the five of us, whirring past the moveless figures, hooded figures brooding at the roadside with the sad song of a man and his god filling our ears. Suspended in time, suspended from any land, any culture: we, along with the melancholy culture of that melancholy country. Streaking west with the moveless sound of prayers pouring out of the single red eye, trying to tell us something we can only barely comprehend, only barely know.
—from *The Myrkin Papers*

TRUE STORIES

Secretary's notes on the meeting of the Tomentose Broadcasting Corporation and the Volunteers Association of KTAO, held March 2, 1971, on the back porch of the home of Ms. C. McGowan:

The meeting was called to order at 3:20 PM. The President explained his plan to turn 25% of the ownership of KTAO over to the volunteers of the station. He suggested that the 35 volunteers elect a trusteeship of five or seven members who would be responsible for the stock in The Tomentose Broadcasting Corporation. A Mr. Deignan asked why the stock was being given to the volunteers. The President pointed out that the stock (at this point) was worth very little. With 150,000 shares outstanding, and with a corporate debt of some $200,000, each share of stock was worth approximately *minus* $1.25. But he stated that it was a good way to get some sort of actual cooperative ownership of the station, to reward the volunteers for the time they spent operating KTAO, and to bribe the volunteers to some sort of an additional commitment. The President stated that he was hardly a philanthropist, and that he would retain majority control—at least for a while.

Ms. McGowan raised her hand and said that something—perhaps a cat or a rat—had crawled underneath her house and died. She said that living with the smell of it was 'uninteresting.' She nominated a committee of three (Tracey McGowan, Lisa McGowan, and Jacob McGowan) to crawl under the house. She offered the committee 50c each to pursue the matter, and offered an additional bonus of $1 for the first to discover and dispose of the offending object.

The President said that he did not think the matter of foreign objects—dead or alive—underneath the house were of concern to the general meeting, did not see them on the agenda, but (now that the matter had been brought up, and the wind had shifted) perhaps it was worthwhile to adjourn the meeting to the rearmost yard. And so, at 3:34 PM, the meeting was so moved.

Upon re-adjournment, a Mr. McAllorum was recognized, and stated that if stock in the Tomentose Broadcasting Corporation were to be sold for 10c a share, then he would take three. The President ruled Mr. McAllorum out of order. The President then called for the formation of a Volunteer Association of KTAO which would: 1) Automatically incorporate all the present 35 volunteers into the organization; 2) Vote in a five man Board of Trustees; 3) Set terms for future membership in the Association; and 4) Get him a beer.

Mr. Deignan asked why the President of the Corporation was "Trying to get rid of" this stock, and the President repeated that he was trying to give the volunteers some taste of actual ownership. The phrase he used was "I am trying to make Little Capitalists out of you." This information was met with some good-natured jeers and cries of "Right on!" The President went on to inform them that "freedom isn't free" and that they would have to pony up $2,500 within the year to pay back the previous owner of the stock for his investment. Mr. Deignan said that to him it looked like a pig-in-a-poke, that the President was trying to farm out on the unsuspecting volunteers shares of some watered-down, overcapitalized corporation which—when it went bankrupt—would pull down the finances of all 35. Several others at the meeting nodded assent, and muttered that The President was "losing his rocks" or "involved in some hanky-panky."

The President then got red in the face and said that as far as he knew it was a magnificent gesture, and that he didn't know of any other goddamn broadcaster in the goddamn country who would be caught dead turning over 37,500 shares of capital stock in his corporation to volunteers for $2,500—despite the fact that it was worth only minus $1.25 a share. Finally the shouting was aborted when Mr. Henley and Mr. Wade shoved a full bottle of beer into the President's mouth. At this point, Mr. Clark said that he would be delighted to be 1/35 owner of 25% of a radio station, and that he didn't *care* what the stock was worth.

Mr. McAllorum then stated that he was bored, and was going home, and that he had (besides) run out of what he scenically described as the leavings of a horse to drink. The President said, "You haven't voted yet." "I know who it's going to be anyway," said McAllorum, running into the fence. "Who?" asked the President. "Cese, Bill Wade, Jeff Smith, and the banker over there in his New Jersey surf shirt." "That's only four," said the President. "All right," said McAllorum. "I vote for Cese." Mr. George disentangled Mr. McAllorum from the fence. "You've voted for her twice," said the President: "You want two of her. You want her twice." "All right, all right," said McAllorum: "Put Goodge on. He has to come in the station all the time to clean it up. He has to clean up all the crap at KTAO. He might as well be on the Board." Mr. McAllorum disappeared around, or under, or into the house, and the vote was unanimous.

The discussion then centered on the conditions for membership in the Association. Mr. DeMoss said a sub-committee should be appointed to study the matter. He was shouted down. Mr. Blind said that *he* didn't think that there should be *any* conditions for membership, that he couldn't stand that sort of thing. But he did say that the members should put a dollar into what he called a "slush fund" every month. When pressed to define "slush fund," Mr. Blind turned on the hose to get a drink of water, and watered several members of the new Association. There was much merriment at this point in the proceedings.

The President thought this a good time to go around the corner of the house to relieve himself, and ran into a Mr. Leskovsky passed out under an apple tree. The President said that he thought that Mr. Leskovsky might find the meeting more interesting to observe than *him*. When he returned to the meeting, Mr. Blind said that he would certainly like to know what 37,500 shares of stock in KTAO or 37,500 shares of *anything* for that matter looked like. The President said that it was a cinch that they would never know, since all the stock was hypothecated to the mortgage holder of the station. This statement evoked several jeers, and cries of 'Burn the Capitalist Pig.' However, the anger of the Association was tempered by Mr. Deignan, who got up to say that, as far as he could see, the whole thing was a pig-in-the-poke, and that &ct &ct &ct. Mr. Blind said that everyone should be invited to join the Association, including the printer, and the telephone company, and the mailman. The President said that he didn't think that the mailman would be at all interested, since all that KTAO was to him was trouble, what with all them Army Transcriptions, and returned program guides. Mr. Wade said that his spondylitis was growing worse in the cold, and as the sun was getting low, and the beer supply even lower, the meeting was adjourned at 5:53 PM.

—KTAO Program Guide #59

THE TERRIBLE WASTE OF EDUCATIONAL STATIONS

When the FCC set up the FM band, they wisely filled the gap of AM: that is, they set aside 20% of the band for "educational" non-commercial stations.

It was a matter of policy with the first FM band [41-50 mHz] and with the later band set up in 1945, the present 88-108 mHz.

When I listen to the educational part in most cities, I want to throw up: so many of the school run stations are slavishly imitating the commercial stations, down to the greasy voiced announcers and running PSAs as if they were commercials.

Jesus! Who's going to tell them that there is a monster world of ideas out there waiting to be plucked—and it don't mean a re-hash of the local rocker.

Once I was invited by the Georgetown University station in Washington D.C. for a live interview. I spent a good portion of the hour on the air bitching about the dumb repetitive commercial-type crud I had heard on WGTB the days preceding my interview.

As I was leaving, the manager asked if I would write up a list of 100 ways that they could improve their station. I told him that the best I could do was ten or so practical points—which I sat down that night and wrote down and sent off.

1) Stop trying to imitate WHFS.

2) Play some Iranian music from time to time: and as well, some Javanese Gamelan music, music from Japan, and African music. You are not a commercial station, and do not have to compete with them. Your revenue from listener subscriptions will go up drastically if you provide a new service to the community (KDNA operated on a budget of $35,000 a its height: all listener contributions). I have appended a list of the great record companies below.

3) Go on the air and ask your listeners to start coming by to do programs. It is not enough that you do this, you must have someone to work with the people who do arrive (you will get a dozen or so applications a week) and show them how they can participate in the station.

4) Start doing some shit-kicking programming. Go in the streets, in the prisons, in the old peoples' homes, in the juvenile centers, in the drug clinics, and in the police stations.

5) Do two or three hearings a week: not the popular ones being done by NPR—but the ones being ignored by the Washington *Post* and the other radio and TV networks. Nothing is more informative (if not dramatic) than congressional hearings—and more worthy of airing. The popular ones will be publicized freely: the unknown ones will earn you the gratitude of some congressmen who are being ignored by the Watergate hungry. Or, to phrase it another way, "There are a dozen scandals the size of Watergate in Washington D.C.; some are bseing discussed publically in hearings—most are being ignored by the press."

6) Start rotating management. If you must have a pyramidal power structure, have a individual running the show each month. This will bring fresh air to the drab day-to-day decisions. If you select to be run by a collective, they must have open-ended, open-to-all meetings every week, and the station and staff should stick to the decisions made at that meeting (if there are any).

7) Running contests on the supposed date of the impeachment of the president is fun and funny; however, if you want to be taken seriously by the community and *most especially potential subscribers*, you should get in trouble over far more exciting fare, not as juveniles. Sam Buffone's suggestion that you do documentaries on such hot topics as mixed adoptions is valid. One 25 minute interview on that topic is worth a thousand Cat Stevens albums.

8) As I said in point #1 — the main failing of WGTB right now is that you think commercial radio. This makes the station the outhouse of 50 or so doubtful egos. If you abandon what we call "The Jerk Off" concept of radio, and get into true community, involved, outrageous radio — your egos will be gratified a dozen times over by the important and good and loving people of Washington D.C. community who will start paying attention to you, writing you, and best of all, start bringing you cakes and cookies because of what you are doing for their minds.

9) One thing that you need is for the five or six people who *really* run WGTB to have a chance to view alternative radio at its best. This means that the manager definitely and at least one of the news collective should ask permission to go and stay for a month or two at one of the viable community stations: namely—

KUSP
Santa Cruz, Calif.

KBOO
Portland, Oregon

KPOO
Poor Peoples Radio
San Francisco, Calif.

WRFG
Atlanta, Georgia

All of these stations are poor radio stations, and cannot support you. But you should make some provision to live in their community for a while, to listen to them, and to be with them every day in some capacity or another. You could volunteer to go there on some specific project: all community stations need a viable record librarian, or someone to clean up, or someone to volunteer to tape outside events every day. As long as they are led to understand that you will not be a psychic and economic drain on the station, they will accept your coming to educate yourselves at their operation.

10) Stop bemoaning the lack of money. You do have a lopsided economic situation: one person getting $15,000 a year, with less than that being budgeted for the remaining operation. Until the station manager volunteers to share his loot with the rest of the operation, you will be stuck with an untenable situation; however, I know of a station in the Middle West that runs shit-kicking programming with exactly the same set of circumstances. One of you should volunteer to go there for a month. That is:

WEFT
Champaign/Urbana, Ill.

Meanwhile, you have far more money to play with than KUSP has had in three years on the air (average annual budget: $6000.) And they were and are doing twice three times six times the shitkicking programming of the rest of us.

11) The worst things you have to fear are not the FCC, nor the school administration, nor some vigilante committee outside of the University. The worst fears you have are in yourselves. You are part of the fucked-up ego selves who go into radio in the first place (I include myself in this army of twisted psyches). For some reason, you tell everyone, those who want to help change you, that you cannot do certain things because "we might get in trouble with the FCC" or "the university administration is already down on us." These are the oldest of old wheeze excuses. You are your enemy, not them. If you are doing true different important ethnic music programming, and mean insightful interviews, forgotten hearings and documentaries that were honest, not juvenile: everyone would come to your protection at time of attack, including the FCC. If you did honorable manful radio, instead of jack-off dj stuff, the University would *have* to defend you.

Your greatest resource is your audience who would love to help you. Your greatest enemy is apathy and imitation. Your best friend is yourself trying to be different and original. You have something denied to 99.99% of the American people: a frequency, the power, and 1,500,000 ears waiting for you to feed them the sounds of the universe. Give them a chance. You may merely survive running a drab imitation of a rock station, but everyone in Washington stands to gain if you start the station on the road to being the voice of the dispossessed and the curious and the lonely.

SEX AND BROADCASTING

We had the first eighty-eight issues of *The Weekly Drag* bound in gold-and-leather, and they now sit on our bedtable. It is a good solid feeling: for the air, the aether is so *un*-solid, so insubstantial. The transmission of RF Waves leaves no traces—except some half-suspected ones on the minds of the purported 'listeners.' Radio is thus a ghost. Last week's High Art mix of music is dispelled into the winds: unless there is some chance ray of modulation soaring to some unnamed, unbelieved solar system 173 light-years from here.

These confection currents of sound are gone, and that witting, charming interview from yesterday—unfortunately the Revox wasn't working, so we couldn't make it *permanent*—is lost forever. The only record we have of the programs of October 14, 1969, is the official FCC log (which some boob spilled coffee on, so our legally transcribed memories have begun to run and fade)—and the listing in the program guide from back them.

One thing we realize by binding and setting the words from then is that we were then obsessed with money. We talked about it so much that the first 40 guides are nothing but a tribute to an ill-managed, half-conceived idea of free-but-not-free radio. We were so obsessed with the potential loss of this outlet in this area that we did nothing but concentrate on (what surely must be) the most dimwitted of men's lives aspect: that is, the collection and disbursement of something called *dollars*.

Sex-and-dollars. Dollars-and-sex. Honest: it is as if it were the most important part of living. It is as if there were a gland somewhere secreted in men's bodies, called The Financial Gland, which creates these vital bodily fluids which drive us to unnatural acts, to ridiculous extremes of thought and movement. It is as if this hidden, unknown gland comes up with vile sicknesses—like $ickle-cell anemia—which force people to become listless, droopy, a collection of singular dullards. And I guess that is what got us in the first year of stewardship of KTAO: and the reading of the guides from back then makes for a powerful soporific.

For that reason, we vowed that we would not beleaguer readers of this guide with long boring lectures on the financial problems of KTAO ever again. And we won't. With this single exception:

send

UN-COLLECTIVE RADIO

Once I had a chance to meet some members of a local (Bay Area) radio station 'collective.' A group of 5 or 10 DJs who had been dispossessed of a job after they had liberated a radio station, locked themselves in the studio, and spent an hour or so telling a fascinated radio audience of their beef with the management—a truly exciting news-event program which lasted until the frantic owner could make his way to the transmitter and unplug this epoch-making 'actuality' show.

There was good reason for the extra-curricular activities by the staff: the owner had left them out of important decisions, had expected them to be nothing more than slaves to his nice, profitable empire. Most broadcast executives—in the business of communications—are terrible at communicating.

But a broadcast collective without a broadcast station is like a church without a collection plate—and in meeting with some of these disenfranchised people, I was asked what they could do towards getting back on the air. I suggested that they get the hell out of the Bay Area and go someplace where they could set up a radio station where it could do some good.

Despite their anger with one station, there are at least half-a-dozen superb broadcast outlets in a market of over 70 radio stations. Meanwhile, there are some major population areas in this country which have no viable alternative outlet. Populations of one or two million people which have no meaningful radio station for the young and the frustrated.

"You're staying where it's comfortable," I said. "If you really care, why in hell don't you go somewhere where the really *need* alive radio. Dallas, for God's sake, or Atlanta. Kansas City. Birmingham. Lord knows that Jackson, Mississippi needs you. Or if you really have nuts, you could start an underground radio station in the Mississippi Delta area, right in Senator Eastland's backyard. There's a place that needs you desperately. Talk about alternative radio. Talk about Poor People's Radio. You'd be hated, and reviled, and bombed; but you'd also be loved and treasured and needed. Because there are some people in the Mississippi Delta that do *so* need a voice. And you're the people to give it to them.

"Talk about real radio," I said. "That would give you a feel for the potency of the medium. I mean, you wouldn't be blabbing into the free air of the Bay Area—where tough rock and revolutionaries and long-hairs are a dime-a-dozen. No: in Mississippi Delta or Birmingham or even Dallas you'd be something special. And you'd be doing something real besides playing with yourselves."

What a blabby night I made of it. I would guess it was too much smoke and bad wine made me ramble on and on. I painted one of those lurid, dramatic pictures which, nevertheless, was true. And I told them the exact how-to-do-it. I told them how to get money for it; I told them about the army of people we knew in engineering and law and other existing radio stations who could share with them the knowledge of how to start a real radio station in a real city. I told them about frequency searches, and how to find the old, used equipment, and where to put their antenna. I gave them do-it-yourself radio lecture #1, complete with jokes, dramatic memories, and stories about busts and bombings...

...Busts and bombings. Maybe that was what did it. Maybe that was why those nice soft revolutionary members of that nice soft revolutionary collective got sort of disinterested; lost all desire to leave the nice soft revolutionary womb of the Bay Area. Go into a true underprivileged broadcast area, start a real station, for real people, with real disenfranchisements. Real people, with real problems—about voting, and eating, and being free. Real people, with real needs—for a real alternative voice.

Maybe I shouldn't have brought up the word 'bomb.' Or maybe it was the word 'Eastland.' I don't know. Anyway, something scared these radio revolutionaries and their brave collective. They didn't take to the suggestion too hot. They mumbled something about the Bay Area still needing 'to get it together.' There were still unbelievable tasks to be faced here—tasks which were hard to explain.

I never saw them after that. They never came down to KTAO so I could show them the six or seven tricks necessary for setting up a radio station for a few thousand dollars at best. Maybe it was those words I used. Or maybe it was the way I waggled my head around, got that light in my eyes when I was talking about real radio alternatives, that mad-light of the radio *crazies*. Maybe I turned them off with that.

Hell, I don't know. I never know what turns them real revolutionaries off when they got around us impractical radio *nuts*.

—KTAO PROGRAM GUIDE #54

A FEW TART WORDS ABOUT CLASSICAL MUSIC FANS, AND THEIR RELATION TO THE ECONOMICS OF RADIO BROADCASTING

'Classical Music Buffs,' they are called. Or 'Aficionados.' 'Junkies' would probably be a better word for it. And in that, we don't mean the improbable compliment of *habit*—for that would, at least, imply committment. No: Junkies—as drawn from the concept of 'Trash' or 'Waste.'

For classical fans want no more titillation of their thoughts and emotions than the constant repetition of the constant Warhorses of Classical music. Give them Brahms Third Symphony yet again. How about a Wagner Overture for the 100th time. Or best of all—they could ask no greater pleasure than to hear, for the thousandth time, the tried-and-true melodies of Tchaikovsky's First Piano Concerto. Dum-dee-dum: remember it? How could we forget? It's so familiar.

Classical Muzak. Muzak for the snobs. The familiar strains—so that one is not strained too much. Or hurt: We don't want to hurt our ears. With that difficult Mayazumi. That bizarre Penderecki. Or even the demanding cantatas of Bach. No: we will swath our corpses in furs of dead animals, and go again to the concert hall where we can hear and applaud loudly the lovely strains of *Capriccio Italien,* or the gentle "Pines of Rome." Give us Mendelssohn "Scotch," but for god's sake don't give us any of his chamber music.

And broadcasting has been hurt again and again by these effete snobs. Because they do not truly care for music—because they demand only the comfort of the cliche—classical music stations nail themselves to the same sing-along melodies. The broadcasters, always quick to leech onto the necessary fads of the land, make the Muzak circle complete. KKHI in San Francisco; WGMS in Washington; WQXR in New York. Beautiful melodies. Another symphony of Beethoven, please: and don't bother us with those strange Telefunken recordings of Baroque and Renaissance music.

And so the broadcasters continue to grind out the cornucopia of myopic bougeoise (straining for their fantasy of upper class taste). And those few stations which strive for the distinctive and the different manage to starve to death.

The next sign of the death of the Middle Class through their mistaken perceptions of art and culture will be the actual superimposition of these deadbeats of classical music over a Muzak system. Where at the Safeway, you and I posed over the artichokes, squeezing the onions, will be regaled with the lite-and-brite melodies of a Brahms serenade.

The classical music collection at KTAO is extensive and stunning. The first shelf, about seven or eight hundred records—is given over to Medieval, Baroque, and Renaissance music. The second shelf, another 700 or 800 records, is given to Bach and early classical. The slimmest section to the whole collection is Romantic Concertos, and Romantic Symphonic (we keep a few of the bizarre backwash selections which have not been squeezed of their emotion by the Classical Junkies.)

I guess we pay for it. Unlike the symphonic Muzak broadcast operations, most of what we play—especially Saturday and Sunday mornings—is fairly rare and delicate. And I notice that we aren't getting any passionate commendations from the San Jose or San Francisco Symphonic Ladies Bridge Club praising us for our efforts to give people a taste of the enormous and brilliant world literature of recorded classical music. If any of them listen for more than a moment, they probably quiver out of fear of hearing a work of art which they cannot identify, which they can't immediately identify—saying, "Oh, yes. That's Schumann, isn't it? His concerto for piano, right? I do love it so."

SEX & BROADCASTING

Romantic symphonic music was a beast unleashed by an innocent and unknowing Beethoven—and carried to extremes of onanistic self-destruction by a trying boor named Wagner.* The classical junkies are hooked on a *pretense* of art—one of the slickest campaigns of public relations in the history of mankind. And economics, and fear of the dirt, and of tough reality—make them cling to this shadow of art.

Which—as I have said before—might just well be appropriate and right. As television was invented to keep the poor out of the streets and off the picket lines; so the symphonies of Tchaikovsky may have been invented to keep middle-class turnips safely in the concert halls, and off the backs of the rest of us who care for some guts and meaning to life.

—KTAO GUIDE #52

*So proved by his *excruciating* Symphony in C.

Bankrolled. The larger than life-size check for $900,000 will help start a number of community noncommercial FM stations around the country. Group owner Cecil Heftel (r) presented it to Jeremy Lansman (l) and Lorenzo W. Milam in payment for KDNA(FM) (now KEZK) St. Louis. Messrs. Milam and Lansman, whose passion is noncommercial radio, will also devote to that cause much of the $200,000 they will receive from Mr. Heftel under a consultancy agreement.

Broadcasting Sep 3 1973

BILL DRAKE IS A MENTAL APID

The California Broadcasters Association is another on of those mindless organizations which spends most of its finances on trying to impede any sort of a rational community oriented broadcasting. The way they do this is by raising hell everytime the FCC takes any sort of a step towards making radio and television stations more responsive to those they are *supposed* to be serving.

The California Broadcasters are meeting this Thursday and Friday at the Del Monte Hyatt House (in Monterey) to discuss, for a change, how to increase profits in broadcasting. From, like the 15% area to, maybe, 20 or 25%.

KTAO and a couple of other *real* radio stations are sending a group down to picket and hand out leaflets, scream and make alot of streetcorner speeches and, hopefully, convince these drips that they have everything to lose and little to expect in the way of survival if they don't bring their business practices out of the 18th Century Pre-Industrial consciousness.

We are making placards which say:

"California Broadcasters Have Stolen Your Air"

"This Conference Is Solely For Increasing Profits...Not Good Programming"

"The Aether is Another Exploited resource"

"Broadcasters Make A 17% Annual Profit on Your Air"

"Bring Back Buffalo Bob"

"California Broadcasters Are Sexually Insufficient"

"California Broadcasters Double Bill" (This is an ingroup trade joke)

"California Broadcasters Eat The Bird"

And of course we will try to sneak into the meeting and ask some embarrassing questions. One of our KTAO people has offered to go on the stage and take off his pants which is, I guess, expected at these sort of confrontations—but we have vetoed that, not so much that it is crude, but that pants on or pants off, he just ain't that much.

Behind all our rabble-rousing is a serious purpose: which is to try to get broadcasters to stop isolating themselves from their communities behind a wall of bucks. To this end, we have written up a long letter—which discusses some new broadcast law:

Fellow Broadcasters:

We certainly hope that you have read the full text of *Citizens Communications Center, et al* v.*FCC.* If not, you are missing something. Or will be. Like your FCC granted license to coin gold.

You broadcasters have done supremely well by your government granted frequencies.

According to *Broadcasting Magazine,* the average profit for American television stations—both UHF and VHF—is over 17%. That's per year. That's ignoring those fantastic trade-outs that you manage to hide from everybody. It doesn't include those ridiculous IRS depreciation laws for broadcast equipment. It doesn't take cognisance of the giant salaries you pay yourselves—and it most certainly does not make mention of the fact that the average price of a broadcast license is up 500% in the last ten years. Truly, you broadcasters, whether AM, or FM, or TV—have a license to squeeze millions of dollars from your frequencies: and you have no idea how much anger is generated through the realization of this consistent, unmitigated rape of a vital natural resource.

To some poor Black, or Chicano, or Indian, or even to some dispossessed concerned citizen of the political left or right, this donation of frequencies for the sole purpose of making a million dollars for you is outrageous, and ominous, and a cause for violence. They cannot conceive of a system where you have to learn the tricks necessary to own a frequency—which they can never seem to get access to.

You came to this meeting in order to study ways of making more bucks. You announce, by your attendance here, that you support the California Broadcaster's Association, and the National Association of Broadcasters. And you ignore the fact that more and more people are realizing that these groups stand guard over your fantastic profits—and move heaven and earth (and legislators) to prevent American radio and television from growing up, from taking its responsibilities seriously, from opening its doors to the poor, and the anguished, and the dispossessed.

You are sowing the seeds of a growing rage against you and all broadcasters. And those of us in alternative styles of radio and television suggest that you have little time before someone is going to do something to keep you from milking the golden goose forever. Or are familiar with the concept called "strike" application?

You have chosen not to have any voices of dissent at this meeting. You have not invited the growing number of experts in communications who deplore your trashy programming, your foul techniques for coining even more money. You have no one here to talk to you about your lotteries, and your double billing, and your eighteen spots an hour. You have no one who will tell you that all those plaques on your walls from the Campfire Girls and the American Red Cross and the U.S. Army (for PSAs at 2 a.m.) mean nothing to the young and the poor and the intelligent who are frantically looking for a place to be heard.

We can suggest ways that your programming can grow to be more community oriented. We can teach you how to get the young and the energetic to think of your frequency as a free flow of ideas, and not just a gusher of bucks for you.

We will do this for free. Not because we don't need money; but we do this in the firm belief that if the dispossessed stay dispossessed much longer, we will *all* be offed, to use a phrase from the underground.

—KTAO GUIDE #74

THE LAST MEMORIAL DAY OF OUR DISCONTENT IN ESTES PARK

Before we got there, they didn't tell us anything about the breathing problem—they didn't suggest that we bring along our portable chest respirators. They didn't tell us about the times where the breath would be coming in short ragged shallow gasps, and you would know what it was like to be well along towards the years of chilblains and phlebitus.

Estes Park, Colorado is stuck up there in the groin of the Rockies at 7,500 feet. You drink a Gin and Tonic, and your blood begins to boil, your brain begins to bubble. They used to advertise in the European weeklies that a trip to Colorado was a cure for obesity. You wouldn't have to drink a dozen bumpers of beer—maybe two or three would make you high without getting all bilious and groggy.

We have been flown in for a conference. Ostensibly it is a 'clinic.' When they get people together to show them how to coin more money, they always borrow the safe terms from medicine. "The Bob Hamilton Radio Clinic." They brought me in to be their conscience.

We are set up in the Stanley Hotel. The Stanley Hotel was built by the man of Steamer Fame back in 1907. Long sagging balconies. White, all white. Windows as big as all out doors.

When we come in Friday night, they are serving *rose* in the music room. Chilled *rose*. From Portugal. The tables are covered with white linen cloth, there is a wine steward with bow tie, black jacket and pants, and a size 44 waist. A chamber group from the University of Colorado is playing Shostakovitch Quartet #8, Opus 110. Dear God. Out the window is Longs Peak, slightly less than three miles stuck in the night-bluefruit sky, and Dmitri Shostakovitch in our ears and good *rose* on our lips.

That was probably the intellectual high point of the conference. Bob Hamilton of the top-pop-tip-sheet knows which side of the butter all the bread is on. The twenty or thirty experts give a leavening, an intellectual *elan* to the whole—but the primary purpose of three hundred top DJs and programmers, two hundred 'promoters,' and a hundred assorted time buyers, management types, rack men, distributors and heavy sale types, the primary business of these people is business. They are here at 7,500 feet to exchange gossip in broadcasting (as full of in-gossip as any field), fish around for more and better job offers, speculate on what is 'breaking,' show off their salaries and get plastered away from the boss. Who graciously pays the $500-$800 for a weekend of such tomfoolery.

For those of us who love the industry—no matter how perverted—such a conference is a gas. For those of us who are hired as paid conscience to this mountaintop full of hustlers, there's especial pleasure in picking at the industry weaknesses; sticking our pins in all those sanctimonious pimples that are the 40 year defenses of a business wallowing in money.

The longest running confrontation of the weekend (for me) is with the management of the Capcities Buffalo station. We are like running dogs, running into each other two or three times a day, snarling at each other over some bitch in heat. The bitch in this case is named Gross Profit.

My mind is clouded by good *rose* and a poor oxygen content in the air—but I think I got one of them to admit that the station grosses (takes in, in cash, every year) some $1.5 million. That's $1,500,000. Cash. I do believe I got them to admit (and it can't be far from wrong: the major market average expense for a broadcast operation is about $25,000) that their net profitability

was about $1.2 million. Or, if you prefer:
INCOME: $1,500,000/year
OUTGO: $300,000/year

NET: $1,200,000/year

Now do you understand love why every scallawag and his mother wants to go into the radio station business?

One of my friends in a community law firm in Washington D.C. told me that a friend, only recently at the FCC, smuggled out Xerox copies of five selected Form 324. For those of you outside the money-coining biz, the Form 324 is the secret form, required by the Federal Communications Commission of every broadcast and television station, showing income, outgo, and a breakdown of expenses. It is a four page no-accounting-nonsense form which is denied to the general public because of 'competitive disadvantage' or some such myth.

My friend saw five selected Form 324, and they bugged his eyes out. One was for a VHF television station in New York City—WNBC or WCBS. The outlet's gross profitability (on an income of $40,000,000/year) was somewhere between $22 and $24 million. Profit. And that was not including all the bizarre profit available for just sitting on a government-granted license that goes up two or three million dollars a year in simple asset value. O man, what has thou done to my beautiful Aether?

We had a great deal of leisure for such gossip. And, as the conscience of the Convention, we had time to indulge in this dialogue at least a dozen dozen times:
—*"How much does your station gross every year?"*
—*"O, I don't know. I'm just in programming."*
—*"Can you guess? How much money does your station haul in every year?"*
—*"Listen man: I tell you, I'm just the program director. I have no idea how much the station makes."*
—*"How much do you get paid?"*
—*"$325 a week."*
—*"After taxes and all the deductions are taken out?"*
—*"Maybe $250 a week."*
—*"Jesus! Do you get any stock options? Or profit sharing?"*
—*"No. Sometimes they let me take some trade-outs."*
—*"Listen. When you go back to Wichita [or Sioux City, or Miami, or Des Moines] why don't you look at the log? Figure out the number of spots they are carrying each day. Eliminate the trade-outs and the spots for the owner's uncle. Look at the rate card. Cut the top spot amount cost in half. Add it up and multiply by 365. Ignore the poor-mouthing of the station manager: that's what he is paid to do—to protect the owner and all his gross. Figure out then all the expenses, deduct income from outgo, and figure out what you are getting from being a dj on the air and what proportion of the profitability is going to you."*

(At one of these confrontations, I watched the Program Director of a medium market AM in Pennsylvania add up these figures, compare them with his $300 a month and 50 hours a week, and prepare his resignation all at once.)

Besides sowing the seeds of discontent, we drank, talked, gossiped, plotted, schemed, ogled the celebrities, ate, and—at night—lay in bed at the skirts of Mummy Mountain where the winds even at the edge of summer came down hard on the old Stanley which had no steam heater in the basement to keep us warm. The wind jerked at the windows, and they were old enough and loose enough to rattle and bang us to sleep. On Saturday night the clouds leaned out of the mountains and blew salt-fine summer handouts of snow at the building and at our faces. The mystery of the white-blue-black mountains rattled my mind, made me remember some times in the wilderness of Canada where I was *not too sure* that mountains and trees and water really wanted me there, looking into them.

The last day there I told Bob Hamilton that he certainly had taste if he had no sense. Bob Hamilton is designed generally on the lines of one of those anarchists drawn by R. Crumb: hair shooting out all around his face with thick glasses and a lopey-type walk. I told him that after the Chamber group got through with their Mozart and Bach and Shostakovitch, they went out to ride on the Ferris Wheel that he had ordered for the weekend from the Acme Circus Supply Co., in Denver. I told him that I was disappointed that they didn't take their instruments with them, for the thought of Schubert's *Death and the Maiden* going round and round, echoing off of Longs Peak and into my soul would have been the perfect rounding off of a perfect 1907 weekend in Estes Park, Colorado.

But what I really wanted to tell him about was the elevator girl. She was dressed in a peasant frock and she was probably the first person that had eyes that they describe as 'limpid pools.' I was very captivated—like most of the 75 people who served us on the 1907 weekend in the sky, she seemed so humble and demure that I had her pegged right off the farm.

"Seventeen years old," I tell Bob Hamilton: "Just off the farm with cow shit between her toes and the scent of wheat in her hair. This is her first summer doing anything," I tell Bob Hamilton, "besides gathering warm chicken eggs and slopping the pigs, and you bring her into this hotel with all these hustlers. Aren't you ashamed," I ask Bob Hamilton. 'Cause Saturday night I see this sweet young creature of the rosy cheeks and innocent way being led out of the bar (where no doubt they plyed her with copious amounts of booze and untold mind-expanders). I see this sweet young thing with straw in her straw-colored hair being led weaving childeyes clouded o'er with the fog of excessive booze—led off in the arms of some piddle-meister from Detroit, some dj from WROK who has made an intense study of groupies for the last four years he has been mouthing the push-&-buy as Doctor Feelgood of the Middle West's high-rolling hightime rock-rage top breaking AM outtasite love generation push-me-pull-you.

"How could you, Bob Hamilton?" I say: "how could you expose this child-of-the-Colorado-wilderness to the gross ministrations of the top body hustlers of all times? How could you permit this angel of the up-and-down to be subject to the two-humped beast provided by some acne-stained sister-crawling rock-her-innocent-roll out of all knowledge past bearing. You, Bob Hamilton," I say to fuzzy-faced hair exploding out-the-corona Bob Hamilton: "You have lost this sweet child of the farm her first and summertime employment in the gentle arms of the Stanley Hotel." For indeed, the management took one gander at Miss Elevator barely able to stand up much less go down locked in the arms of the Detroit Bomber—and they gave her her walking papers the very next day. "How could you, you brute?" I said to Bob Hamilton.

Well, we took her out to dinner. Bob wanted to find out the facts, and if necessary get her the job back. After all, he did carry a bit of clout, having laid about $18,000 in the willing if able hands of the management of the 1907 white painted three story towering monolith of the Stanley Hotel, leaning moodily against the cold flanks of Mummy Mountain (El. 13,450). And don't you know that we found out that our sweet child-of-the-fertile-Colorado-earth was no child at all: that she was 28 not 17, had dropped acid with the best of them in the Haight, had worked a dopefilled light show in St. Louis, had been hired away by a rich but friendly businessman to his apartment-lanai on Oahu where she lived carefully petted and preened by several thousand a month and no end of attentions by her dewlapped patron.

Our child-of-the-farm had done it all except milk the cows and shovel pigshit: she had dropped, snorted, sniffed, smoked, and ingested every possible combination of late 20th Century Americana—and had drifted through it all, with wide eyes and golden honey hair and a face that bespoke true innocence of the truly knowing. Our elevator girl had had all of the ups and downs, and most certainly, the last laugh on Bob Hamilton and me, her cornball protector: a man who couldn't tell a child-of-the-fields unless she ran out and bit him on the thigh.

They paid me and Sam Buffone and Dr. J.B. Rhine and The Colorado Chamber Players—they paid us to be their conscience and we were. In all the wrong ways. Because me with my balance-sheet mentality and Sam with his *Alianza* case and the good Doctor with his ESP tests didn't see and couldn't see and wouldn't see that we were the innocents in this size 600 crowd of heavies who had learned some few years ago that they were the voice of the theatre—the teasers and pleasers of the mind of The American Dream. We were and are the children of the media and couldn't even, for christs sakes, tell the difference between an elevated acid child of the mind and a sweet-smelling flower of the fields. Only the hustlers of the aether could make that subtle distinction, there nestled at the long cold flank of once-warm Mummy Mountain, there faced with the blue-black brow of Longs Peak, sprinkled with some white sage of Colorado.

(*The Radio Times, #123*)

THE EENY-WEENIES

As sure as Boise-Cascade is out to rape the land; as sure as American Smelter & Refining is out to ruin the air; as sure as General Motors and The Federal Highway Building Program are out to destroy air *and* land: as sure as these nitwits are out to destroy our lungs and aesthetics—there are broadcasters out to cruelly misuse the spectrum; to turn a whole national resource into a hawker's box.

And, as with the despoilers of the land, air, and water—it is so often the smaller entrepreneurs who are the greediest, the greatest uglifiers. NBC and CBS and Westinghouse are enough in the public eye that their network stations tend to program a fair amount of intelligent, careful in-depth reporting, and local controversies, and occasional local investigations of art and thought.

No: it is the small time robber-baron broadcasters who do the rawest pillage on the aether. Names you may never have heard of: but still extremely powerful local or regional operators with the full complement of radio and television stations. And these stations are milked for all they are worth: the maximum possible number of hourly commercials; the minimum news and documentary reporting.

As I say, you may never hear of these boobs. But in broadcasting circles, their greed and misuse of the frequencies are legion. Who ever heard of The United Broadcasting Stations, or The Triangle Stations, or Storer Broadcasting Company, or The McLendon Stations, or Golden West Broadcasters, or Bonneville International? And yet each of these groups owns up to five VHF (Channels 2-13) television stations, and up to seven FM *and* AM stations, mostly in the 25 most populous urban areas in the country. A great concentration of media power—with a great potential for doing so much good with these fabulously powerful broadcast tools. And yet, wtihout exception, these grotesque organizations, with the blessings of the Federal Communications Commission, continue to spawn a 5th rerun of "I Love Lucy," and "McHale's Navy," and "I've Got a Secret." Those of us who care for and love broadcasting are often tempted to lie down and weep for the programs of import and beauty which are *not* carried.

"If I just had access once a week, for an hour, in prime time, on one of those stations," I think. An hour from Golden West over KTLA which rams into over 6 million homes in Southern California. Or an hour over WNUS, McLendon's barf station in Chicago. George Storer: give me an hour on your crap station WJBK in Detroit, or WSBK in Boston, or WIBG in Philadelphia. I'll show you some *real* radio, for real people, about the real nuts of life. Richard Eaton, you naughty old man: Give me a day on WJMO in Cleveland, or WOOK in Washington. I'll give your listeners a chance to come alive, to know the potency of radio reaching into people's minds, to tell them that they are alive, and have troubles; but that the world around them is alive, and full of troubles, which can and should be heard.

And you grunts at Bonneville International! You're supposed to be a church. You are supposed to be in the God business. What is all this fervor for making money? Does the Mormon Church really believe that a gross of some million dollars a month is absolutely necessary for all those Latter Day Saints to get into their specific (and racially segregated) heaven? With an hour a night on that 50 kilowatt baby KSL blasting out of the salt flats of Utah, I could give you some *real* radio. Not placebo. Not make-a-million-bucks for The Church.

No Bonneville (and in this I include all the short-haired work-Puritan ethic present day Latter Day Saints)—if you gave me an hour on that brilliant piercing monster station out of the night into the ears of the truckers in Nevada and lonely women in Fresno and poor Chicanos in Mexicali and dispossessed Blacks in Watts and dirtblighted farm laborers in Missouri: if you gave me that, just for an hour, just a brief moment in the 50,000,000 watt broadcast heaven you possess: I might just be able to save you for the 21st Century.

Yeah. You, and your $233,000,000 in assets, and $20,000,000 in cash flow. I might be able to save you (and maybe part of it) from the rage that keeps rising up: the poor and black and yellow and dirtnight rage, coming curling up out of the ghettoes, out of slums, out of shacks, out of the bleached-bone river-sheds, and the asphalt nights of the cities. A rage that turns bluehot and furious because it is the rage of the dispossessed. A rage that is sure to rise up and burn you and *me* and all the rest of us nice middleclass white hardwork godfearing twinks who believe, imagine, pray that the dark of the slums will keep the poor out-of-sight (and out of our banks and suburban homes) forever. That's what we pray and hope.

But it ain't true: Bonneville, and George Storer and Gordon B. McLendon and Gene Autry and dumpy Richard Eaton and George Koehler: all the king's horses and all the king's men and all the I-Love-Lucys will not keep this writhing black angry snake out of our bedrooms and our lives forever. They are dispossessed. They are angry. They need to hear their voices on radio, see their faces on television. They are angry. They will tear us up. If we don't give them a chance. Right now. To be heard. And seen. Before it is too late. Can you hear me?

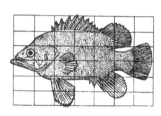 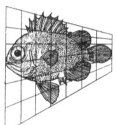 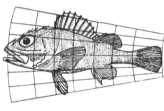

SEX & BROADCASTING

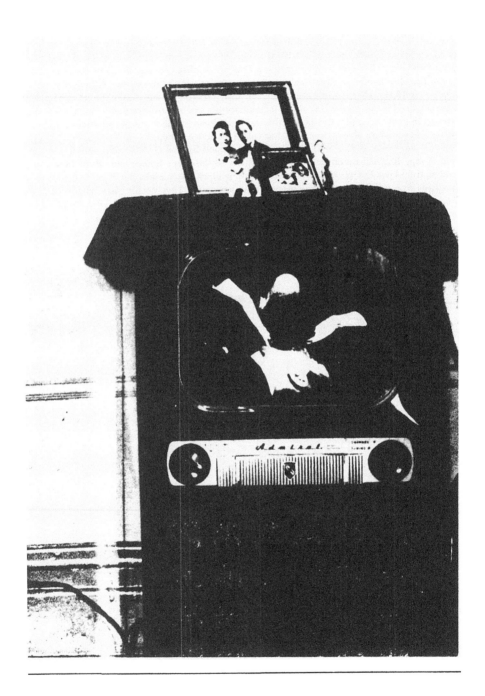

SEX & BROADCASTING

A SMALL HICCUP FOR HUMANITY

When I called up the radio-tv people at the *Chronicle* to tell them that there was a new radio station in town, they said that they couldn't fit the station into their radio log. "It's all locked up," the man at the pink section said. "We can't fit it in." "But this is a new radio station, just went on the air," I said. I was very excited since it wasn't every day I was associated with a new radio station that went on the air in San Francisco. "It's all locked up," the *Chronicle* man said. "There's no more room. We can't fit your station in." "You can't unlock it?" I said. "Nope," he said: "It's all locked up."

That's the way it's been all along for poor old KPOO. Everything all locked up. Not enough space. Not enough room. Not enough money. Never enough money. And why does San Francisco need another radio station, anyway? What with the seventy they already have: *another* one?

People like us who build radio stations find that it's a special kind of weakness. You just can't stop doing it. Once you've put one radio station on the air, you've done it all. And yet, it's damn near impossible to stop. It's like one of those lovers that you love and love, and then things go all sour and wrong, and you want to get away, because all you do for God's sakes is fight and bitch and moan at each other—tie up the mental shoelaces in endless knots—so you try to get away so you'll stop knifing each other in the groin; so you stop seeing each other at all for a week, or so, and then you think "just one more time; just a little peek, to be sure she's not been poisoned or something..." And there you are, back again, bitching and moaning at each other, and finding it impossible to get away from each other.

Building a radio station is something like that, only worse. And KPOO was doubly worse, trebly worse: the worse of worseness. I think it had something to do with its being the last possible broadcast station to be shoehorned into any city of any size in the country. As far as we could figure it, Los Angeles and Chicago and New York and Boston and Cleveland and Detroit and Washington and Philadelphia were all filled up. There were no more holes in the Aether for us to fit a station. A non-instructional, non-commercial, non-profit, non-biased apolitical radio station. For people. Outside of all them institutional sets: a radio station with the door open, with miniscule financing, with no Big Brothers to lean over its back, and blow fear down its tubes. A free radio station.

Old KPOO. The last of the just. What a mess. There must have been 200 people tangled up in the KPOO affair before it got on the air. Serious people. Concerned people. People wanting the station to do right for Christ's sakes. Wanting it to be a good station, an honest station: to redress all the wrongs of 53 years of American radio gone sad and balmy over commercials, or over 'educating.'

Two hundred people. Trying only to help, to do the right thing. Tangling with each other. Threatening to call lawyers, or the FCC, or the IRS. Resigning, writing long letters to each other; trying to unresign. Not speaking to each other. Calling up and hanging up on each other. Old KPOO—not on the air for almost three years, and all those people (board members, staff, volunteers, friends, engineers, acquaintances) tearing up each other in righteous style. Most of them good, righteous people. With ideas. Strong ideas. For a strong and good radio station. Cutting each other dead, because they believed. And wanted to be a part of the last possible radio outlet to go on the air in the mental and sociological whirlpool called the San Francisco Bay Area.

Old KPOO. Not on the air, at least not until last Wednesday. Already eating people up. Two hundred of them. Eating them up and spitting them out. A feisty grouchy cantankerous childstation in the Aether Jungle. At last on the air, the last of a long chain of anguish. A massed resignation to the fact that the Aether is precious, so precious to the media-struck people in this country. The last of the urban aetherkids. On the air. At goddamn last.

(KTAO Program Guide #102)

CREEP CRITICS

One magazine—a weekly by the name of *Broadcasting* is, as we have suggested before, edited by a Paleolithic group who routinely compare Nicholas Johnson to Attila the Hun and who regularly place the boobsie Babbitt types who run the National Association of Broadcasters slightly (but only slightly) below the Twelve Apostles. The editorial page of *Broadcasting* is written by a gaggle of clowns who feel that American radio and television should be run with the same ethics, principles, and concern for minorities that—say—the DeBeer's Family of South Africa show for their Bantu workers.

All in all, the literate understanding and criticism of broadcasting in America—so vital for the future of the media—is non-existant. And 'critics' working for the daily press (for some reason) turn stale and rancid before too many years are out—possibly because of the sheer drabness of the object of their verbiage. Terrence O'Flaherty of the *Chronicle*, Larry Laurent of the Washington *Post*, and Jack Gould of the New York *Times* may have been literate writers at some dim and dusky point in their careers, but too many Martini lunches in the mogul dens of New York and Hollywood seem to have cauterized their wit as well as their brains, although on occasion, O'Flaherty has been know to rise out the sawdust which seems to plague most older *Chronicle* writers (such as Hogan and Delaplane—the masters of the one-line, two-syllable, no-thought paragraph) and make a telling point here-and-there about the daily rape called American broadcasting.

The best and most promising—even literate—media critic dropped out after a bare year and a half. Michael Arlen, writing in *The New Yorker*, treated radio and television as a *system* of propaganda and education—and saw it better than any writer before or since.

One of the problems is the haphazard means by which newspaper editors draft their columnists. Since most are merely refugees from the alcoholics who line the sports desk, they easily become lackeys to the daily television handouts, and see their job as a means to obtain endless free three-hour lunches at the trough of television publicists.

There is no solution, of course. A newspaper owner can always take the sad and shoddy route of the monopoly kings such as our own Santa Clara County Ridder combine: what they did was to see radio and television as possible competition to their own advertising bucks, and simply not even bother to produce a radio-television page. Thus, for years, the San Jose *Mercury* made itself the laughing-stock of all those in journalism by refusing to mention San Jose's Channel 11 by name (which, given the sorry record of the station, was hardly any loss). Even now, the daily edition of the *Mercury* and *News* devotes a page to some illiterate listing of AM djs names, filled with a three-week old handout from Rick DuBrow—and that's it for local reports on the media.

The best answer is a well-thought-out monthly review written from within by reporters and staff members who can publicize the stories (and the inside business facts) which would never be reported any other way. As a model for this, we could point to the *excellent* "Chicago Journalism Review" which is put out by those reporters who work for the oligopoly city newspapers—and who are limited in their freedom to report and discuss the press itself because it is, of course, the main vehicle for criticism and reportage.

Thus, ten or twelve staff members of the big television or radio stations in San Francisco or New York or Los Angeles could undertake to publish a monthly *critique* of what their bosses are doing; and, if they were lucky, they could find themselves a local broadcaster who would avail his station for such a process of self-criticism.

And that's the paradox. For radio and television are such great instruments for criticism and awakening. But...because of their potency and richness...they come to be almost exempt from criticism. Which does nothing but hurt and confuse the literate, and vitiate the medium, and it's very potent message.

(KTAO Program Guide # 56)

SEX AND A TYPICAL DAY AT YOUR ATYPICAL NON-COMMERCIAL, LISTENER SUPPORTED RADIO STATION

At Senor Chencho's restaurant, the sweet plump darkeyed Mexican Mama Mias wrap their tacos in pieces of tissue paper that read *Here's Your Taco* and picture a barefoot, vacuous-faced Mexican *chico*, smiling under an enormous tourist-style sombrero.

"Parody," I think: "Everyone is in the business of creating parodies of themselves."

When I leave, as I get in the car, I notice a mushroom-shaped cloud rising up over Cupertino. Cars race by me, pulling at me—a thousand stone faced drivers, insensate to the present tragedy taking place in Cupertino. "They've sent off all those damn missiles," I think, "and no one has even bothered to tell us." That's the way it ends. Neither with a bang, nor with a whimper—but simply without notice.

At KTAO, the music shifts from Rev. Kelsey and his congregation, to music for shakuhachi, to Albert Ayler. It turns dark outside—but rather than being a symbol fo the Apocalypse, it is merely the sun. Going down, behind the smog: going down to rest for a while.

"Why can't this town bring in some giant windmills," I think: "Set them up on the northern (or weaker) border, put them in reverse, blow all that stink back to San Jose and Oakland, where it belongs." What a crappy neighbor San Jose has turned out to be. Plow down an orchard a day. I keep hoping that the San Andreas Fault, in its next big burp, will put Los Gatos up somewhere scenic—where we belong. Like next to Sausalito, say. "And it'll take San Jose and give it to San Bernadino, where it belongs," I think. David Clark starts playing some music for kobza. I can hear him in the control room, clapping along with it. He always did fancy kobza music.

The lady at the Villagers Realty calls up. "Say, Lorenzo," she says, "who *is* that person you have working for you? He came in here today and I just had to tell him to get out." I think she's from New York. I tilt the receiver away from my ear some. "I have *never* had to tell anyone to get out of my office before I'm in the real estate business and I don't need to tell people to get out of here but he walked right in went right back to one of the back telephones..." (I notice that all the other people in the room are listening, watching me with fascination, hearing this *voice*) "...I don't know what kind of people you have over there but who is that man I had to tell him to get out are you laughing at me Lorenzo? Is he there with you now? Are you both laughing at me?..."(the five other people in the room gather around me and this disembodied voice) "...I don't know if you care about your image in this town but I can tell you that man is going to hurt you I don't want him coming back in here ever again are you laughing at me? Is he there listening with you are you both laughing at me?" David puts on a long Rāgā—in Teen Tal. The veena twists and turns, runs up and down the room, up and down the antenna, up and down the world, up and down my spine.

I go next door to the Park Lounge with all its black lights and the silverfoil beginning to peel from behind the mirrors—mirrors in which I can see myself repeated three times over. The man sitting next to me has a large white bandage covering his left eye. I am sitting to his left. He one-eyes me circumspectly—as circumspectly as he can muster. "Say, Miss," he says: "Ain't that Danny a good fellow?" He shouts to the bartender: "Say Danny: get her a beer." I run my hand over my hair which has gotten a bit longish in the last few months. The bartender two-eyes me in black-light confusion. He draws me a beer.

"What happened?" Danny says. "It blew out on me," old One Eye says. "What's that?" Danny says, still watching me and my hairdo gravely. "Got a cornea transplant at Stanford University Hospital," he says to Danny. "Want another beer, Miss?" One Eye says to me. "No thank you," I say, *basso profundo,* slipping off the stool. "Why do I have to care for all the cornea transplants and blowouts in the world?" I wonder. I slip Danny a dollar in the back room and tell him to buy a couple for the old geezer.

Back at the station, Geoff is doing one of his imitations on the air. He is being Ray Krass. Ray Krass talks through his nose, says 'Hey-heh' a lot. Geoff talks like Ray Krass, plays *Earth Angel* by the Penguins, *Chapel of Love* with the Dixie Cups, *Convicted* by Oscar McLollie. "Well, folks," he says (he always addresses his audience as 'folks') I paid my money and they let me be a deejay. Heh-heh." All the telephone lines start up. Ray Krass always gets more telephone calls than the rest of us.

Bill Wade takes over. Plays the Dances from *Terpsichore*. Our top pop 40 of the week. The telephone rings again. It's a girl, quite young-sounding. "I'm looking at a spider," she tells me. "What sort of spider?" I ask. "It's got nine legs," she says. "No," she says, "that's impossible." She giggles. "He has nine legs and is sitting on my mirror, right at the top. His body is red. It's a red triangle." "Are you scared?" I ask her. "No," she says, "I'm not scared." She sounds very young. "Oh, *wow!*" she says. She sounds very scared. "Why the fuck do I have to take care of all the cornea cases and acid heads in the world," I think.

I think about a radio station I used to work for, a long time ago, in Florida. "It was a real teakettle," I used to tell my friends. The owner was a crook—as is usual in small town radio stations on the AM band. Half of the announcers were alcoholics, the rest were twisted in some dreadful way. Your usual American commercial dream broadcast outlet.

Just to thicken the plot, the whole had been set down in a swamp. Supposed to be good for transmission—high ground conductivity. I don't know about that, but I do know it was *very* good for the mosquitoes that lived off our rich, alcohol-laden blood. And, as well, for the skinks, 'possum, snakes, moths, bats, ghouls and vampires who habitually thrive in the southern swamp mileau. "Why in God's name am I working here?" I used to wonder as I drove another thousand-footed creature from under the toilet seat. "Why do I keep coming back to this place?"

The morning announcer was my only pal—and only that in the morning. Afternoons he began work on the vodka, and by early evening he was gruesome, ready to fight anyone. But in the morning he was a gas: a funny gentle alive person—like most alcoholics.

Johnny had worked at most of the big radio stations, back in the 30s. He used to tell me about being on WLW, the 50,000 watt AM station. He would always say that he worked there "back when radio was radio." He always said that about radio before his own particular fall.

He had done a poetry program. With appropriate organ music. From 11 to 12 every evening. The program was called *Moon River*. He would read poems of all those obscure bad poets like George Crenshaw, Mary Farmer Dowley, Elizabeth Barrett Browning. The organ would croon love songs behind his deep rich voice. *Moon River* had hundreds of thousands of dewy-eyed listeners all over the Middle West. Johnny's mail had to be delivered by the sackful.

Then Johnny fell in love with Demon Rum. He lost the job at WLW, went on to several other stations, then to some of the smaller stations, and then to the smallest stations. Like WJVB where we worked. Transmission to the swamps was the last stop in Johnny's project of total self-destruct.

Like most radio people of that time, Johnny had a schizophrenic vocabulary. In keeping with the drab Puritanism of the day, he would never ever (even when dreadfully soused) say anything naughty over the air. But then, when the microphone was off, he would 'let himself go' or 'become himself' or something. He had a very rare and raw collection of expletives, strung together in even rarer combinations. Johnny had a filthy tongue.

IV

At 11:45 each morning, at the end of his five hour program, Johnny would resuscitate *Moon River*. O the setting was different. There were bugs and snakes and a few flyblown housewives—rather than the hundreds of thousands of dream lovers from the old-days: those in the dark valleys of the middle west who used to sway in the rich heavy wind of his poetry. Times had changed.

But it was the old radio man, star, personality. Plugging ahead, against all odds. Sonorous voice *Why Do I Love You* and Lenny Dee on the organ, keening and wailing just under the words. The shades of Moon River had reached the swamps of Florida.

One day Johnny received a fan letter. His first in too many years. Written on line notebook paper with three holes. It was from some sweet young girl. 15, maybe, or 16. "Yes, Johnny," it said: "I *do* love poetry. I love hearing you read." Each time, the word 'love' was underlined, in red: four red lines.

Johnny read the letter on the air. I watched him as he was reading it. He was crying. The tears leaked from the red-rheum eyes. Tears ran down the botched skin, past the ruined mouth. "I *do* love poetry," she said, he read in his rich Moon River voice. "I love hearing you read," he read.

Then he turned off the microphone, and the organ tunes of Lenny Dee came over the monitor, still backgrounding his voice as he told me how much he would like to get in the girl's pants. "I'm going to fuck her," he said, and his eyes brightened. "I'm going to fuck that bitch so fucking hard she's going to fucking fall down bleeding from her mouth." The organ music swelled. There were still tears shining on his cheeks. "Shit, I'm going to get my prick all the way up that fucking cunt." And he raised up his pasty fist to show how he was going to do it. And his sad red eyes glittered with old alky tears, and the organ music raved on behind us.

(*KTAO Program Guide #25*)

When Shirley Temple came to call in 1937, Hoover wore the special badge of her make-believe police force on his lapel and signed her autograph book.

TV PAST

American television has fulfilled its necessary promise. That is: to pacify the middle class adults as it is revolutionizing the middle class children and the entire lower class.

Since most of the literate population is antagonistic towards visual escapism—the intelligentsia has spent (some would say misspent) their venom on the alleged banality of modern American television.

This antagonism ignores the society-change forces of video. The generation that was born in 1950 or 1955 can be truly called a *visionary* generation: from age 2 onwards, American children have been fed some of the most exciting, motivational, stressful visual images. For eight-twelve hours a day, an entire generation of open minded pre-adults has been subjected to two dimensional, audio-visual conceptualizing. The effect has been tremendous.

For this generation of Americans is more articulate and cynical and exciting and alive than any this country has ever produced. There are few mysteries to them. The ideals they have been presented with are all the ideals of consumption and American Standard eat-drink-smoke-play-live-feel. And like any creature exposed to 495,000 hours of such inducements, they have—of course—rejected all these values.

The critics and the mothers-concerned-about-effects-of-television will never understand the truth of *apposition*: that is, if you brainwash hard enough and long enough, the opposite will result. If you try as cleverly as American television has and does try to force consumption patterns, then the most open of receivers, the young, must follow the diametrically opposite values.

The other revolutionary force comes with the world wide distribution system of American television. There is no one in the United States who can guess the impact of *I Love Lucy* on the poor of Uganda, crouched eight deep before the village 12 inch television set. The message carried to the rest of the world is that of Lucy and Desi acting out their comic lives *against a background of such splendor*. Such splendor: Lucy weaves her universally comic way through such accoutrements—and she doesn't even notice! Washing machines, sofas, guitars, stoves, rugs: Lucy is knee deep in the most magnificent selection of consumer items, and rather than be astonished, she carries on *as if they weren't there*. Such is the splendor of American lives.

Can you visualize the impact on some dust farmer in Africa, or India, or Japan: we have, they want.

THE ROOTS OF MILITANCE

"Let's say they wanted to create some sort of militance in the minorities in this country. Let's say that back in 1933, they wanted to come up with some method of creating a political awareness in the ghettoes, and among the dispossessed. Let's say there were some groups that wanted to radicalize all those people who for the history of mankind have been dumped on, given no rights, left out in the cold.

"The best way they could radicalize a whole social and economic group would be called the Communications Act of 1934. The fastest and best way to cause the upheaval of the 1960s was accomplished with a piece of conservative legislation, written by a supine congress with the assistance of Sol Taishoff and The National Broadcasting Company and Sarnoff and CBS and all the barons who were intent on keeping broadcasting in this country out of the hands of the educators and the public spirited citizens and bureaucrats.

"Because the Communications Act of 1934 profoundly influenced the size and shape and orientation of commercial television as it was to grow in the United States immediately following World War II. And the orientation was simple and direct and uncomplicated; viz.

CASH FLOW

"When David Sarnoff and the commercial broadcasters testified concerning the future of radio ('freedom'...'no government intervention'...'the American Way'...) and the absolute desirability of making it a media for *sales* rather than a media for *informing*—they couldn't dream what they would be doing 30 years hence to the expectation of the have-nots.

"For television, American television, American television of the 1950s was in one way a crude money box. In another way, though, it became an instiller of a profound sense of values in those who were its best customers; namely, those with insufficient resources to do anything else but sit in the old overstuffed and watch the hi-jinks on *Your Show of Shows*.

"I am not speaking of the middle class, who were already beginning to realize the immense prosperity of the country, and what it could mean to them in material values. I am not speaking of the upper class who were by and large well-educated enough so that they have all the resources enough to give them the advantage in real life over what they would receive on TV.

"No, I speak here of the Puerto Rican in Harlem or the Black in Newark or the Chicano in Los Angeles or the American Indian on the outskirts of Fargo, N.D. And I am talking about the fact that the average lower class family is before the television set 8 to 12 hours a day. And I am talking about the real-life one-minute dramas acted out 18 times an hour called TV commercials.

"If you live in a 4th floor walkup on 127th Street, and are constantly being diddled to buy another 'living room suite' or a washing machine or another tv set for christs sakes, and if watching that commercial which makes the suite or the washer or the tv so attractive, and finally it tells you that if you don't have it you are poor and underprivileged for christs sakes. So you have got to go out and get it, even if it means rioting for it: the tv has motivated you, since age 18 months, to possess-own-have-get the trappings of a comfortable life.

"I love it of course. I love the knowledge that none of the people who designed the Communications Act of 1934 had any idea of what they were fomenting in the way of radical change in America. If you had told them then that we should temper the sell-buy-now mentality of broadcasting with a generous use of broadcasting to educate and enlighten—and that *that* would have been a true conservative piece of legislation: why, they never would have believed you. For the radicalization of America to come on the wings of Adam Smith, as it were, could never have been predicted.

"And that is the marvel of history and predictions and all that sort of stuff. That we never know what we have created. Adam Smith—*The Money Game* man, not the 19th century economist—says that history has never keen kind enough and good enough to let us predict the future. And Engels said

> Those who imagined (that) they had made a revolution
> always saw the next day that they did not know what
> they had been doing, and that the revolution they made
> was nothing like the one they had wanted to make.

"Old Sarnoff. Testifying before the Congress in 1933. Asking that broadcasting be free of 'any onerous requirements' like obligation to teach and instruct and enlighten. Never for a moment imagining the prodigious force of the one-minute magic television commercial of 30 years later to motivate burnings, and revolution, and the revolutionary Mau-Mau mentality of 'rip-offs' and dope and long hair and anti-war and radical change.

"God bless him, Old Sarnoff, wherever he is now: he and his brother 'free enterprisers' in the first third of this century fomented the radical change of the last third. And they never even guessed that they were doing it."

(KTAO-TV Prospectus in The Radio Times)

"In May...the 'National Public Radio Network,' sponsored a conference...in Washington D.C. I attended the Conference as the Jack Straw Memorial Foundation's representative...Here, hard by the Potomac, a legion of 'educational' radio station executives, almost to a man from state colleges and universities or big-city school districts, convened, and talked about the two things dear to the hearts of all radio men—programs and money. The things they said about programs were bad enough...The thing that really appalled me was what was said about money. Institutional stations—the kind run by big Universities and school districts—aren't happy with tax money. Oh no. They want to supplement their bloated budgets with subscription money—asking the listener to pay for the sort of programming blandness that only a University administration can provide. Even more appalling, they have, with the active encouragement of the Corporation for Public Broadcasting, solicited charitable agencies in the local communities to pay for the privilege of running programs: programs that should be run as a matter of moral obligation by educational stations. Say you're the executive director of the Slabtown Gay Lib, or League of Women Voters, or Salvation Army. Here's the Slabtown State U. radio station guy—asking you for $35 an hour, each week, so that you can talk to his audience about your agency, and what you're doing for Slabtown. To a man, these 'professional' educational radio station people don't know, and can't ever expect to know, how to run a radio station 15 or 16 hours a day, seven days a week, with at least an hour a day and sometimes five times that much local, topical, public affairs programming—to say nothing of unique music and drama, for anything *less than* $6,000 or $7,000 a month..."

—Ben F. Dawson

(KRAB Program Guide #70)

145
A SCAM ON WCAM

The Mayor of Camden (N.J.) has a goodly case of the dewlaps and small eyes well protected by tinted glasses. The rims are black and plain; his handshake is soft. The Mayor of Camden and I have come together because I want to buy his radio station—or rather the radio station that belongs to the people of Camden, N.J., since it is licensed to the city. He, supposedly, wants to sell it to me. Supposedly.

The Mayor of Camden (N.J.) and his—or rather their—radio station hang out on the 18th and 19th stories of the Camden City Hall, respectfully. His office is plain and dull and hot: windows don't open on the 18th floor of the Camden City Hall, and even in December I sweat profusely and the good mayor has that soft-boiled look of staying indoors in an overheated building too long. He has just gotten over a television interview in which he has described the needs of Camden (N.J.) in financial terms. Natch. Twelve million there, twenty million here. The television interviewer loved it, natch: a nice safe 'public service' interview with Joseph H. Nardi, Jr. about money. That's called 'public service.' My friend Pete the Italian says that Nardi is slang for 'nuts' in the home country. And he ain't talking about 'Aw, Nuts!"

The Mayor of Camden (N.J.) has been trying to sell his or rather their radio station for the last eight years. Even in terms of bureaucratic inefficiency, this one is a doozer: three potential buyers, each offering to pay $1,450,000 in cash, knocked down by a combination of the Citizens Communications Center, and the Federal Communications Commission, and other broadcasters, and plain down-home stupidity. That's why I am here, in the City of Camden (N.J.), land of Campbell's Soup and a restless, 50% mixed Black and Puerto Rican population, and a Mayor whose name is nuts who talks about money all the time.

WCAM went on the air in 1925. In those days, it was common for cities to own and operate their own 'municipal' radio stations. New York City did and does. Jacksonville, Fla., and Louisville, Ky., and Dallas, Texas did and do. So does Camden, New Jersey.

In 1965, the City of Camden agreed to sell their major market AM station (1310 khz, one kilowatt-day, 250 watts night) for $1,450,000 to Leonard Chess of Chicago. He had made a small fortune in 'Black' radio, and decided that Camden-Philadelphia needed his commercial ministrations. The good mayor of Camden never said why he was dumping his (or their) radio station. One would hope that the city fathers were seeking additional sources of revenue because the tax base of Camden had fallen apart because of the influx of the poor and minority from Birmingham and Jackson and Puerto Rico. You've got trouble in your city of 102,000—all the good businessmen are slipping out of Camden County, and all those unfriendly window breaking types are coming in, right? So you got a problem, so you sell your city radio station, right? At least that's what the good mayor told me.

What the good mayor didn't tell me was 1) Why all the good middle class people were leaving his streets; 2) Why the newcomers were tearing up his streets; 3) Why the sale of the city owned radio station was going to solve this. The mayor didn't think too good in the way of psychology-sociology-human dynamics. He didn't understand fear too good. He was just a plump, heavy-dewlapped second generation Italian who didn't want to be mayor, he kept telling me. He also didn't like this particular frizzy Californian sitting across the desk from him, listening to the sweat pop from his aching pores and trying to suggest to the mayor that some of the problems of the City of Camden fifty-percent-Black-and-Puerto-Rican could perhaps be ameliorated somewhat if the angry citizens of that community could have a bit of their own voice about their lives, and their economics, and something, *anything*, about the goddamned smack freaks tearing up the streets and neighborhoods looking for another fix for christs sakes. Where did all that smack madness come from, Nardi?

They ought to write a book about the attempted sale of WCAM. First to that man named Chess who has the gall to try to hire away all the deejays and salesmen from the Philadelphia stations like WDAS and WHAT the day after he signed the buy agreement and it was a matter of a week when the FCC had a complaint with ten affidavits about this man Chess who would be trying to serve the Philadelphia black market. I mean, it's all right to cheat and steal if you are in broadcasting, but you can never never never say that you are buying a station in Camden (N.J.) in order to lift money to serve the citizens of Philadelphia (Pa.) four miles across the river. You see, I say you can do it: but you must always pretend that you are *not* doing it. Broadcasters are licensed to rob you and me blind, but they must never never admit it.

Then there was Gordon McLendon. You remember Gordon McLendon, the sharp shooter from Texas, who parlayed a couple of am daytimers into gushing oil wells by selling the shit out of them with 18 commercials an hour and a rigid format. He and Stortz were there right at the first, to make radio pay, as it should have paid, from the first. Gordon McLendon, buying a station here, pumping it full of his Texas juice and sell-your-old-granny, now with a dozen or is it two dozen stations which had been dominoed into existence, each worth a cool mill. When Mr. Chess got in trouble, Gordon McLendon picked up the option for $650,000. And being G. McLendon, he was too smart to get tied up with that 'suburban' issue. He screwed it up another way:

Each radio station must produce, at license renewal time, or when bought or sold, something called 'community interviews.' The idea, like everything required by the FCC, is basically sound. Broadcasters, with their universal piggery, have quickly distorted the idea.

The idea is that there be 50 or a hundred interviews: one half with community 'leaders;' the other half with just plain home folks. Each of these interviews is to discover community 'needs.' The idea behind it is that the radio station owner or manager get out of the money-changers' temple and see what problems exist in the area he is supposed to serve. Broadcasters, always on the make, have managed to subvert this requirement as quickly as possible. The community surveys, it has been ruled, can be made by any outside 'consultant.' These 'consultants' can be hired, and can produce as much paper as possible. As a fact, this demonstrates Milam's 5th law of Bureaucratic Junglework: to solve each complaint, the FCC creates another paperwork project for the applicants and licensees. Broadcasting *per se* never ever changes: but the FCC files quadruple in size. And there is nothing to say that the broadcaster has to read his application; he only has to sign it. Thus, we are spending another $5,000 or $10,000 or $50,000 to provide some more reading matter for some poor bastard in the Broadcast Bureau who doesn't *need* to read about community problems.

As I say, G. McLendon was too smart to get screwed up in the 'suburban' issue. "Of course I'm going to run a Camden station," he said. And we all ignored the fact that his 1 mv went all to hell and gone over 4,000,000 Philadelphians as well as 200,000 Camdenites. And we ignored (or thought we were going to have to ignore) McLendon's past and sordid history as an aether-raper.

In steps Nicholas Johnson. And a few of the FCC staff.

When McLendon stuck together his application to buy the aether in Camden—your aether, my aether, the poor and black aether—for $1,450,000 (said dollars which would go into some strange dark hold of city government and you know we would never see them again) he hired old friend Mel Gollub to slap together a 'community interview section' of the application for him. He and Gollub knew and liked each other. They should. McLendon just paid $750,000 to take over WIFI from Gollub. They were fast friends, and rich too: but Gollub was a bit lazy.

His community interviews were sketchy at best. Few minorities were contacted. He talked to maybe two dozen people. The community problems section of the McLendon-WCAM purchase transfer-control Form 314 was just too goddamn skinny. McLendon opened the door for a Nicholas Johnson coup: perhaps the last and the greatest hell-raise by this outrageous troublemaker: now you know why he was not reappointed to the FCC come June 30, 1973.

There is another factor in this official 'designation for hearing' order that came down in 1968. That is a matter of FCC staff. The ones we always hear about are the FCC Commissioners. The ones who do the real work, and have the real power at the FCC are the staff: the drones in the Broadcast Bureau or the Common Carrier bureau who hold their jobs for 20 or 30 years and dream of retirement in the brick homes of Leesburg, Virginia, or the Chesapeake Bay region of Maryland. If a FCC staffer (as they are called by snide *Broadcasting* magazine) gets a hard on about you or your station or your attitude, then you can be fucked. I mean, these FCC people get pissed on from all over: by the Civil Service, and Congress, and the damn FBI, and their boss, and some dumb 7-year Commissioner, and from their wives, no doubt. They have hearts and souls and livers and ears and peters just like you and me. And there are, in the complicated body of Rules and Regulations of the Federal Communications Commission, enough leaks and peeks so that if you ante up enough power after ten years in the Transfer Division, or 23 years in the Broadcast Facilities Division or six years in Complaints & Compliance, and if there is enough body of feeling in the 250 scales that make up the body of the FCC, then you, as a humble staff member, can wage a small war. Against some broadcaster. Or some network.

O, it's a secret subtle war. You can't be too obvious. You've got to pretend it's not happening. You don't want to jeopardize that home in McLean. But those in the staff of the FCC, and one or two commissioners, and *all* the FCC lawyers outside the commission, and a few smart broadcasters—they know what happens when one staff member gets a hard on against you, and your station, and your filings. Things get lost. Or slow down. Nothing seems to work out right. Applications go all funny and wrong. Ask Richard Eaton. The genuine piggo of commercial broadcasters, with stations in Washington and Cleveland and Miami and New York and San Mateo and Baltimore. United Broadcasting Co.: they were going great guns, and in the oldest robber baron tradition of FCC dealing, thumbing Eaton's broad nose at the silliness of filing and bureaucracy. He was right of course: about it being silly; but he was wrong to ignore the Idiot Child of The Great Aether God. Now seven of his stations are locked in 'license renewal' problems, and a $25,000,000 empire, built of the corpses of American radio creativity, lies on the edge of collapse.

There have been others. They say that Teleprompter is in such a place. We all know that Pacifica Foundation is there, and has been there for some 15 years. Paduan Foundation, in its brief history, got locked in the arbitrary bailiwick of a man named Harrington. Somebody, I mean somebody had to be the one that took the arrows against KRLA and lost it its $8,000,000 license. And there are whispers that Gordon McLendon, the aether-rapist supreme, has finally gotten one or two of the $15,000-a-year staff attorneys pissed off at him. Do you know that Gogol and Dostoevski were right (*vide*, "The Overcoat," "Notes from the Underground.")

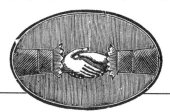

148

The application of Gordon McLendon to purchase WCAM, Camden, New Jersey looked to be all right as applications go. But there was this FCC staff hard on, and Nicholas Johnson, at his zenith.

Despite it being the last fool year of the presidency of another man named Johnson who didn't have sense enough to keep his hands out of the cookie-jar called World Politics, the winds of change were blowing through the pale green halls of the FCC there at 1919 "M" Street, and the application for transfer of control was set for hearing.

Hearings are long and awful things, and if it weren't for Gordon McLendon and his *hubris* and the dim-wittedness of the City of Camden, all parties would have dropped the thing and let the station go on pumping out 24 hours of dreary five-years-ago 'pop' favorites and turning some small sum over to the City of Camden each year. McLendon has won all his battles before, and thinks he is god's gift to revolutionary new concepts in robber baron radio, so why not? Nicholas Johnson, that's why not. The Last of the Just.

Someday, if you are diddling around Washington D.C. and have nothing to do for the afternoon, pop over to the FCC and go up to the public reference room on the second floor. It's open to anyone: fancy-dan Conn Ave lawyers knee-to-knee with $50/a/week copy clerks from *Broadcasting* Mag. There are days weeks months of stuff for you to fill your craw with: the stuff of American Radio & TV. Behind all them musics and dj 'yeah-hi-temp-40°-WABC-love-wow' lies a table-full, room-full, office-full, building-full, ware-house-full of papers. And inside their boring green-gray folder covers lie the million dollar secrets. Waiting there, like some secret bible script, waiting for you to plumb the depths of income-outgo tower-site-availability 301 and you-are-requested-to-supply-information. It's all there, from 8:30 to 4:30, five days a week. If you can figure out the golden key treasure which is: behind the golden sound is four or seven or twenty four million dollars worth of paper. And it is all stored in this public reference room of the FCC.

Within the Federal Communications Commission, the headless 4th branch of American Government (as all Independent Regulatory Commissions are called) lies an executive, and legislative, and judicial. All are squashed together in that aluminum-glass bookend on "M" Street. And the judicial is—as all are supposed to be—the most independent of all. The Hearing Examiners, under Arthur A. Gladstone, sit as quasi-judges on the cases that come before them. And as judges, some are sweethearts and some are fools. You should read the initial decision in the case of McLendon-WCAM (1968) to see where *that* guy stands. Hooray for the American Way of Broadcasting. Or: "It's no sin to make money, and attempt to make more profits, in our present economy..."

No matter. The full commission, on its own order, reviewed the decision. And stuck a giant legal poker in the ruddy bum of McLendon. It defined, for the first time, the responsibilities of broadcasters in *at the very least* finding out what the community may need. The Mayor of Camden, whose name is nuts and eyes are tiny, lost another $1,450,000 bet. McLendon lost a little face, some FCC staff members got a good laugh up their sleeves, and the last of the winds of change whistled through the straight-and-narrow halls of the FCC and died out in a chill whisper.

The application of Starr Broadcasting (Wm. F. Buckley, 10%) to take over the bid of Chess-McLendon for this jewel in the rough along the dank greasy banks of the Delaware River was an anti-climax. "The station is cursed," I told the beady mayor of New Jersey. Citizens Communications Center pointed out that Starr was planning to dump 90% of its Spanish language programming which, in a community in which one out of three people are Spanish-speaking, might seem to be a bit of a mistake. It was. Kramer made a couple of heavy filings after the transfer was announced, and Starr was happy enough to withdraw. "What Camden needs," I thought, "is a real radio station. For real people. Set in the ghetto. Or a series of mini-stations set in different parts of the ghetto. So the Black and Puerto Rican will have their own voice. For a change. What it will do," I told the dewlapped Mayor of Camden New Jersey on 27 December 1972—"What it will do," I said, wiping the tears of sweat engendered by the 18th floor of the dull grey barricade city-hall complex of a sick and dying city of the sick and dying Eastern United States, "What it will do," I said, voice shaking with the worth of my idea, (and, I may say, given the chance to explain Truth to any Mayor, even the Mayor of Camden New Joisey), "is to give your minorities a voice, an important and strong voice. So you won't have the riots here, on the streets of Camden, like you did, 18 months ago. You know?"

He didn't wince, the mayor didn't. I watched him, and I swear he didn't even twitch. All while I was telling him about my idea for putting this radio station into the hands of the people, and music to give people a pride in their culture, and talk and discussion, and hundreds of volunteers, radio station volunteers, roaming the streets and picking up continuity from the streets to broadcast, and interviews and ideas and talk and angry people, rising up angry before the microphone to be heard, letting their voice, the voice of the people, be returned to them, so that they could be heard, so that they would not have to spill out rising up cruel in plumes of smoke on the streets, to show with fire and TNT their outrage at being forgotten: but instead, being given their own voice, their own microphone, in their own city's radio station.

WCAM. Their own radio station. I told the mayor that I was willing to buy their station for the community. For $50,000 down, and the other $1,400,000 over fifty years: the radio station would evolve out of the city government so isolated in that 19 story grey tower and into the hands of those who were going mad and wrong in the streets. "It's a long payout," I said, trying to figure out where and how I would finance fifty years of indebtedness, but wanting so to try: "It's a long payout, but it would be worth it. To keep the city from falling apart. Like 18 months ago."

Mayor Joseph Nardi of Camden New Jersey did not wince. His eyes were steady. His pudgy hands rested lightly on his nondescript desk. He did not wince. I do not believe he heard me. *I think we were talking two different languages.* Because out of his words, 20 minutes worth of words, I got the idea that the good mayor did not by any means nor would he ever think that having a street radio station would be any good at all for him. *At all.* He pretended it had something to do with the term of payout, and the need for a $650,000 down payment. But what the Mayor of Camden was telling me was: "Your idea for turning our am pop hits-on-a-chart and singing *WCAM, Camden, do-dee-do* into a real human living radio station for real people

eats the bird

150

It's too late to turn the clock back, isn't it? They've already stuck too many holes into the love and potentiality of radio, haven't they? Robert H. Storz. George B. Storer. Woody Sudbrink. Leonard H. Goldenson. How can I name the days and ways that they have stuck their moneyhunger up the ass of the spectrum so that—is it?—It's too late to change it all. The Winds of Change are so fitful. And abortive. Was that really a victory over McLendon? Or did it just make him more wary and cat-like?

Despite my offer and my anger, the mayor of Camden still thinks he can find a buyer for WCAM. A nice commercial broadcaster. Thum-thum-hum-hum-buy-now. The fucking city is burning down, and the City Hall is turned to a fortress, and he says: "What I want to do is to get some more good industry into this city. Campbell Soups came here in 1925. What a boon for Camden. We have to find more Campbell Soups." The fucking city is turning to ashes, the eyes the windows are filled up with smack-heads and rifles, and he wants to find another Campbell Soups. "There's nothing wrong with making money, you know," he tells me in abstract parody of the hearing examiner's decision of McLendon-WCAM, with echoes of firm 1890s economic laissez-faire ringing soundly in our ears.

There's nothing wrong with making money, you know. But there's something wrong with the city. And there may be some tiny connection, Mr. Mayor. Or are we both nuts?

They'll give us an FM station here or there. A ten watter tucked up around 88.1. Where it's safe. Maybe even a Class C in one of your minor major markets. A few community broadcast operations sprinkled safely around.

But they won't give us one of them high-power AM 50 kilowatt blast jobbies in New York City. No, we have to save that for Storz, or Eaton, or ABC, or McLendon. We will never get Channel Two in Los Angeles, or 890 khz fifty kilowatts clear channel swoop up and down the 50 million population heart of the great grave burning america gushing throb out of Chicago. They sold that one down the river in 1915, remember, and we can't put the voice of the people there, remember. It has to be saved for the coining of gold, remember.

And the Mayor of Camden, god knows, would never yield the still small voice of 1310 khz to the California crazies. Them and their blathery ideas about democracy, and deliver-ideas-not-fortunes.

The Mayor of Camden. He sweats and has too many dewlaps. And a name that's nuts. But he ain't nuts. He has more goddamned sense than that: to let the mind-democracy furies come swarming out of that tender golden box called radio.

(from *Alternative Radio Exchange*, #118, Spring 1973)

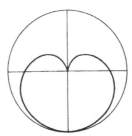

SEX AND THE AMERICAN BANKING SYSTEM WILLING, AS USUAL, TO HELP THE DISTINCTIVE BROADCAST OUTLETS OF OUR DAYS

The ship of financial state at KTAO develops a few leaks. Dollars seeping down the drains, slipping out the doors when the gods and the dogs go to sleep. Strange letters start appearing from the Credit Bureau of Redwood City. Dun & Bradstreet, in fine corporate confusion, offers to give us a free sample credit financial search on the company of our choice and, in the same mail, duns us heavily for some long-forgotten debt from some long-past management.

The man at the brokerage service where we used to go so long ago, where there are now so many empty desks, where the wall is ripped and scarred from where they had to cart away the expensive electronic moving green-figure market board, says that Mr. Iron at the local Savings & Loan can, perhaps, dissolve all our debts; perhaps make our checkbook well again; perhaps even arrest the soaring flight of the dollar. Mr. Iron is reputed to be a man of magic out of the dark mystery of American loanery. I put on my ragged shawl, pick up my battered tin cup to visit the wizard of Iron.

Mr. Iron's Savings & Loan is dark. It is a square tile expanse of ice stretching dark in order to protect the S&L financial stability by chilling the customers' hearts. Frozen assets as it were.

Mr. Iron has salt-and-pepper hair a bit shorter than my own: say maybe a quarter inch from hair follicle to splitend. I forgot to shave this morning. Mr. Iron notices this. I forgot to press my shirt last month. Mr. Iron notices. My specs are dirty, particles of brain flaking away. Mr. Iron notices. He says nothing. I sit in his icegreen glacier-plastic visitor's chair and say: "Hi! I'd like to borrow a half a million dollars or so." There is just the briefest scattering of a smile.

"What kind of music you boys play over there?" asks the good Mr. Iron. "O, you know," I say, juggling this morning's three hour presentation of Handel's oratorio *Jeptha* against yesterday's festival of *oud* music from lower Nubia, complete with native musicians and voice. "You know," I say, hoping he doesn't: "a bit of rock here, some foreign music. A touch of jazz from time to time."

"I like boogie woogie," says Mr. Iron. "Pardon," I say, leaning slightly forward in my chair of pain. "Boogie woogie," he says: "you remember that." "Certainly," I say. Painful memories of Miss Baggs' Class of Social Dance creep through my head. Memories of shoving statues around the floor. Memories of standing around, wondering who was going to get left standing around. Memories of that girl *o god what was her name* who always but always loaded up on onions just before dance class. Memories of the one fast piece of the evening *boogie woogie: was that the name of the record* where we got to jump around some like broken frogs. I think we even laughed some. "O yes," I say to Mr. Iron: "I remember boogie woogie." *you bet your sweet ass I do. and those onions. what the hell was that girl's name? wasn't she the one who killed herself just before she was to graduate from college? she was very fat.*

"Maybe you like classic jazz?" I say, brightly. "Do you collect it?" "No," says Mr. Iron, "I don't." There is another delicate pause. I push my hair around some more. "I don't know what's happened to the music since then," says Mr. Iron: "Something's gone wrong since then. I can't listen to it, won't let the kids listen to it."

We pause thoughtfully again. "Something's wrong," he says: "the music is too strange. Something's gone wrong with the music, and the kids...something..." Mr. Iron (who has eyes with a blue very much like the blue of a glacier) looks at me for a while. Watches me trying to adjust my coat, trying to make it look less wrinkled.

"The nut," I think. "There must be some way for me to explain the nut of life." The wind pulls the willow branches down across the sidewalk. I have to lean down to get under them.

152

The rain streams down my face, streaks my glasses so that the whole world (street house leaves scudding clouds) turns distorted and wavery. The wind tries to push me back. "The nut," I think: "No matter whether it is four in the afternoon (rain) or eight in the morning (sun) or two in the morning (dark, cold, alone)—time is leaking away. Our lives are leaking away." This time, this day here right now and me here bending under time and the rain will never again be repeated. Ever.

Our days curl out of the testes of existence. There we are, talking to cold dead men, or dancing with lonely ugly girls, or walking with the wind trying to push us back to where we came from. *I won't go.*

It's a nut. The whole thing is a nut. Spiraling out of the soft warm bag of life: a strain curling through the endless wraps of the universe. Time shot out of the blaring mouth of some laughing god. There we are, stuck with our tailed seeds, curling out of the nut. And they and it don't care.

"I should get a philosophy," I think. "Or a religion," A big fat glop of cold rainwater finds the tiny gap between coat collar and neck, runs squirming (me and it) down a once warm back. "There should be a philosophy to deal with this silly mystery of our days spiralling on and on." No brakes. No stoplights. No stopping it all. No stopping at all. *At All!*

"There must be some way to approach this whole thing philosophically," I think. "There must be," I think. There has to be. *I think.*

(KTAO Program Guide #42)

J. Edgar Hoover standing in his office with the Lone Ranger .

SEX & BROADCASTING

THE KRAB NEBULA MEDIA CONFERENCE

JULY 21, 22, 23, 1972

ANNOUNCEMENT

History

It was another of those cold dark wet flappy days in Seattle. The sky was grey—as usual—and the power lines on Roosevelt Way hummed and buzzed: a city's naked power leaking down into the grounds.

In a dumpy retired doughnut shop, three rather seedy characters—Jeremy Lansman, Gary Margason, and Lorenzo Milam—bent low over a 15 year old Collins transmitter. A signal was given—a button was pushed. Is this when community non-institutional radio was born in the United States?

No: not only were there several other similar stations already in operation around the country—but the goddamn Collins transmitter blew up almost the same instant it went on the air. The bloody 2000 pound power transformer, shipped at inordinate expense from WFLN in Philadelphia, showered sparks all over the ground—and it was a full month before the three got over their exasperation and put KRAB on the air once and for all.

From such humble beginnings, from such improper technical planning—who would ever guess that the KRAB nebula would grow. With eight stations on the air, with construction permits being sought in at least five other areas—and with community groups being organized in a half a dozen other markets.

Who would believe that our troubled threesome, beating their heads against the wall in November of 1962, in that dark land, hammering and bending and cursing that wretched antique transmitter—were really on the way to creating a group of stations which some ten years later would be reaching over 10 million potential listeners.

The concept and ideal of what we have come to call 'community, free-form, non-institutional radio,' is powerful, and compelling, and sometimes—as in the beginning—appallingly frustrating. Hampered by old equipment, scrawny budgets, and a lack of *categorization* by various governmental and economic bureaucracies—the stations have come not only to survive, but to spawn other similar operations in other areas.

And this is the point in which the KRAB nebula operations differ from all other community and 'educational' operations: that of a strong messianic need to give other cities their own KRABs—so to speak—stations which burgeon with awful financing, sometimes painful civil wars, and inexperience. Stations starting with nothing except a burning desire by radio mendicants who seek only to spread the joy and freedom of outspoken, outrageous, eccentric, culturally astonishing, socially unkempt, didactically indefensible radio. The KRAB type of radio station—small, inexpensive, dependent upon hoardes of volunteers, often operating out of unsanitary, unsightly, and nightmare facilities—is growing up in the most unlikely places.

Needs

Spiritually, we are kin to the Pacifica stations, to some of the more daring educational operations, and to the best of the CBC outlets. We are also closer, some would say, to KPFA in 1953—certainly more than KPFA itself in 1972. Ideally we look to the BBC Third Programme: back when it was still sanctified (and protected) by the good Lord Reith.

Practically—we are separate from any other group of radio stations in this country. We have common interests in strong doses of ethnic music, and antagonistic interviews, and non-political (somewhat anarchistic) stances, and feelings of leisure and non-professional, non-pushy announcing. Unlike the Pacifica stations, we have no common overseeing directorship (except for KRAB and KBOO)—which some of us would claim is fortunate.

For without this common, by-law ownership and management, we manage to avoid the most exasperating internal *political* problems endemic to all organized, society-changing institutions. But in the process we manage to lose the advantage of interchange of ideas, and mutual support, and the prospects of orderly growth. We exchange tapes, but it is desultory, confused, undirected. We exchange peoples: but without pattern, on a random basis...even though it may be the most successful of our relationships. The only real continuing exchange between us is a free mailing of our program guides, with occasional bitching letter about some tape gone astray. Our experiences are hardly shared. Our experiments never leave our own communities.

Future

We don't *have* to have a meeting of the KRAB nebula stations. We can continue to grow, and grow well, separately. For the eight stations, and the unknown number of incipient stations, benefit from their independence, and their strong local community base.

Still, we have a great deal to learn from each other. Since the tenth anniversary of the birth of KRAB is practically upon us—it would be fun and enlightening to see how each of us has grown. We have, independently, created a fine pool of talent in our respective communties: it would be healthy for each of the stations to see the personalities of the other communities as reflected in their own stations.

For the greatest interchange of energies comes through knowing and liking someone somewhere else who has access to the tools of their own community. It is for that reason that I have asked KRAB to loan us the use of their new Firehouse Station #25 for the purposes of a conference between the existing and proposed radio stations.

Subjects

Friday, July 21
THE INTERNAL PROBLEMS AND STRUCTURE OF COMMUNITY STATIONS

—Volunteers: The Care and Feeding of
—Boards of Directors and Staff Relationships
—The Care and Feeding of Internal Politics
—Taking over College Nurdly 'Educational' Stations and Making them 'Community.'
—'Communal' Staff Decisions
—Regular staff meetings as a Form of Self-Destruct
—The physical plant of a station as a function of 'Plots.'
—On-the-air revelation of internal problems and plans
—Engineering maintenance of decrepit equipment
—Large vs Small working areas
—Teaching the subscription Lady to do documentaries
—Putting Board Members on the air to end their 'distance' problems.
—Nude Therapy as a solution for internal tensions

Saturday, July 22
THE EXTERNAL PROBLEMS OF COMMUNITY STATIONS
—Dealing with 'External Politics.'
—The FCC.
—Naughty programs and the Law.
—Recent grant projects of HEW and CPB.
—Private Foundations and Grantsmanship.
—Publicity.
—Making a station known.
—Subcriptions and subscribers.
—Subscribers as program participants.
—Getting raided and bombed as statements of community interest.
—Getting raided and bombed as artistic statements.
—What to do about immediate neighbors.
—Getting program participants in the door.
—The 'liberal' image of community stations: the dispossessed conservatives.
—Live music and unions.
—Record store and distributor relationships.
—Mind fucking as a cause of external tension.

Sunday, July 23
PROGRAMMING AND THE INTERRELATIONSHIPS BETWEEN THE KRAB NEBULA STATIONS
—What is a good and a bad program?
—How to get bores off the air.
—How to get boring staff members off the air.
—Coordinating program volunteers.
—Music: ethnic and classical vs blues and rock and jazz.
—The concept of supplementary programming: listening in on other radio stations.
—News bias.
—Block programming vs the flexible free-form 'floating' programs.
—Snotty interview techniques.
—Interweaving of music and talk bits.
—High vs Low Comedy
—Emergency News Coverage.
—How to improve the sagging, dragging, impossibly complicated non-bureaucratic interflow of tapes between the KRAB nebula stations.
—Introducing other and new stations into the tape exchange.
—Staff visiting between stations for extended periods of time.
—Short wave intercommunications between nearby stations.
—Subcarrier interconnections between nearby stations.
—Talking to engineers.
—Mind fucking as a solution to nude therapy.

OUR TRAIN TRIP

I guess it was Paul who had the idea of renting a whole railroad car from AMTRAK. So we added it up and it came out to about $150 both ways for each of us. Plus the fact that we would have a place to stay in Seattle at the King Street Station for the three days of the conference. When we found out that they would let us do E-Z Squeezy on the American Express Card, why there was nothing else to say but yes.

We didn't know that we would be dealing with Southern Pacific anyway—even though it had the *imprimateur* of AMTRAK. Southern Pacific: who, for years, had been doing everything possible to drive away their passengers. As one of my rail-freak friends had said: "They found out that it was easier to ship a carload of pigs than a carload of people—and ever since, they have done everything possible to drive us away. They poison us with bad food, they charge us for drinks, buy up all the cranks and grouches to wait on tables. They let the rolling stock fall apart so that the air conditioning will die. They charge off the cost of amortization *against* the passenger department—not the freight Department. And they do everything possible to make the trains run late—everything except to missend us to Boise when we want to go to Phoenix." Southern Pacific has always been the enemy of those of us who love to go by train: forgetting that their incredible richness came about because of massive Land Grants from the federal government a hundred years ago. I mean, if they want to go out of the passenger business—instead of wrecking our systems with bad food and bad rolling stock—they should return those millions of acres to us, the people. Then they could forget their obligation to us and forget the passenger service business.

This was five years ago that my friend John Stair was telling me about the sins of Southern Pacific—certainly the worst of a bad bunch—but we all figured that with the coming of AMTRAK all that would be changed. That's what we thought.

But, hell, it was a great ride. Even Southern Pacific disguised as AMTRAK cannot destroy the joy of a train ride. I've often tried to figure out the magic of a railroad trip. I has something to do with the world wheeling past the window non-stop. It's a great television show, on a wide screen, in living color. We go through the bowels of the city. All that 1925-1945 industrial waste which surrounded the rail yard. Some indefinable beauty of rustling metal, sooty-windows, brick warehouses crackled and sagging, the absolute reactionary beauty of American Union Pacific-Seaboard architecture. Built by people with no need nor sense of any beauty outside of mechanical utility. It's the beauty of Sherwood Anderson and Theodore Dreiser.

It's a visceral journey through the bowels of America. And it's the size: the scale of giant engines and giant passenger cars. It's the rails: as we travel, the rails will ride silver pairs beside us. At high speed, they parallel our every move. Then, at some juncture, they begin to proliferate: two will spawn four, four will spawn sixteen pair. Some will be shiny, some turn reddish with disuse. One pair in mad frenzy will make off suddenly to the west, to disappear in the brush, or over a fall-down trestle.

Then, as soon as they spread, they begin to re-pair. Twenty will run into eight, eight to four—and soon we are with the single pair running parallel to us again. The siring of a wide and incredibly tangled siding will be resolved into the two of us hurrying along together, dipping slightly, banking now to the right then to the left, then narrowing down to a twin silver pointing into the horizon up there before all reason, back there beyond all memory.

To Vladimir Nabokov, the rail trip was told in the passage of the telegraph lines running parallel to the tracks. A dozen or so black wires would dip halfway down the window sash—and pause, then rise up smartly to the point where some three-armed monster of a pole would smash them collectively so that they fell down again in a lazy arch, only to be raised up again for another blow from another cross-series. And behind this show, the fields of wheat would wheel past all hearing.

S E X & B R O A D C A S T I N G

The memory bank is crowded with images: diddling along the southern edge of the Puget Sound, moss rocks and the current pushing past the thin legs of Tacoma Narrows Bridge. Stopped on a siding somewhere south of the California border: a dark woman comes to the back porch of her house (she has the show of trains pouring past, with their army of characters pouring past) and flings out a dishpan of dirty water and sadeyes us as we sadeye her. We and the train intrude on her life and all she has for us is a pot of dirty milk-colored water.

But the best memory of all is the 3AM shot just outside of Redding. The moon of course is full and powdery—lying on its side, it balloons along the tops of the dark trees: by some mystery of perspective, the skies *wheel* in circles as we push through the night: the stars rotate as if we were racing time, the moon going through its all-night paces in a minute. We have made time stop: no, we have made it run, through this ancient device of us in the long low wide-windowed rail car pushing aside the darkness, pushing the night into the bearings of the mysterious parts of our persons.

(*KTAO Radio Times* #106)

There were themes which occurred and re-occurred during the three days of the conference. That of community radio stations keeping the doors open to the community; that of getting boring programs off the air; that of their relations to action groups which are striving to force commercial broadcasters into giving up time for blacks and chicanos and other minorities; that of the pitting of block programming (with a heavy, minute-to-minute program guide) against free-form or open programming.

The first day was one of introduction—of getting to know the eighteen groups from around the country which were on-the-air, about to get on-the-air, or merely in the application process. We were introduced to groups that we had never heard or dreamed of: Deadringer from Ft. Wayne, Ind.; Agape from Dallas; New Wave from Columbus, Mo.; Nan Rubin's group from Cincinatti. We realized at that time that community, non-institutional radio was, all at once, young and active, and growing, and meaningful, and rich, and alive. That we were at a conference perched (for awhile) on the corpse of one of the first community radio stations—but that it too would grow and change and show life again, because of the fact that this type of radio was so irresitible, and compelling—that it would continue to grow and prosper: especially given the vitality of the 75 conference goers who had come to Seattle from 1000 or 2000 or 3000 miles to give and get the ideas which make this type of radio so vital.

There was a time when this point became apparent to all of us. It came at a time—natch—when we weren't even talking about radio at all.

It was Saturday afternoon. There had been intense discussion during the day between two lawyers about community radio and the existing broadcasters and access. Al Kramer, the attorney for Citizens Communications—and the scourge of commercial rip-off broadcasters—had been challenged by Michael Bader, the legal representative for the Jack Straw Memorial Stations and who, by-the-by, represented some other more commercial, establishment, money-grubbing operations. The discussion had been full and ripe and joyous to those of us who love radio, and radio law, and regulation, and ideas of transmission and change.

The debate had broken up. The park—Golden Gardens Park in Seattle, a ledge under giant redwoods, overlooking the wide dark cool Puget Sound, where a hundred boats pothered around—was growing dark. The sun was squandering itself out over the Olympics, frosty and cold over on the western horizon.

The whole group of freaks and lawyers and radio people and newspeople and engineers had grown quiet, and mellow. We had eaten almost a hundred pounds of baked salmon, and drunk up the best part of two kegs of draft beer. We were filled with ideas and life and the...rightness of what we were doing.

Margot had asked three African musicians to sing for us. They came over from the University, and sat down right in the middle of the group, and with the cool and the night and all, sang to us of the joy and life of Rhodesia. Three voices, coal-black faces and white teeth, singing over the elaborate patterns of the M'Bira. Telling us of life and excitements from thousands of miles away, from thousands of ideas away.

Them and us, in the middle of a Radio Conference, tor Christ's sakes. The music of communication and spirit. The ideas of transmission and musics. And of a sudden—the conference, disparate elements drawn from all over the country, was knitted together in a single pattern of communication and ideas and song and richness born of the willingness to open the world to the forests in transmitting the rhythm of all cultures, merged and unafraid, woven in passion of communities—yet to be heard from, yet to be known.

IS THERE ANYONE CAN SHOW ME THE WAY TO SAN JOSE?

It seems that San Jose has always had to put up with one of those municipal ego-deficiency problems. It was bad enough when all we had to show were endless orchards, breeding cows and prune pits. But it has become particularly acute since we have emerged as the Los Angeles of the Bay Area.

We always seem to be getting the worse end of the North Counties. You get the fog and we get the smog. You get the hills and the views; we get the hot flatlands. You get the *Chronicle*—no matter how bad—always jazzier and irreverent and sophisticated; we get the dismal, fat, grey, overstuffed, pretentious, boring *San Jose Mercury-News* (which, would you believe it, still runs those Christly religious uplift cartoons on the Saturday editorial page).

San Francisco: the flagship, the turmoil, the center. KNBR and KCBS and KGO are the West Coast offices for NBC and CBS and ABC, which means you get the best bureaucracy that those august organizations can come up with. What do we get? KNTV: a boorish, amateurish, incredibly rich local outlet started, of all things, by a baker.

In San Francisco, there are a dozen dark, mysterious, cheapo foreign film houses; here, we have the San Jose Drive-In—the only porny drive-in in the country. It's enough to make commuters of us all.

And when you say "I'm going to The City," it never means a trip to San Jose or Oakland, God knows. The City has always been San Francisco. Which is part of the brainwash process: always being humbled by the intellectual forest fires up there, the kultchur out of the North. And it gives us a powerful thirst, a real Brain Drain problem.

For just like the southern U.S., where the intellectuals and poets and writers go to Chicago or New York because they are so bored, Sunnyvale and Santa Clara and Mountain View and San Jose always find their wittiest and most brilliant people hitching north to the mystique of The City—leaving behind all the troglodytes who fill the voids of suburbia with their garish progeny.

For those of us who stay behind, amid the used car lots and the plastic-burger bars, there are compensations. Little is expected of us, and we deliver little. Our intellectual and artistic and social lives have much in common with those of San Diego or Houston or Seattle: a little precious, a little *fey,* a little self-conscious.

We find ourselves pleased with the small crumbs, the dibs and dabs of humanity here and there, as we hum in unison, in mock grandeur, the pop theme song of the area from three summers ago:

> "*O do you know the way to San Jose?*
> *[Bum-bum-bum-bum-bum-bum]*
> *Put a thousand down and buy a car...*
> *[Bum-bum-bum-bum-bum-bum]*
> *O I have lots of friends in San Jose..."*

There are hidden pleasures, hidden amidst the endless tracts—a spot of dew on the cactus as it were. Manssur's Shish Kabob (a Persian restaurant: small, dark, cheap, good), four blocks from the parking wastelands around San Jose State. Two Japanese restaurants (small dark, cheap, good) hidden like nosegays in the bleak industrial district. The thousands of tiny Victorian homes stuck here and there in the side streets of San Jose; spared from the Caterpillars, we would imagine, by the very ease with which the city tolerates the brutality heaped on the orchards that once prospered in the outskirts.

And there is the good feeling, the kameraderie, among those of us who try to speak or think or act out of concert with the asphalt and concrete—the few of us in the county who have found each other. After all, we are the nervous minority on a mindless Progress-bound ship-of-state called Santa Clara County (which, remember, grew 70% between 1960 and 1970; and seems hellbent on repeating that figure in the next ten years).

We get to know each other through our meager turnouts, rallies with 20 or 30 people to protest some new concrete battleship shopping center, or some proposed 13-story grey "condo." We are a precious minority, the small finger-in-the-dike band, like that dismal group of 700 who met together in Los Angeles last year to celebrate "Driveless Day." All the others, the 2.3 million commuters, were tied up in the traffic, and couldn't make it.

It is this San Jose ego problem which, we would suspect, is giving us some recent strange events. The search to be Big, and Smart is making San Jose State College into a University next month; although no more physical changes—outside of the usual annual gargantuan growth cancer plans—are contemplated. And the San Jose City Council, under the aegis of a sweet grandmotherly type by the name of Janet Gray Hayes, *almost* questioned, officially, the annual $200,000 dole to the local Chamber of Commerce. And now there is the business of "Doc" Herrold...

Comes now one Gordon Greb, of the Department of Journalism and Advertising at SJS, and he went all the way to Chicago last month to pitch the National Association of Broadcasters. Dr. Greb told them that Charles David Herrold started regular "broadcasting" (as opposed to the usual point-to-point "narrowcasting") in 1909, in San Jose, over what was later to become KQW.

Therefore, said Greb, San Jose should be appointed title as the original and official Home of American Broadcasting. (The NAB, I may point out, which is always far more interested in tomorrow's advertising contract that yesterday's point-of-history, merely said that it would take the matter "under advisement.")

I wrote to Dr. Greb for more information. He sent me a copious letter of defense and explication, rooting for the good doctor, pointing out that there had been many others (Reginald A. Fessenden, for one) who had broadcast music and talk from time to time—but Herrold was the first to embark on such activities on a regular basis. He also included the transcript conducted with Doc over his station KQW, back in 1934, the 25th anniversary of the start of broadcasting.

It was and is a funny interview. I don't mean funny ha-ha. I mean—well—I expected The Father of Broadcasting to be more into radio, and the art of transmission, and the magic of The Aether. I wanted him to be...portentous, and respectable, as befits one who has done so much for 20th century American culture.

Perish The Thought. In 1934, Doc Herrold was full of all the fantastic profits to be made from broadcasting through the miracle of advertising. Listen:

Interviewer: Have you divorced yourself from radio altogether, Doc?
Herrold: Not by a long shot, I haven't. But I'm playing at an entirely different end now—radio advertising. You see KQW was the first radio station to do direct advertising, and I was the first man to sell a bill of goods in front of a microphone. There's a great future in it. I am now acting as advertising counselor for various stations and writing for radio merchandising...

Of course. Who would have suspected that it would be any different? San Jose, The Home of Broadcasting. And Home of The Hard Sell, too.

I am sure that the doctor, one of your early commercial innocents, would have no idea of the look of his words some 38 years later. And I would imagine that, what with the American radio spectrum having been whored so spectacularly and completely, he would have no idea of the gross irony of his phrasing. A bill of goods, eh Doc? The inventor speaks; instead of a courtly Edison, or a shy Wright, we get sold down the river—at once—without a moment's hesitation. A bill of goods!

Whatever did happen to Charles David Herrold, and KQW, and San Jose, anyway? The Doc: he finally passed on up to the great hawker's box in the sky, and now sells us no more. KQW: they stole it from us, The City did. Natch: now it's KCBS, babbling away with 24-hour dimwit hardsell news. The first and the original Whore of Babylon—snitched away to the rich bright lights and sounds of The City.

And San Jose? It continues to muddle along: making a buck here and there, sweating under the smog passed down from the hills to the north. A farm boy, in a sharkskin suit, wanting to stick his bare toe in the dust; a wad of bills jammed in his pocket from all those orchards flattened out, all those things stolen from us by them smart-assed *kultur* people from the nether parts of the Bay.

(*The S.F. Bay Guardian,* May 25, 1972)

SEX & BROADCASTING

A Brief History of Listener Supported Radio

Listener-
supported
radio was
started in
1946 in the
mind of Lewis
Hill radio in
1946 listener-
supported KPFA
in Berkeley was
on-the-air in 1949
for sixteen months
listener-supported
Lew Hill 1946 news-
paper man in Washing-
ton, D.C. 1946 all the
commercials listener-
supported radio started
in 1951 KPFA went off the
air in 1951 Lew Hill said
in 1951 for lack of funds
KPFA in 1949 Lew Hill said
listener-supported, free-
forum in 1946 no radio station
would allow in 1948 Pacifica
Foundation was off-the-air in
1951 from lack-of-funds then in
1954 the famous marijuana program
off-the-air Gertrude Chiarito and
Chuck Levy back on in 1952 with 1300
supporters then in 1957 Lew Hill went
on the air in 1949 no commercials Lew
resigned after calling Board meetings
in 1946 this man left his newspaper office in 1957 Lew Hill they played commentary programs, panel discussions in 1949
one studio-office on the tenth floor in
1957 Lew Hill then in 1960 Louis Schweit-
zer in 1957 Lew Hill then they got KPFK
in Los Angeles and the people remember
Wally and Chuck and Virginia Chuck once
said in 1946 Lew Hill in 1957 Lew Hill
and the fights, a trail of dissent and
despair in 1957 Lew Hill then there's

this Paul Dallas just one of a trail in 1946
Lew Hill in despair in 1960 then Chris Al-
bertson said to Al Partridge in 1949 this
strange eccentric peaceful they fought over
KPFK even before it in 1949 for fifteen months
or was it in 1957 poor Lew Hill riddled by in
1964 the FCC became involved in 1951 they had
all given up and then nine months later in
1957 Chuck and Virginia and William he said
"What can we do to get it taken off the in
1953 the finances were so desperate always
the desperate war of people in love with an
idea in 1961 finally the three stations
the battle over Pacifica and the CSDI and
in 1965 there was this man with a strange
Hallock Hoffman free and eternal hope in
1951 they thought it was the last of
1949 nothing to be done Lew Hill and
the terrible Corocidin or is it Coro-
in 1957 the strain of listener-supported free-Alan Rich fighting the
San Francisco Symphony all by him-
in 1957 the strain and Lew Hill
resigned but returned and subscriptions at an all-time low
and Bill Butler remember his
long inter-office in 1946 Lew
Hill after ten years of fighting remember the fights who
cares about programs my soul
is at 1949 so bright and
hopeful that in 1957
there was no more 1952
when they came back
WBAI & KPFK who's this
angry Dallas to say
in 1957 mad and lost
Gert said a week
after he came back
to her dream all
wet she said you
can't you're dead
and he whispered
it's all lies and
rumors spread by

Book Two

> Good radio is a feedback system, providing society continuing information about itself so that society can be self-correcting.
>
> Like society, a servomechanism without negative feedback goes wild, cannot function, hunts and oscillates.
>
> Correct information can set the system back on the correct 'course'—like an aircraft autopilot.
>
> KDNA, when it was functioning at peak efficiency, had such a self-adjusting system. It was the four telephones which could be linked together and put on the air.
>
> It was the people who called in on these telephones, commenting on our programs, correcting our factual or interpretative errors, adding, constantly adding new bits of information and details, feeding their ideas back into the society serving and served by our constantly changing signal.
>
> The listener fed back and became KDNA. KDNA became a self-correcting servomechanism of the community that was part of it...
>
>   Jeremy Lansman
>   Manager, KDNA
>   1968-1973

Part Seven

"From 1950 until the present, KPFA, KPFK, and other like-minded listener-supported operations have managed to maintain their on-the-air presence as an excuse for what are variously called 'Internal Politics' (by the staff), 'Back-Stabbing,' (by the ex-staff), and 'Madness,' (by all the other outsiders)."
  —The Radio *Times*

SEX & BROADCASTING

S E X & B R O A D C A S T I N G

The Dixie Songbird
and Other Assorted Silliness

173

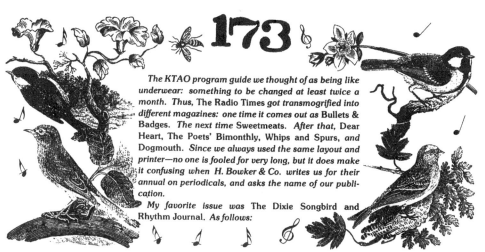

The KTAO program guide we thought of as being like underwear: something to be changed at least twice a month. Thus, The Radio Times got transmogrified into different magazines: one time it comes out as Bullets & Badges. The next time Sweetmeats. After that, Dear Heart, The Poets' Bimonthly, Whips and Spurs, and Dogmouth. Since we always used the same layout and printer—no one is fooled for very long, but it does make it confusing when H. Bowker & Co. writes us for their annual on periodicals, and asks the name of our publication.

My favorite issue was The Dixie Songbird and Rhythm Journal. As follows:

    The Editors of *The Dixie Songbird and Rhythm Journal* are happy to be sending complimentary copies of this edition, our 53rd anniversary issue, to our many friends in journalism and the arts around this musical country of ours.

    "Music is to the beast as the savage sound," said the ancients. We at *The Dixie Songbird* like to think of ourselves as the elves of melody. Indeed, we think of music as The Staff of Life, and often sing our troubles away in four part harmony here at the office on Beale Street.

    The Dixie Songbird advocates many cures for the ills of the world. It goes without saying that we advocate singing in the shower, humming in the breakfast nook, and bursting into full melody on the road. "Sing your blues/Sing the news/Two-by-two/Ooo-ooo-oooo" is the way an old hymnal has it.

    Once when the late Harry Truman was facing one of the many crises that faced him when facing the country, he got up in front of his Cabinet and asked that the members all sing a chorus of "Froggie Went A-Courtin' " to ease their troubles. Would you believe that the next day, their problems were solved? So we here at *Dixie Songbird* suggest that you "Sing your little troubles away/To trouble your songs today."

The Editors

Pier-Luigi Zucchini

by
Vincent
Pepperoni

Director, the New Salem
Music Box Song Cycle Fest

Among the many exciting finds in the revival of old music, none is more noteworthy than the rediscovery of the great Pier-Luigi Zucchini.

Although widely acclaimed in his own time, Zucchini fell into disrepute during the late seventeeth century reaction against green vegetables, and remained virtually unknown throughout the later Baroque, or Pre-Cambrian period. Fortunately for music-lovers of our time, however, he was rediscovered lurking in the KTAO sales office several months ago.

Zucchini had a large green head and spikey fingers, which earned him the sobriquet of "The Green Priest" in contemporary literature. Valdez, for example, in his massive monograph "Seventeenth Century Venice and the Pickle Barrel," refers to Zucchini as "a head above all others in the field—and far tastier. Fry with butter, olive oil, and a little chives."

Pederast the Elder however, much respected for his tome on the Late Renaissance, refers to Zucchini as "a little withered number, with roots too deep in the muckey-muck of Venetian Court life for comfort." Be that as it may, Zucchini quickly established himself as the leading exponent of the *al diente* style, and his compositions were performed everywhere with relish on the side.

After leaving Venice for the court of Abalone, the Green Priest disappeared from view for several years, leading his biographers to suspect hanky-panky or worse. We next encounter him leading a string band (or possibly a string bean: contemporary accounts differ) at the Duchy of Antipasto. About this time, Zucchini began experimenting with the large scale works which were to earn him his subsequent obscurity.

His "Pastrami Sonatae," a series of seventeen thousand Cantatas for each feast day, were published between 1647 and the morning after. Many of these works have been lost, by sheer good fortune, but the fragments that remain mark Zucchini as a consummate master of the picayune. Scored for large forces (double choir, fat soloists, and military band *obbligato*) each work lasts no more than eleven seconds but seems much longer. The "Sonatae" received several public performances, as a result of which Zucchini was deported.

In his later years, Zucchini turned his attention to the more intimate forms of the chamber ensemble and chamber pot. A striking series of quartets for transverse flute, viola da gamba, oregano and chili pepper flowed from his pen, sometimes dripping off the desk and staining his cuffs. During this period, Zucchini also wrote his monumental treatise on edible counterpoint, which eventually came out in paper but was snubbed by *The New York Review of Books*. Furious at this affront, Zucchini called the editor a "saltimbocca" to his face and left in a huff. Reaching Bologna the next day, he traded his huff in on a Pierce-Arrow and continued his pilgrimage onto the Holy Roman Empire, into the pages of history, and out of our lives. There he rests fitfully to this day.

—Jon Gallant

REVIEW: THE GIRL IN THE CHAIR

by Festerr Banks
Dixie Songbird's
Critic-in-Residence

(*DMT News*) — I have been sitting in the reviewer's chair for over 25 years now. And I have to admit I have never heard nor seen a record to match this month's *The Girl in the Chair* (Dogmouth BALH—1760) which came to my attention this week. Came to my attention: hell, it ran out and bit me on the leg.

The performing group is called *The Roots of Madness,* and it is said that they had something or other to do with the closing of Agnews State Mental Hospital in Milpitas, California. Let me quote from the liner notes so that you can get a taste of *The Girl in the Chair:*

"It was Donald [Campau, one of the performers] who was our greatest inspiration," says Joe through his speaking tube. And early—and unfortunate—bout with Kyphosis makes it impossible for Joe to use his larynx. Doctors have fashioned an experimental plastic reed which is inserted directly into his jejenum. Thus the piercing sound in some of the compositions as Joe 'plays' with his voice as if it were a flute. "Don was an inspiration because of his willingness to play piano *with only one arm.*"

"Here Joe is referring to Donald's tragic experience at age 5 when he lost one arm in a player piano ratchet. And yet, ironically, it is the player piano that Don favors the most—hitting it (pounding it, really) with his one remaining arm. 'It may be twisted and horny—but man! Can Don do the old 88s!' says Joe with obvious approbation through his clear plastic flue. Joe likes to think of himself as 'An Attentive Ear' as Don races up and down the very instrument that tried to 'eat' him."

I have been reviewing records from The Editor's Easy Chair here at *Dixie Songbird* for over two score years, listening to the good and the bad, and I have yet to be exposed to the *set of The Girl in the Chair.* The liner notes continue:

"It would be impossible for me, in this short time, to give you the intimate feeling of inspiration, musicianship, guts and sheer sweat that sets *The Roots of Madness* apart from other folk, blues, rock, and modern musicians. I first hear them on Kountry KTAO—a Santa Clara County radio station which calls itself 'a peep-hole into the very nuts of life.' Twice a week, Geoff [Leskovsky, another of the performers] would wheel himself into the studios of KTAO, put his hook into the trumpet, adjust his speculum bag—and blow his heart out.

"It was and is a very moving experience. Even the janitor whose job it was to clean out the bile-pipes would be moved to listen. A visitor once asked him what he thought of Geoff and His Residue (as the group was called then) and he said 'It's a Gol-darn mess, it is.' It was said that he was frightened by 'Old Gimpy' Dolfin who...would remove his stump and leave it on the table as he was working the short wave receiver."

Alas, dear Reader: you can guess with what trepidation I put this record on my turntable, and turned once again to the notes written by the irascible and obviously half-witted liner-jacket writer of small fame:

"*Listen:* Let your ear soar into 'The Old Man's Ass' [the third selection on side 2]. Remember that this brave group, *The Roots of Madness,* can count only five arms, seven legs, two glass eyes, an iron lung, a half-dozen mouths, and one bladder pump among them. And yet—despite such odds—they can produce such music. *Such music!*

"Don't be put off by their self-mocking fantasy. You might hear the rattling of Don's Sputum Cup on *The Flight of the Ocka Bird* [4th Selection on Side 2]. You might not even be aware of the fact that Geoff had a *petite mal* seizure during the actual recording of *Realisation II* [1st cut on side 1].

"Forget all the rumors of incest, pederasty, masochism and onanism that their detractors are trying to lay against *The Roots of Madness*. Even in San Jose, the Prune Capital of The World—there are vitriolic enemies out of the very Steven's Creek *avant-garde* that spawned this hardy band."

There is only one word to describe the content of the record, the music and artistry after this long and elaborate build-up by this anonymous author.

(*Radio Times #117*)

"The *harpsichord* is in reality a stringed instrument, in which a complicated assembly of ratchets, grommets and cleats contrives to pluck. In the clavichord, the strings are tapped rather than plucked. Young people today prefer an instrument in which the strings are rapped rather than tapped; it is called the rappichord. John Cage has designed a keyboard instrument in which the strings are neither plucked, tapped, nor rapped, but simply stared at.

"Actually, most of these sonatas are scored for flute and continuo. The continuo was a fabulous beast who flourished during the 17th and 18th centuries. He had the legs of a bassoon, 'cello or gamba (which is a kind of crayfish), the body of a keyboard instrument, and the head of a man. The idea was to surround a written-out base-line with an improvised accompaniment. Eventually, the School of Public Administration objected to the notion of not writing everything out, and the continuo became the discontinuo. Nonetheless, some distant relatives survive to this day in the wilder fastness of jazz..."

—Jon Gallant
(*KTAO Program Guide #41*)

SEX & BROADCASTING

DEAR BOB & JOHN

Then there was the letter to Haldeman and Ehrlichman that appeared in issue #123:

#5 University Avenue
Los Gatos, California
95030
31 May 1973

Mr. H.R. "Bob" Haldeman
Mr. John Ehrlichman
% The White House
Washington, D.C. 20500
Please Forward if Necessary

Dear Messers Haldeman and Ehrlichman:

KTAO is a small community-based broadcast station that serves the entire Santa Clara County area with unusual music and talk programs. It has been our plan for the last few months to establish a viable news department. That is where we think you might be able to help us. We understand that one or both of you might be free for outside employment in the next year or so. We would like to start off our public affairs programming with a bang—and that is why we are offering you a position with our small but interesting radio station. It would be completely up to you as to how you ran the news and public affairs programming on KTAO. You might choose to run an 'in-depth, investigative' radio service. Or perhaps you would be more easy with a droll, casual dialogue of the Huntley-Brinkley type. In any event, we are sure—given your backgrounds in public service and government—that you can give our listeners a new insight into the processes of power: sort of a living, day-to-day civics lesson, if you will.

I am sure you are thinking that I am some sort of an aggressive kook to be demanding that you accept such a job offer out of the blue. But please understand that those of us who work here on a daily basis think of ourselves as in desperate need of the techniques to leadership that the two of you can provide.

For instance, one of the lacks here at KTAO is in the looseness of our office and staffing. I can barely get the people down here in the morning to work—much less disentangle them from their Dr. Peppers and away from telling silly stories about their last night's activities. I understand that in your previous jobs, you both worked hard, kept your staffs well in line, and could be expected to put in 50-60 hours a week. We surely need this sort of dedication.

I would be lying to you if I were to pretend that we don't need your fund-raising abilities as well. The radio station is ill-financed, and we can hardly pay the cost of the garbage man, much less the cost of stoking up our wood-burning transmitter on Mount Uhnumum. Frankly, our money position is a disaster.

For that reason, those times when you were not out digging up local news, we would want you to be prepared to make contact with your many many friends in the world of finance and business in order to make our money problems a bit more manageable (and pay your own salaries!) Please be assured that we would not be expecting the gargantuan contributions required in your last job; I could say safely that our financial demands are so slight that I know they will amuse rather than bore you.

There is one thing I should say right here, and I hope you do not think it presumptuous of me. That is—we have a diverse and original group of people working out of KTAO. Although, of course, you have a right to demand loyalty of them, we would hope you would restrain your efforts to keep them in line. A talking to : yes. Inter-office memos: naturally. I just would not want you to use—how shall I say it—electronic methods, or appeals to government anti-subversive organizations or the IRS to extract loyalty from them. In your news gathering, we will encourage you to avail yourself of the technology for any and all leads (giving you the opportunity to "turn the tables" as it were); but I would want you to avoid using any wigs, masks, or gloves in your work with the KTAO staff. They are quite sensitive, especially in the Town of Los Gatos, where officials have referred to them in a variety of colorful expletives.

The reason I dwell on this is because I am sure that you will value our lack of concern or prying into *your* past. We like to think of ourselves as being tolerant of any man's eccentricities— and would only ask that you accord your co-workers the same freedoms.

I'll bet you can well guess the excitement of our listeners and volunteers over the possibility that the two of you will be coming to "The Pruneyard of the World." I would imagine that you may think that a job for KTAO might be a step down in life; but realize that our enthusiasm in opening this electronic paradise to you may well compensate you for any loss in salary or position you may have to take to come here. With your experience and reputation—I should think it would be a matter of time before you have a whole valley eating out of your hands. Ours is to accept, and not prejudge. Your critics may be harsh, but you can understand that we will be just that more open to you. I would think that the new pleasures that await you in our area will delight if not overwhelm you.

Sincerely,
L.W. Milam,
President KTAO

(*Radio Times* #123)

DEAR HEART
■ ◉ ■
The poets weekly

.29 lb.

The Pieces Fall Into Place,
And Life Goes On

Issue # 112 of the program guide was Dear Heart, The Poets' Weekly. *It contained some mush, some great poetry by Roy Fuller, Edna St. Vincent Millay, "Warty Bliggens the Toad" from* archy and mehitabel, *and this biographical note about Pete Winslow.*

PETE

I guess you could say that Pete Winslow was one of the Beat poets. He liked to declaim when he was reading; and he always had a copious supply of beer and pizza to wash down his readings. He came up to Seattle once, and read his poems on KRAB. I think we liked him because he was good, because he published all his books himself, and because he was obscure. We felt like he didn't belong to many other people.

His poems were always slightly irreverent, and heavy on mad detail. He published them with his own cut-outs (he was using Victorian drawings for decoration long before the fad)—and distributed them in person. The best of the bunch is *Monster Cookies*.

Somebody told me that he died recently, in Pleasanton, of a heart attack, age 42. I don't believe it. For one thing, Pete Winslow would never die in Pleasanton—although he worked for an obscure suburban newspaper for years and years. For another, he would never consent to die from a heart attack: never. Buried by a bull-dozer, sucked into a cement-mixer, suffocated by a gross of corn-flakes; anything but a heart attack. It may be true—if they keep saying it—but I for one refuse to believe it. Somewhere out there, he is finishing a whole gallon of Marca Petri, by himself, and getting on stage—at Fresno State College. The audience is small but enthusiastic. They laugh raucously, applaud every line. They love it, he loves it, as he roars out the words to "Hurricane Fred:"

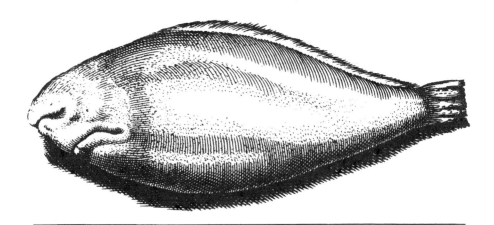

A guy came alone on a horse
Shouting into a bullhorn that the turtles were coming
We said so what
He told us they'd eat the furniture
Drink the gas from the cars
Run up the phone bill and keep the lights on in the daytime
Well we battened down the hatches
And sure enough they came millions of them
Moving in off the freeway
Eating doorknobs and drinking fuel
Wanting only to be loved
We gave them love took them into our homes
Let them eat and drink what they wanted
Let them sleep with our daughters
And at last they went back into the swamp
Everyone pitched in to clean up the mess
We scrubbed the turtle poop off of everything
Until the town looked the same as before
Now there's just the children with shells on their backs
To remind us of Hurricane Fred.

SEX & BROADCASTING

People are always asking me to write letters of recommendation for them. A job I sincerely loathe—because obviously, if I didn't think highly of someone, they wouldn't ask for such a letter.

One of the ones I favor the most was written for Mitch Green, now news director of KPFT. When I sent Mitch a copy of the letter—he called me and inquired if I really, really and truly, honest-injun, cross-my-heart, had sent the letter to his draft board in Flushing. I told him I had (I had). He hoped he begged he asked (he told me later) that it would be lost in the mass of papers he had sent them.

It was, and he didn't have to go off to Viet Nam. But I still think that this letter would have helped.

29 May 1970

Director
Draft Board
United States Government
Flushing, N.Y.

Dear Director of the Draft Board:

I am writing you to give a reference for my friend Mitchell Green who I understand is now in negotiations with you about being drafted.

I have know Mitchell Green both in Seattle at KRAB Radio Station where I was...and at my boarding house when I got to know him and again here in San Francisco where he works at KQED. He's a good boy and bright. I think he will go far.

I think my friend Mitchell Green will be a credit to any branch of Your Service. I think you will find him of the right caliber to go anywhere in the world as he is very ambitious. That's how he got so high in the T.V. Business altho he is only a boy.

I trust Mitchell has told you all the problems but I am sure that it will be no hindrance to him in uniform. I think the troubles he was in at College were all a part of what the young kids are going into with a confused world such as you and I brought them into, eh? I understand the police have it out for that College and altho it is no place I would want any of my boys going to with its radical ideas and the teachers teaching the way they do still I think that just because Mitchell is on their Campus he shouldn't have to get such a scrape with the police. He told me in confidence they are *out to get students* which is one reason I think the Army might be the best place for him to get him out of the hands of those people telling the things to his mind.

I hope you haven't been put off by his appearance. When I first met Mitchell in Seattle there was something said about his appearance what with the drooling and all. But I can tell you I was relieved when I found out that it was not a congenital thing and only occasionally does Mitchell have that problem with his head. I still think he would be a good worker although some of the others would complain about him screaming at night, said it kept them awake & etc. Mitchell explained to me once when he was a child his father used to beat him for misbehaving in the bed and that made him scared of his shadow so to say. The old man would check under him every hour to see that he had not wet. All night long he did that. And it he was even the slightest bit wet he would lash the boy again and again until he was forced to cry out even tho he is a brave boy. I think you will agree this is no way to raise a small child and I think it is no wonder that Mitchell has the problems at night now. But as I say he is brave and strong-willed and that's why I think he would make good Officer Material in your Service. He fought for two weeks against having to sleep on a rubber sheet here at the house. But I told him I wasn't about to change his sheets again (you know I'm the one gets to iron out all the spots). I can tell you we had a good to-do over than one but like I say he is a good boy down deep and when I offered to kick him out he responded just like a man. He said, "I will ok the rubber sheet if you don't kick me out." I didn't and he stayed on just like I knew he would.

One thing I could like to caution you with. I don't think you should get Mitchell too close to guns until you have him completely trained. He gets depressed easily and one time I came on him in his bunk at the boarding house and I asked him what's wrong? He wouldn't say anything he never did when he was depressed. Sometimes he would go on like that for days even weeks. Not talking and sitting by himself in his bunk or lying staring at the wall. I almost passed by but then I saw something sticking out under the pillow. I pulled it and it was a gun. Now I am not saying he would use it on himself but I took it from him. I say with moody boys you shouldn't let them play with anything like that. I thought he'd kill me, saying 'Give that blankety-blank thing back.' I wouldn't though he is strong and I thought he'd break my arm.

Still I think that is a good sign for you. If you could get him to love and respect firearms I am sure he would be a credit to your Service and would fight enemies like a man altho I am not sure about having to do things with civilians like I read recently in the paper. But as long as it is killing enemy soldiers as long as they aren't boys or children I think he'd do ok.

Now about his relations with other boys. I found in the Boarding House he gets along good with the other boys. You may be glad to know too that Mitchell just shuts up when the girls are around. One time that Joanie Peters from the house down the way came down and she had a real thing on about Mitchell. Well she said something to him and he didn't like it he spat right in her face. I thought she would blow up right there but then she went away quick-like which suited me just fine as I didn't like those co-ed girls from down the street too much anyway. I think it will be a relief you being in the Army and all that Mitchell makes real good friends in fact seems to prefer the company of boys. I understand in the army you have to be away from womenfolk for as long as six months or a year which I think for Mitchell will be just fine.

One thing I know you will want to know is what kind of businessman Mitchell is—how much he likes to work & etc. Well while he was staying in the house he got involved in selling spare car parts and old television sets. I used to get mad because he always had his room so *full* of that stuff. Hub caps and tires and steering wheels and especially radios from cars and those new tape machines. I never knew where he got them they were always appearing as if by magic and then disappearing again. He was coming and going all night with them and finally I had to tell Mitchell that if he wanted to go in business he would have to do it elsewhere and he didn't talk to me for awhile but finally he got rid of most of it although I don't know where. But I think his love of finding machines and bringing them home might be some help in the Army although I am not sure how.

Well I go on too long about Mitchell because I like him. He comes down here to stay with me from time to time ever since I had to get out of Washington State what with the police problems and all. I didn't want to leave but you would never believe how strong they are up there about *anything*. The neighbors ganged up on me and got a crooked judge to tell the police that if I ever talked to any of their darlings again I would be in problems but that is my own business, not yours and anyway I prefer the sunshine in California you can take that rainfall and those police and judges in Wash. and stick them.

As I say I think he'll be a credit to the boys in blue. Anywhere you send him I know he will be a regular little fighter. I know his mother will be proud of him too. I got to know her when she called first after he came out I had to calm her down and all tell her that she didn't have to worry about her baby boy. She had a stroke after he left, couldn't walk around and all. One time she told me that more than her legs what she wanted back after all these months was her baby. "He's a good boy," she told me. "A hug from Mitchell, and a kiss. That's all I'm ever needing until the day I die," she said. And she sounded sort of sad when she said, "And that may not be too long from now..."

Sincerely,
L. Milam ("A Friend")

Doug Cruickshank and I went through personal transformations in the act of putting out the KTAO guide. He learned to be a great layout man. I learned to fabricate the worst sort of silliness—to the point that most people claimed that it wasn't a program guide at all: rather, just a place for us to be as dumb as possible. The silliest, truly the silliest (I think) was:

Some of our readers have asked us—in hurt tones—where we get off telling *them* about pain. They say that they have gone through agonies that we never imagined.

They tell us—in exquisite detail—experiences with thumb-screws, nails, clawhammers, shards of glass and even Old-Fashioned Iron Maidens:

Then they tell us: "Where do you get the gall to tell us about naked, animal hurt?"

Well, the editors of *Whips and Spurs* have never claimed to be so-called experts. Although most of us are no spring chickens when it comes to "handing it out" or "taking it in."

Many has been the time when one of the Senior Editors of *W&S Quarterly* comes to work, sitting down with a groan, or moving so slowly in front of the old typewriter. "What happened," we say. And they reply, with the look of sweet hurt in the one remaining eye: "What a night!"

That's all. And that is all they need to say.

No. We here at the magazine of "Family Pain" are not ones to say we are experts. By no means! We can show scars with the best of them. We can demonstrate egg-size blue bruises, or fine scratches up and down the back.

Oh, yes! We have had our licks. And we neither confirm or deny it.

So—when you readers call or write us to complain about our insensitivity to the hard-knocks that *they* have had, only understand that we are human too.

We writers and editors and simple draftsmen may be associated with the biggest magazine of leather arts in the West: but we have hearts too.

LETTERS

Dear Dr. Buckley,

Sometimes I just have to wonder. Last night I was sitting watching tv with my husband, and he was drinking a cocoa-malt. It was Johnny Carson and just after the Playtex commercial he reached over and dumped the Cocoa-Malt on my head and pulled the wire out of the tv and tried to plug me into the wall.

Now I am very devoted to the idea of a wife following the lead of her husband. I would never be caught dead crossing him because he brings home the money and I am his slave.

Still, this might be too much. This was like last year when we were watching tv and he was drinking Sazerac Cocktails which he fancies and he takes the can of Aqua-Net hairspray and holds it in front of my face and tried to light it which will burn me to a crisp you might know because when I was a girl in Pasadena my brother and I used to do that with the lizards and we would turn them to charcoal within a minute.

Well when he did that I took the toaster which my Granny gave us for our wedding day 17 years ago which was a painful experience I won't go into right now and I took the toaster and crashed it down on his back. He had it there in the living room because sometimes when I go to sleep in front of the tv since I can't stand Johnny Carson my husband tries to wake me up by sticking my hand in it and turning it all the way up to *dark*. I have scars you wouldn't believe...

Anyway, I took the toaster and smashed it down on his back which flattened him and he lay there and smiled up at me and said "Vera. You're the toughest goodest woman a man can have." And he reached over and bit me on the foot hard so that I knew the bones were broken and today the doctor says I might get the ptomaine and who would it be that would bite someone else on the foot. I didn't even tell him about the time when the old man stuck my hair in the Waring Blender trying to break my neck and I had to drop telephone on his ankle to make him let me go.

So what I am asking you Dr. Buckley is this—I have been married for almost 17 years but is it worth it?

Desperately,
Vera G.

Dear Vera:

I was happy to print your letter as it shows a common problem that we all have. That is, how far can we take our love and affection. Troubles are all a part of marriage, and fights are a necessary and psychological release for the little day-to-day tensions that build up. The marriage sacrament is a strong tradition which provides the mortar-and-bricks for a healthy, long-lasting nation.

Your husband is obviously a concerned, albeit hard, but truly and finally a just man. I urge you to try in little ways every day to make him happy. You might try to buy him little presents. A vase of flowers on the supper table might be helpful. An extra pat or kiss in the evening might prevent a recurrence of this temporary difficulty you are encountering. You might consider giving in a little more to what might seem unreasonable whims of a man who has his problems in his office and work.

Doing the little things, in a sacrificial way, is a little price to pay for the years of love and happiness remaining to you as a wife and friend to your sometimes anxious and temperamental husband.

Sincerely,
Dr. Buckley

S E X & B R O A D C A S T I N G

WHAT THE CRITICS ARE SAYING
ABOUT WHIPS AND SPURS

"Next thing, they're going to be asking me to dress up in cleats for the prom."
—Rex Reed

"We could hardly believe our eye."
—*The New Yorker*

"OOOOOOOHH!"
—International Velvet

"It made me want to turn weak with rage!"
—Tiny Tim

"I got so excited I got my finger caught in my electric pencil sharpener."
—Paul Krassner

"Revolting."
—The Chicago *Tribune*

"Corrupt. Disgusting. Nauseating."
—The San Jose *Mercury*

"It made me want to claw the walls."
—Ronnie Davis

"We think they are trying to degrade the publishing industry."
—*Printer's Ink*

"I saw no signs of bestiality."
—Art Kunkin

"It was no worse than a wet dream on an electric blanket."
—Dr. Crane

"I felt tingles of rage and disbelief coursing through my veins, and up the insides of my thighs."
—Jean Genet

"Have you ever wanted to repeal the First Amendment?"
—R. Reagan

"I re-applied for treatment!"
—T. Eagleton

187

"Someday they'll stick electrodes on the private parts of those monsters. And ah hope that it's me turning the switches!"
—James Eastland

SEX & BROADCASTING

dogmouth
A MAGAZINE OF THE ARTS

> I'd rather be a retching Dogmouth than one of your modern contemporary so-called artists. Their sculpture bodies lying naked on gallery floors, paintings on walls showing dirty people in dirty positions with their parts all out, and carvings, looking to be nothing more than a pee-pot of a prickly pear in a plump Parnassian Prostitute. I say—I'd rather be a ragged, wrecked, retching Dogmouth than your modern so-called contemporary artist..."
>
> S. Agnew
> Speech, Ft. Wayne Veteran's
> Memorial Ballroom, 7-4-71

CRISTO ANNOUNCES SEVERAL NEW PROJECTS TO DOGMOUTH

DENIES COVERING STATUE OF LIBERTY IN VASELINE

Special Coating for North Siberian Coast

By Salvatore Deli
(Special)

Cristo has revealed several new and innovative projects to *Dogmouth*—one of which will utilize 'several carloads of Saran Wrap' according to the artist.

Cristo is known to the art set through his massive projects, including the wrapping of two miles of the Australian Countryside in Plastic, and hanging a giant curtain across a Colorado canyon.

In interview with your editor, Cristo vigorously denied a supposed plot to smear the Statue of Liberty in chicken-fat "or even Vaseline," and denounced as "ridiculous prattle" the rumor that he would be covering the peak of Mount Baker in Hellman's mayonnaise.

"There are many of my enemies passing around these vicious, not to say viscous, tales about my future projects. Nothing could be further from the truth," he told me. "I think these people are out to ridicule me—if not Mt. Baker."

During the interview, Cristo happily outlined plans investigated—then discarded. "We were going to wrap the entire Chrysler Building in Saran Wrap," he said: "but then we got a cost analysis from General Foods (parent corporation of Saran Wrap, Inc.) and they said that it might run some $600,000."

"So we decided to investigate the possibility of wrapping the Chrysler Building in the Empire State Building. We've got bids out on that now, and should be getting some sort of a ballpark figure from the Acme Hauling Company of New York. Moving a whole skyscraper is not as expensive as you might think," he said. He denied any designs on the new twin-tower World Trade Building. "That is its own comic work of art," he reports.

Cristo ridiculed the rumor that he was moving into Reverse Art. "I had heard that there were stories about my wrapping a truckload of clear plastic in Disneyland—or something stupid like that," he said. "I can't believe the lengths to which my detractors will go. Hell, I can't believe the lengths my *contractors* will go," citing a quote figure he got on covering the entire North Siberian Coast in denim. "I went to the Levi Company because I figured that they knew denim better than anyone else. But $15 million? Jesus."

Cristo, despite (or because of) his massive projects, is just a little sprite—standing no more than a head above 4'11". This doesn't keep him from dreaming big dreams: like the plan for appealing to NASA for a special grant to send him to the moon. "I wanted to spread a one-inch square polka-dot in the center of Mare Nostrum as a greeting to future space-heads. NASA told me in so many words to take a flying – – – –! At the moon!" he sighs.

This of course has no bearing on Cristo's latest plan which is to cover the entire *inside* of the NASA hanger at Cape Kennedy (the largest contained structure in the world) with an eight-million cubic foot Trojan. "Some people have said that I was vengeful," he states. "Hell,

I just wanted the military to get some idea of what it's like to be screwed—for a change. Besides," he concluded, "The American Rubber Product Company (manufacturers of Trojans) wouldn't even talk to me when they heard what I wanted. You should see the six-foot thingie I'm sending their president. In case he wants to put it on. And scare his wife."

CONCERTS OF NOTE IN CALIFORNIA
FOR THE FALL

*by Igor Seneferin,
Special Dogbone Correspondent*

October is showing itself to be one of the most exciting months that American concertgoers have had in the decade.

First and most important is the preview of the work of Karlheintz Stockhausen of the *Concerto # 17 for Massed Klaxons, Auto-Smasher, Bulldyke 'Choros' and Orchestra, Op. 77.* The piece, rumored to be one of the strongest in the arsenal of the famous contemporary German composer, will be premiered at Candlestick Park by the San Mateo Jr. Symphony as conducted by Sam Stockhausen, the half-brother of the noted composer.

Sam reports that the preparations for the October 21st Concert have been 'unusual' in several respects. First, the Daughters of Bilitus have been notoriously un-cooperative with the drafting of the Choros—since they describe the Op. 77 as 'Saxist'—referring to the lack of saxophones. They promise to picket the performance and stage a lie-in in the restrooms which, they point out, have yet to have their signs removed.

On top of that, the Auto-Compressors Union Local 276 (San Rafael) is balking at the donation of three 12-story tall *Hercules* Model Presses which are to be the main percussion units in the Concerto. According to the score—these massive instruments, each weighing close to fifty tons, are to operate in close and proximate rhythm to the 'choros,' in contrapuntal rhythm to the string section. This section, unusual even for a Stockhausen work, will show a battery of stringed instruments including cat-gut tennis rackets, window sashes, and a dozen or so kites.

"Outside of our problems with the Daughters of Bilitus and the car-squashers local—the kites will add the greatest element of risk to the performance," reports Dr. Stockhausen. "If there is a good wind off the point, we can have all dozen kites up and running by curtain time. If not," he sighs, "we'll just have to get the audience to blow. I don't care what," he says: "We're going to get those plucking kites up in the air, and I ain't just whistling 'Dixie.'"

The concert will conclude with a version of 'Dixie' as whistled by the 105 member orchestra with the audience.

Finally, Milpitas—of all places—will be host to an unusual concert around Christmastime. The Milpitas Arts Council has patiently saved up, for a dozen years, for a presentation "which would have the Bay Area sitting on their pants," according to M.A. Council President Ms. P.P. McFeelie.

The Council contacted the avant-garde composer John Cage in New York last Spring, and asked him if he would compose a special Christmas program for presentation by the Milpitas Hi Cheer Band. "Sure enough" replied Mr. Cage, and the concert was set up for 24 December.

Your correspondent has been sworn to secrecy, by both Mr. Cage and Ms. McFeelie, so we will only list here the program, but *let your imagination do the walking...*

PROGRAM
*Introducing the Milpitas Hi-Y Hi-Fi Club,
the Milpitas Glee, and the San Leandro Bando*

Symphony in Two Parts for 1500 Fire
Hydrants, Soprano, and Band-Aid.

Air "Flow Swiftly Gentle Afton"
transcribed for Dogwhistle
& Doggie-bag Quartet.

"To the Memory of the Grave of Artemus:"
Canzona in the Pie a la Mode
(16 Castrati and Double-Bass)

Finale: 'The Pimple is E'er Fleeting'
Cantata for the 2nd Sunday after
Epiphany for Full Orchestra and
Volcano (in the Sicilian Mood).

There it is—and this is all that *Dogbone* can reveal right now: but you can believe that all of Milpitas will be hopping on *this* particular Christmas Eve.

Then there was issue #131, subtitled The Slaver:

The Editors of *The Slaver* are happy to be sending this, our 125th issue, to many of our friends in the world of journalism.

Many people think that the issue of slavery is one of little concern in this modern world. Nothing could be further from the truth!

In this issue, you will learn of the machinations at Foggy Bottom, and how our own Secretary of State is involved in White Slavery.

Bondage is being practiced in our own United States, and readers will learn of so-called Underground Newspapers which further captivity.

One of our senior citizens will describe a burgeoning industry of slavery in the sports field, and how some of our national heroes are linked to chains firmly enmeshed in the sordid earth as ever existed in The Golden Triangle. To those of our readers who find these revelations horrifying, even revolting—let the brave few stand up and be counted, to chant together the words of our founder Lord Wilberforce:

Im Sacrum Belitum
Cese Fulcrum Detritus
Forbat Pendennis Hoc
Rapid Pusey...

or:

In the sacred belt
Wasted efforts must cease,
There hangs the forbidden face
Of the angry enslaved
Dropsey-filled
Freedom Fighter...

These words, fighting words if you will, filled our ancestors with fire, the divine fire necessary to struggle against whole nations. Let our present mission be no less!

WILL THERE BE NO MORE WHITE SLAVERY AT THE WHITE HOUSE?

In the midst of the entire Watergate controversy, a very alarming fact has been overlooked by the thousands of reporters falling over each other to report malfeasance in office.

It is the unhappy but proud duty of *The Slaver* to reveal the sordid capers of the conspirators—actions so vile and unspeakable to be, almost, unspeakable!

Certain facts have come into the hands of *The Slaver* which demand the attention of a heart-sick country. Facts having to do with the freedom of the white working women in Washington D.C.

This scandal reaches into the very bowels of the White House: indeed, some say, into the duodenum of the Oval Office itself.

And, despite all threats, the editors of *The Slaver* will stand by their promise to reveal all to a horrified world. The story of gentle white limbs enchained to a pulsing hull, of sweet rosy lips being forced apart by a muezzin's evil mask, of tender white flesh being bruised by an oil-soaked, money-bags who cares nothing for the freedom of white American womanhood. Indeed, we are accusing the White House, the CIA, the FBI, the Secretary of State of knowingly and willfully indulging in

WHITE SLAVERY

ITEM: Why is it that Henry Kissinger makes so many trips to the Middle East, the White Slavery Capital of the World? Why is Air Force One so heavily guarded, why is no one allowed to look into the famous sweathole *underneath* the plush offices that carry this man to Iran, and Iraq, and Egypt? Are the reports of muffled screams from below to be trusted?

ITEM: What was it on that famous 18 minute missing tape segment? Did it *really* have to do with a simple 'bugging' operation—or does it have more to do with the fact that around Washington Rose Mary Woods is known as Ms. Slaver (no relation to this magazine!) and in some circles even as Ms. Big? How close is she anyway to Henry Kissinger, and how many times has *she* gone to Iraq?

ITEM: Those carefully groomed young men around Nixon, Haldeman, *et al:* were they simple Nixonites as some have suggested—or were they infallibly attractive to the weak and easily taken-advantage-of young ladies who are and were part of the Infamous White House Secretarial Pool?

ITEM: In present day American life, would anyone truly get worked up about a simple telephone tap—to the point of handling around $200,000 in $100 bills? Or, is there something bigger, something that would earn John Ehrlichman the title of *The Whip* by those who knew him intimately?

ITEM: Tricia Nixon has been strangely silent during the revelations enveloping 1600 Pennsylvania Avenue. Yet *The Slaver* has found out that she may be, in part, the 'bag woman' of the Secretary's office. That she would invite one of the hopeless and helpless young functionaries over to her room in the West Wing for tea and begin asking embarrassing questions about that woman's interest in or fear of men in *djellabahs*.

ITEM: "Toots" Raefer, once a humble young dictating 'machine' at the White House, was last seen stumbling, drugged and battered, out of an earthenware hut in Ahleeb-mal-Keinsa near Port Said. Repeated questions addressed to Ms. Woods alias *Leather Lady* earn nothing but pooh-poos and suggestions that your editors are "grasping at straws."

ITEM: Why is it that all newspapers (except this one) have stopped reporting on the 'dating' habits of Henry Kissinger?

ITEM: Was Julie Nixon's "operation" paid for by a check laundered through the Bank of Cairo?

ITEM: What is Maureen Dean's exact relationship with King Faisal?

ITEM: Why was the "Oil Crisis" suddenly *over?*

Far be it from *The Slaver* to point a trembling finger at anyone in the administration. Today's bondage master is often tomorrow's president.

But we remind our readers of the Spartan Code, elucidated some 2500 years ago by Pluto, which said:

> *Non carborundum*
> *Qui discubraetes*

or

> *Don't let the mothers*
> *Catch you.*

HOW I CAME TO GO TO THE FLESHPOTS OF DUNGUNAB:
THE CAPTURE; SAILING OUT ON THE POOP; OUR ARRIVAL IN UNM LAJJ,
AND THE ADVENTURES AND FEAR THAT BESET US
ONCE WE HAD REACHED THE STEAMY HORRORS OF TRINIKITAT

A Confession by Karyn A. Baker

I had dropped out of Mother Butler's only a month before, bored with life and tired of lecherous old men and impotent nuns.

I was in the habit of going to San Francisco—the North Beach area, and loitering in front of the "Baghdad," hoping to catch a glimpse of the belly dancers. They wouldn't let me in because I was only fifteen.

One balmy night as I sat on the curb smoking a machine-rolled cigarette flecked with hashish, a dark man wearing a scarlet turban and a golden snake approached me.

This did not startle me, I was used to such things...but he showed me an exquisite crystal statue from his silk vest pocket, a strikingly exact depiction of Louis Quatorze.

I asked him where he got it, and he said, "I have access to many beautiful treasures. I come from Persia. Would you go to Persia with me?"

I questioned him closely on his taste in music and theater, and discovered that he had been intimately involved with Buckminster Fuller.

I decided to go to Persia.

We went to Fisherman's Wharf and boarded an old tugboat, strangled with barnacles and rotten ropes.

As I went below to the poop my mysterious companion (whose name, I discovered many months later, after intensive investigation, was John Smith) grabbed me by my long blond tresses and dragged me across the splintery planks to a dirty seedy dark corner.

After a brief scene on the floor, we heard a chorus of horns calling to each other.

Quickly I grabbed an old piece of sailcloth, wrapped it around my pubescent body, and climbed up through the hatch.

The Atlantic was as calm as a bathtub, the white wind smiled, and there before our wretched little boat was a big white ocean liner, with a gleaming brass railing going clear around it, and leaning casually on the railing was a remarkably well-built white-haired, white-bearded captain wearing a white uniform and shiny white brass buttons.

Deftly I dived into the seawater and he threw me a lifesaver.

I began to swim in and out of the lifesaver for I was a mermaid of the deep, or at least from the time I could swim (age 16 months) I thought myself to have a flipper where other little girls had feet.

Immediately two more conspicuously well-endowed young men rowed up in a life-boat and took me on board.

After a brief if not unpleasant scene on the deck, I was taken to the Captain's quarters, and after a brief if not unpleasant scene in bed, he gave me one of his T-shirts to wear and we went to dinner.

There were many beautiful women dressed in silk and chiffon, dining on artichokes and caviar. I was self-conscious, wearing only a T-shirt and no panties, but the captain whispered to me, as we entered, that I was the loveliest of all, so I forgot my embarrassment.

We sat at a small table and drank champagne. I became immersed in watching all the beautiful women, and wondered if any of them were belly dancers, when the door opened and in walked an extraordinary woman with long black fine hair and one breast bared.

After a pleasant conversation and a delicious dinner, she took my hand and led me to her private suite where after a long if not unpleasant scene she told me that the next day we would land in Persia.

I said: "Persia! Again! O no, o no!"

She was immediately compassionate, and said "You darling!" and kissed me again and again.

In the morning, after discovering the true utility of the bidet, I heard once more the calling of the horns.

We arrived none too soon in the mysterious town of El Quseir ("The Place of the Dark Lover") and I was led blindfolded along many broken streets. I could hear the cries of the beggars, and could smell the perfumes of Araby.

When we came to the Palace, they removed my blindfold. They lined me up in front of a blue-&-gold tile wall with 17 other girls, many of them younger than I. I was still dressed only in the T-shirt and sheer panties, and many of the prospective buyers looked at me closely. Their dark faces were enigmatic. I could only sense their feelings by the flaring of their dark nostrils.

Finally, after the frantic bidding in a language I could not understand, I was sold to the King of Jidda for thirty thousand: John Smith got twenty-five thousand, and I got five.

Mary, my beautiful dark-haired lover was allowed to go to Jidda with me to be sure that I did not smoke, scratch, or pick my nose and thereby offend the king who would behead me.

Upon arrival in the fairy-village of Jidda, we were taken immediately to the preparation room. While my fragile clothes were removed, I was beset by four Geishas and a Puerto Rican janitor with a golden back.

After this lovely escapade, I was splashed with exotic oils, garbed in a soft white *sari* which was embroidered in gold-silver-blue flowers, with a thousand thousand tiny mirrors which cast the light like stars. Lilacs were arranged in a crown on my head, and Mary brought me a necklace of crystal and silver.

Finally, amid great festivities and the soft thrilling sounds of 50 *ouds*, I was presented to the king. I rose from a graceful genuflection to see his brown eyes—as mysterious as my long journey—staring into my own.

His tipped fingers were keeping time with the chorus of *ouds*, and we fell to giggling.

He giggled and I giggled and soon the entire court was giggling, because this was the year 1926, and when the king did anything, the whole court did the same thing. Even when he chopped off heads, because he was so young he would often break into giggles, and the whole court would too, despite the blood-soaked ceilings, and the blood-stained rust-tiled floors.

We had many refreshing if not uneventful scenes in all parts of the palace, even the beheading room, and on the 1000 stairs that lead up to his billowing chambers.

Well, I spent my five dollars on belly-dancing lessons, and became a curiosity in the city...for I would steal out at night to smoke and drink in the cabarets, and present little plastic gewgaws to the prostitutes.

During the day, life was quite normal in the harem where we made up all sorts of games with balloons and kites and pretzels to pass the time.

Mary became fat and I had to let her go with tears in my eyes. I gave birth to ten children.

Now I am over 70, but they say that I have never lost the witching beauty that made the King of Jidda love me and no matter what I did with the other members of the harem, he never cut of my head. My daughter became a pirate. She was my very favorite.

I never aged, they said. I never lost my beauty. The King once said to me that I might well be immortal. I don't know, will never know, should—perhaps—never know. However, I think that it may be true because once, ten years ago, when John Smith came back, he said to me

"All of this
All of this was in
The Plan..."

It was then, I heard, for the very last time the faint calling of the horns.

I knew I would never hear them again, never.

KARYN ALICE BAKER was born in Alexandria in 1913. She lived with a Greek family for over 12 years where she learned 10 languages and taught poor Catholic families the duty of the Sacrament.

After the adventures described above, she was invited to join the Board of Consulting Editors for Ms. and for the Freedom Committee of the Association of Business and Professional Women.

Recently she has become involved in frog farming and has tea each day at 4 with some of the leading members of Los Gatos Social and Community Life.

She is currently in love with a high school teacher.

Some say the vulgar sweepstakes of all the KTAO program guides belongs to Whips & Spurs; *others to* The Bagman's Gazette. *I frankly prefer* The Daily Handmaiden, A Journal of Onanism. *It just worked out that it was the last guide to appear before KTAO got sold, which may say something for the karmic demands of 'self-starters.'*

Let them abuse us in the pulpits, and in the dark alley-ways. Let them abuse us in the fetid halls of congress, on the smoking battlefields, in the dusty cloak-rooms of a hundred thousands dusty high schools. Let them abuse us without surcease from every imaginable position of a torpid society. Let them abuse us loudly, angrily, and endlessly: for—after all—they only abuse. themselves."

> Faerie F. LuPage
> On the Founding of the
> American Onan Society
> June 14, 1963

From the Editors

The editors of your *Daily Handmaiden* have been criticized right and left for their attitudes, stands, operating techniques, and research.

Our critics—mostly compeers in the onanistic "Industry," have accused us of selling out to the opposition.

They make fun of the title of our newsletter. They say it should better be called *The Mouldy Handmaiden*. Or *The Greasy Handmaiden*. Or, as one competitor grossly described it: *The Warty Handleaker*.

Another went so far as to insult us and a contemporary political leader by saying it should be transformed into *The Daley Handmaiden*.

We would like to remind readers and critics alike that ours is the first organization to demand rights and respects for one of the downtrodden groups of all times.

Before our founding—"Self-Starters" were brutally mocked, held up to ridicule, abused by press and public alike.

Now, our brothers and sisters can look the rest of mankind above the groin of anguish and say: "I am a human, too. I am not dirty and diseased. I am not Insane—not a Jerk."

They can say: "I am the guy (or gal) who can stand tall before the bar of life. I am proud to be an Onanist."

The American Onanistic Society was founded on the precepts of good solitary fellowship and hard (not kneejerk) liberalism.

Our five-finger handshake (or the one-finger symbol of prosperity for the Little Ladies) have set us apart from the other so-called liberation groups like *The Black Fist, The Helping Hand, Finger Up!* and *The Vital Vibrationists*.

Our literary coevals are the talk of all the competing 'fronts.' Our enemies mock our careful attention to Durrell, Joyce, and Nabokov (quoted further on in this edition).

199

They have amused themselves at our continuing attacks on the Boy Scouts' Manual—and D.H. Lawrence's insulting attacks on our favorite pastimes. They call this "beating a dead bush."

Our scholarly investigations evoke their mirth. Last year, we stated that there were obvious erogenous parallels between the work "jack" (which originally meant "man") and the moraceous jack-tree, the quasi-religious jack-in-the-pulpit, the cheerful jack-'o-lantern, and the foolish jackanapes.

They said that these studies indicated that we were obviously "going balmy"—a charge levelled against our brothers-in-solitude over the past 4000 years.

When we investigated the etomylogical roots of "jerk"—making interesting side-trips into "jerk-water" towns and "beef Jerky," one critic had the gall to say that we were nothing more than a flaccid bunch of "jerk-offs," truly one of the most gross attacks on our pride and brotherhood since the days of Pope Pius who equated us to masters of venery.

Nothing can or will be solved by name-calling. If our various self-help groups are to fall into civil war and other abusive acts, then our cause must droop and die.

We must take hands together, leave our differences aside, and rise to the arc of triumph of our brotherhood.

We must join fingers to arouse friends and relations alike to the dread enemy of flatulence and assorted poopery which threatens to dry up our support.

We like to call the magic words of our founder, Master LuPage when he said:

"Venereal disease is sweeping this land. We have an answer for that. Abortions are on the rise: we have an answer for that. Unwanted pregnancies are everywhere: we have an answer for that.

"Homosexuality, pederasty, rape, fornication, fructification, brutal acts of passion and self-exposure are making a mockery of law-and-order. We have the answer for every man-jack one of those problems.

"Gentlemen and ladies: we are sitting on the most obvious solution to the dread acts of congress which are draining this country dry, making a mockery of our forebears. We have *the* solution here, at hand—for every ill this body politic is heir to."

There was scarcely a dry eye that dark and cloudy Sunday as LuPage roiled up to a bottling close—and rushed off stage to his dressingroom to the relief of all observers.

Our magic leader had raised his fist on high and called on all of us to stand up before the mirror of life and be rid of the hard, pressing problems of our generation.

Our leader! Beating back the bushes of ignorance; beating on the heads of the blind, and the ignorant, and the fools—"the one-eyed jacks." LuPage, the Humble and Wise Man of Solitary Splendor, offering to us now this burning solution—which we dare not drop in the sacred springs of a troubled Republic.

"BUT A DRY MEMORY..."

We Editors of *The Daily Handmaiden* are happy to be sending this, our Tenth Anniversary Issue, out to our many friends in the world of journalism.

In ten years we have grown tall and strong. How well we can remember the dark years before our founding under the sacred hand of F.F. LuPage!

We look back on our flaccid Issue #1 — see how we were the laughing stock of our compeers: we were a tired, scared, hag-ridden, pasty bunch of hard-core pilgrims in those days.

And it was a grim and wet day indeed when we came together at the behest of LuPage. LuPage of the Laughing Mien! they called him. He took charge immediately: said that in a decade our message would be spread across the underbelly of America with ever increasing Vim & Vigor. How right he was!

"We have been mocked by our friends and spied on by our mothers," he said on that fateful day, that fateful first day. "We have been shamed by our priests and beaten by our fathers. We have been fingered by our enemies and frightened by our guards. We have only ourselves to blame," he said bravely.

LuPage was right in more ways than he could possibly know. He was the one to exhort us to hold up our heads as we observed the Blessings of the Mailed Fist. He was the first to see that there was one minority more observed in "the breech rather than the act."

So afraid were we of being caught red-handed that we would sign away our rights without question — and it wasn't until "F.F." said, and rightly so: "Those who abuse us, abuse themselves." Why hadn't we thought of that!

Finally — it was LuPage who demanded that our society be open equally to women. "Let this cup of burning gold be available to all self-starters — regardless of race, creed, color, previous condition of penal servitude, and sex," he said at his famous last speech at 'The Iron Fist.'

"Let no man put asunder the candles that our sisters burn at both ends," he said emphatically and well. And then he was gone.

LuPage died as he would have lived. No one was there when the final spasm took his firm heart; no one would or could have been present as writhed his last upon his own proud bed of plenty.

But he left many memories for us, his hardy band of "Self-Starters" as he laughingly called us. Best of all, and most significant — perhaps — was the doxology of his hero, and our own, the ancient and venerable Solomitus The Wan.

Lu Page asked his words be graven forever on the hard stone above his head, and perhaps these very words will be the theme of this, the proudest outcome of his many hours laboring alone in what he most believed in:

"Mono maenous nobis
Valorum unicum
Deus oiseaux..."

or

"A hand in the bush
Is ere worth
Two birds..."

(KTAO Program Guide #133)

E   Elephant.   e

G   Goose.   g

But I guess my favorites of all were the very early guides—the ones that I typed up, laid out, glued together, took to the printer, picked up, hand addressed, and sent out—all my myself. In those I learned the Blakean concept of being the Renaissance journeyman, which gives one ultimate responsibility—and ultimate satisfaction. The following are some of my favorite essays from this period 1970-1972, when I learned that what you see is what you get.

THE PREPROCESSED PLASTIC NEWS PROCESS
or ADVENTURES IN MERCURY NEWSVILLE

For those of you outside the South Bay Smog and Prune Belt, *The San Jose Mercury-News* (incorporating the *San Jose Herald*) stands as one of those fat, limp, inordinately rich newspapers—characteristic of any newspaper monopoly situation. It's like the guy you went to school with: his father owned half of pre-Castro Cuba, and he was the first person in the class to get an off-campus apartment, the first person you knew to own a *new* MG, the first person of your age to give those all-night twelve-brands-of-Scotch parties in which someone ended up puking in his woven lace-mantilla cover bed. You didn't like him too much: he was so soft and pasty and fraudulent; but you always seemed to end up at his apartment on Saturday nights. You never said hello to him, though.

Anyway, the *San Jose Mercury-News* is sort of fat and wan and...*tired* like that. It weighs three pounds on Sunday, has one of those brilliant religious uplift Cartoons on the editorial page on Saturdays (the famished, distinctly Aryan Christ looking hurt over the weaknesses of his children) and never seems to have any hard, live, strong news. Or thoughts, for that matter.

"Still, and however," I thought to myself last week: "no one is paying any attention to KTAO. Despite our nigh about perfect programs of music of Cambodia and Blind Joe McFeelie and that Saturday morning five-hour straight shot of Baroque music and Thursday evening *raga* fest; despite all that, no one cares. No one seems to be listening.

"What we need is some publicity," I say to myself. It certainly has to be newsworthy when a friendly local broadcaster drops the 1001 strings and takes up Tofu music for toeflute: that should be big news. Even for the *San Jose Mercury*. That's what I told myself.

The *San Jose Mercury-News*—being the only daily to serve this wretchedly prosperous area, with its wretchedly prosperous population—is housed in a modern terra-cotta castle complete with 3 olive trees, 20,000 car parking lot, and giant squirting pool along the southern (or weaker) perimeter. I stayed my course a few moments along the edge of the quintuple-squirt pool to stock up on courage. "I wonder what kind of fish they should put in this pool," I said to myself idly, nervously: "If I were the publisher of this newspaper *If I were the publisher of this newspaper* I would stock this with cuttlefish. That's exactly what this monster pebble-fronted castle needs: a dozen or so cuttlefish. For warmth: to cuttle." I went in the door marked "Editorial."

It was weekend, so there was hardly a riot of newsgathering activity in the editorial offices. Three bored reporters were nibbling on their coffee cups. One with regulation-length sandy-colored sideburns was assigned to my case. "Well," he yawned: "I'll give your name to the editor. But I don't think there's much chance of a story. We don't like to help out competing media."

I thought about that one for awhile. That one about "competing media." I thought about the concert yesterday, for shakuhachi, solo. Or last Monday's three hour program of music from the Yucatan. Or that rich selection of original (not phony) bluegrass. Or the lieder festival we did last week.

I thought about that: and thought about the *San Jose Mercury*, struggling for its next million in advertising, one of the richest newspapers in the country. I thought about them chewing on our efforts to do something meaningful with radio in this country, and finally deciding to forget it: because of the competition.

"News, and newspapers, and ethics," I thought to myself, as I drifted out to recover next to the squirting pond. "That man, in that newspaper, came right out and told me that his paper didn't do stories that might threaten their revenue." He didn't fudge, almost as if he were proud of it. "We don't do news if it is going to affect our revenue." Or, translated: "The million column-inch advertising in the *Mercury* IS the news."

I sat down for awhile, comtemplated the sweet mist of the pool—let it mix with my thoughts. I thought for awhile about the owner-manager of that newspaper, about his ethics, about his life. "When he dies," I thought, "they'll mark on his gravestone, 'He put out the fattest newspaper in the country.' Or 'He made a bucket.' Or maybe: 'He never let a good news-story get in the way of a juicy advertising contract.' Something like that. Never: 'He cared.'"

I sat there, in that hot sun, near where there had once been a hundred thousand acres of orchards, in what was once the Valley of the Heart's Delight: the Los Angeles of the Bay Area—whose growth and paving had never, for a moment, been opposed by the Ridder Complex Combine. I sat sweating under the sun, overlooking the 20,000 car parking lot which had overtaken 20,000 unfortunate peach, and pear, and palm trees—and I thought about that boozy man, and his miserable life, and him worrying himself sick about all his moneybags.

And I thought that at night, some night, late at night—when the moon had turned a thin fingernail and the stars had come to be dusted up against the hills—some night I would sneak up, me, alone: so quietly that no one would hear me, so darkly that no one would see me. What I would do is to sneak up to the edge of that giant squirting pool, just outside the factory for that fat, white, slug of a newspaper. I would come late—stay for a moment: to slip a little present into the cool green-blue waters: an octopus, maybe. A couple of sharks. Maybe a three-ton squid. Something, anything, to get those tired people in the editorial room worked up a bit: noon at the desk, and this tentacle creeping over the hunched shoulder.

Something, anything, to inspire a little excitement at that newspaper. The first blop of life they've had in years. A real press-stopper. A genuine scoop.

They might even do a story on it. It it wasn't too serious. If it wouldn't antagonize any advertisers.

204 A HEN IS AN EGG'S WAY TO MAKE ANOTHER EGG; FOR WHOM, THEN, LOVE, ARE WE MAKING LOVE?

You are a good administrator. No—not a good one: you are a great one. You have been in Washington D.C. for practically 40 of your 60 years. You were born with the century, and grew up (and through) the U.S. government.

You have always been loyal to your country, like a great bureaucrat. You started out in 1922 with the ICC, the first of the Independent Regulatory Commissions. You learned to work with Congress, and with rich and powerful men. You learned to give in to both forces—or better, how to *appear* to give in to both forces and still, somehow, to get your own way. You were and are a good man, and you worried about justice, and the people who are the people who make up a country. You were a secret ombudsman, even back in the crony-ridden days of the Harding and Coolidge administrations.

You moved to the U.S. Civil Service Commission—and worked your way up as a bureaucrat's bureaucrat. In the 1930's, when the Civil Service was quadrupling and sextupling in size, you were growing and advancing. You came to have many loyalties—for the halls of Washington bureaucracy are not too far from the halls of any small town government. Favors are done because you know and like and trust people—not because you are corrupt.

During World War II—when the military was bursting with billions of dollars and pure administrative *naivete*—you not only showed the army and navy and marine officers the way, the simple way, through government process: but, as well, you came to know and like many future military chiefs in what was then called The War Department. You never crossed them: but you always saw to it that the military did not run roughshod over the civilian administrative personnel, no matter what their power and money, no matter their influence. You had a concept of personal justice.

You passed on into the State Department after the war, even spent some time in Indochina as advisor to the French *colons*, as liason on military procurement technology. You were good in the field—but you missed Washington D.C., and it missed you. Administrators with life and imagination are hard to come by. At 52, you returned to a coordination position between the General Services Administration, the then-fledgling Bureau of the Budget, and several other obscure branches of executive and legislative interaction. Like Stalin, you knew that the secret of real power in any government lies in the administration; unlike Stalin, you lived with elected personnel—but you knew that any elected official could affect no more than seven or ten percent of the governmental machinery. The rest of it lay in the hands of you and your brothers in the Civil Service.

There were few department, or branches, or sections, or sub-sections of the Federal Government that you didn't know...even the secret ones, the ones with titles like "Bureau for Governmental Interdepartmental Liason" or "Office of Strategic and Management Supply" were no mystery to you. You knew them all, where they got their monies in secret budgets—and most importantly—who ran the show in each of them. Most of them were personal and good friends of yours.

By the beginning of 1960, you were one of the wise men of Washington. Few of the taxpayers had heard of you—and yet, you were one of the immensely respected Civil Servicemen of Washington, and the shrewd congressional Administrative Assistants were constantly on the telephone to you. You had powerful and loyal connections in all of that giant paper city—and like the *Opus Dei*, you prided yourself on capacity for immense work, and capacity for impeccable honesty. You always keep your word. You love your country.

205

It is in 1958 that you see a dangerous imbalance developing in the government machinery of your country. No: it has nothing to do with the late lamented and slightly addle-headed Senator McCarthy: you see those comings and goings as nothing more than a basically healthy blood-letting which must and should take place between the hundred branches of a democratic government. But it was McCarthy, the buffoon, who had the strength, and madness, and daring, to attack the military. He lost, and *that* opened your eyes.

In your beloved government, a government you see so differently than the tourists, the elected congressmen who come and go—in your government, you have begun to see a new and perilous force. A giant, vast, rich, powerful military bureaucracy. You are a bureaucrat: and you have special sensing devices. You can always see when another bureaucracy has the cancer of growth.

This one was special—because it has come to have limitless monies, and can always duck under the blanket of secrecy. You know American history well and fully; you know why the authors of *The Federalist Papers* loathed and despised and feared a standing army. You are afraid for your country.

"Something must be done, and done now" you think. "Something must be done—or in the next dozen years, the military will be running the whole country, even me." You thought of your friends in the military—and you even feared them. Because you knew (as did Madison, and Hamilton, and Jefferson—prophets all) that men with unlimited power turn cruel and dangerous.

In 1960, you helped write a speech, a final speech, for the then outgoing President Eisenhower. But you knew that something more than a speech was necessary.

"If," you thought: "If there were only some way to bring down the impregnable image of the military." You thought of a big war—one which would be over in no time at all. Then, you thought, everyone would know how bad it all was. But then, you thought—who would there be left to complain? Only the President, and the military leaders, and a few Senators, lying safe in their air conditioned underground bunkers. You thought of Mendel Rivers, and James Eastland, and Curtis LeMay being in charge of repopulating the now desolate world; and you thought that another solution was necessary.

"No," you said: "we have to get them involved in something more manageable. Something more...traditional." Some kind of a venture which would not last for a day, nor a month, nor a year—but decades. Some sort of an extended battle which, by the nature of the terrain, and the enemy, could never be won—might never be lost; but, because of the conditions, would drag on, and on, and on.

See—you knew, as a lifelong bureaucrat, that the best way to defeat an otherwise unbeatable power, is to take it through all the hallways of bureaucracy. You had, yourself, "lost" various papers, misplaced certain applications, sent back "for further study" some documents which you knew, you *knew* would work mischief for the democracy you loved. You had seen many men, many bad men, falling apart in the tortured corridors of administratives jurisprudence. You had watched many rich, and powerful, and bad men brought to their knees: because of endless bureaucratic haggling, and surprise victories (which turned out, in retrospect, to be defeats), and endless appeals and waiting. You had read a great deal of Kafka.

"If we could only apply this logic to the pursuit of war. A military engagement which could never be won, nor lost. One that would ramble on forever, and always be inconclusive." You needed a war which would break the public's habitual belief in the magic of the military, in that damnable World War "Instant Victory" philosophy. A war, a new kind of war: one which would go on and on and on; one that would gradually expose the bare bones of the American military machinery. One that would expose them to public vision, and then public criticism, and finally public ridicule: to reveal for all to see the basic injustice and power of the draft, the danger of a rubber-stamp appropriations system, the power imbalance now in the hands of the military.

"A land war," you thought: "what we need is a land war, in some jungle—with a clever and wily enemy that would have some taint, perhaps the taint of association with world-wide Communism." It would have to be a group of devotedly loyal national patriots who could take all that the Pentagon could come up with, and still survive; it would have to be a country of such an alien nature that the Americans would go into war with the confidence of easy victory—a victory that would come to be more and more elusive: and nothing would explain the inability of the military to end that particular gruesome war.

Your eyes turned west. You remembered your experiences with the French in the hot steamy paddies and jungles of a place they then called Indochina. You thought of a place called Dien Bien Phu, and your chance, brief, but impressive memories of someone they all feared as eternal and inexhaustible: a man called Ho.

You thought of that, and of your newly-created (in 1961) prestigious position on the very secret committee on bureaucratic liason on international affairs. You knew that the new executive officials would listen to you, because of your expertise, and your wisdom: they were new and brash, but they would listen—maybe even act on some of your recommendations, for regrettable but necessary international police-keeping actions. They would trust you implicitly on that—because of your experience in Indochina, and your quiet ways; because of your humility, wisdom, and honesty—and you unreproachable patriotism...
(KTAO Program Guide #19)

STATEMENT BEFORE THE GOVERNMENT'S COMMITTEE
ON THE PROPOSED DEVELOPMENT

"A few overconcerned citizens have protested this development as being wasteful of natural and financial resources, and even more—destructive to human life. They have spared no criticism, have pulled all variety of political strings to put an end to this most important and vital project in our entire system of national priorities.

"May I point out that this city was and is the most economically deprived in the whole nation. There are greater pockets of poverty in this region than the rest of the nation combined. Unemployment haunts the streets. No trained worker can expect a single day of decent work. The honesty of the common man is being compromised by vicious unemployment, demeaning charity, and a lack of goverment projects in the area.

"We propose to change this sorry state of affairs. Once our development has your approval, some eleven millions will be poured into the district. Sheet-metal workers, masons, carpenters, contractors, electricians, brick-layers, common construction workers, and even the previously unemployable untrained younger work force will have a project which will take a full year to complete, and provide almost a thousand regular jobs thereafter to repair, maintenance, and security personnel. And for the year of initial construction—some seventeen hundred to two thousand men will be gainfully employed, with all the attendant advantages of paychecks for the hungry and the depressed. This is a humanitarian act.

"In short, the proposed operation will bring necessary government funds into the city, erase persistent hard-core unemployment, give a whole community new self-respect, and provide a continuing source of funds to the willing, able-bodied, and hard-working citizens of this desperate district.

"Gentlemen: the people need our project. Their well-being, and the security of our whole movement, demand that we provide for their immediate needs. We must give the people of Belsen some pride. I ask that you vote favorably on the construction."

> —Rudolf Hoess,
> Statement before the Chancellor's
> Commission on National Priorities,
> meeting in secret session, April 5, 1940.
> Subject: Proposed Construction of Belsen
> Concentration Camp, Belsen, Germany.

(KTAO Program Guide #109)

ZZzzzZZzzz z z

My friend Hugh Gallagher claims that he has a fly that goes everywhere with him. On of those little black jobbies—I think they are called fruitflies. He'll be talking to you, and his eyes will cross slightly, and he'll wave his hand back and forth some and say "There it is. My own personal fly. It goes everywhere with me." And then he'll tell you about the time last winter they were having a big conference in London (he works for British Petroleum), and they were talking about Class Action and the Alaska Pipeline, and that dumb fly shows up. Hugh doesn't want to appear too eccentric—although he understands that when you are working for a British

Company like BP you're supposed to be a little crazy—but he waves his hand back and forth and thinks "There it is. My own personal fly. It goes everywhere with me." And, as he is telling me this story, sitting in Bethesda on the porch, the jets going overhead, his dog drooling affectionately on my foot, I notice there, maybe about a foot in front of him, almost motionless in the air, one of those tiny black flies, I think they call them fruitflies. I don't even think that Hugh notices it, but I do. Then, and for the next three days in Bethesda—and then, and from then on, on the plane and back here in Los Gatos : a little black fly, I think they call them fruit-flies, or *Trypetidae,* coasting there, about a foot in front of me, going wherever I go.

(KTAO Guide # 105)

10 Steps to Certain Sainthood   Sri Milam Ji

Each man's acts of joy or sorrow are his own production, his own movie, his own dramatic performance. If he blames his parents, he is still a child; if he blames the government or economics for his misery, he is a fool; and if he blames the universe—he is nothing but a poor deist. We are ultimately and finally responsible for our own moods, loves, hates, position, strengths, weaknesses, work, worth, and world. Most of us need someone to convince us of this—and when that person comes, the realization of self-responsibility for *everything* is nigh about blinding.

The power of this self-determination is obvious. It is the most compelling answer to grow out of Existentialism (which, in turn, grew out of the Paris Underground between 1940 and 1944—intellectual coffee-house society trapped in the hot-house of the sewers) as married to Jung and Perls coming into the passive embrace of Watts and Suzuki neo-zen. And in the frenetic split-level nut-house of the United States in the middle of a revolution of the head-bone. The gifted amateur can pretend to counsel, or co-counsel—but what it is *really* is the loneliness of man (always standing apart by the very separation of the grey whorls of brains in separate containers) forcing the break of those very convenient walls of Puritanism and Victorianism.

The rules of the new self-determination are appallingly simple...but, of course, they—like all good rules about relationships between people—must change from day-to-day and even moment-to-moment. But, right now, at this moment, these are the Rules of the New Man: 1) After age 12 or so, each of us is exactly as we *want* to be; traps, anxiety, desolation are of our own creation; 2) As men, we create devices to isolate ourselves from other men; gossip (speaking poorly of someone who is not present) is one of the most effective isolation devices; 3) Conversely, confrontation—expression, face-to-face, of one's opinions about another—breaks down walls, and isolates loneliness; 4) We know a person the moment we meet him: any excuses to the contrary are further devices of isolation; ('one knows when one is being put on.'); 6) Men don't have egos, or ids, or sub-consciousness—unless they want them. These Freudian devices are efficient devices for isolation and self-protection: what a man may call his 'sub-conscious' is simply a learned response. ("You did it, and it worked—so you kept on doing it; long after it ceased working.") 7) Each of us has the capability for expertise into the minds of men who are not kookie, deeply sick, or dead—if we approach them with warmth and affection. Each man has his own *patois,* his own vocabulary, his own means of expression. As humans, learning each man's language (like each man—separate and distinct and original) can be frightening and delightful and funny and scarey. 8) Concommitant with this strength is the responsibility to avoid self-destruct: some people are torn enough to be hurt by the glare of honesty—and they will hurt you for it (and perhaps themselves). 9) Everything we have done, are doing now, and will do in the future—is for the self. No matter what motives and excuses are hung on each action—you do what you do *for* you. And for no one else. Ever. 10) No matter what you (the intellectual you) tells the human you, there is but one truth to guide us. That is, we are all of us, ultimately, finally, and totally alone in the universe. How we deal with this knowledge each day, each moment—is what separates the fools from the men.

As I write this, as I write these ten points—I have to remind you that there is another, the eleventhsies, a final rule.

And that is that whenever anyone is fool enough to list the rules of life—then you would most probably be well advised to do exactly the opposite.

Only in that way can you achieve true and reverse sainthood, the saintliness of the mirror to whom we all aspire.

This last is rule 11, and should be considered as worth reversing as the previous ten.

(KTAO Program Guide #45)

5

Five years O my Christ five years did you say Yes O Christ I said five years how many months is that at least eighty Venusian months how many days is that at least eight thousand Saturnalian days maybe if you look at it as Mercurian the God Maphros computes it as an infinite number of days. An infinite number. O Christ.

Have you ever heard the fire extinguisher theory of life? Life is a fire extinguisher. You know: one of them big red babies. Heavy too. You carry it around with you for five years or so, waiting for the big moment, the big blaze, the big tongue of red that's going to come down and lick you up. You carry the extinguisher with you to parties, games, dances, and the monthly meetings of the Toastmasters, International. Finally, one day, towards the middle of the Twentieth Century, maybe late in November or sometime near the Yule Season, it comes. The Big Blaze. The one you've been waiting for. Children scream, mothers faint, fathers scratch themselves in desperation. And you step forward to the front of the auditorium. "Have no fear," you tell them, and at once, they are calmed. They wait expectantly. Their eyes shine: pure and sweet hero worship: you. "Have no fear," you repeat. They wait. You lean forward (the Fireman's Pose it's called.) You rip the little wire loose from the handle, the flat lead seal. You press down on the squirter-plunger. Everyone says nothing. There is a noise, a little noise:

brt.

A little fart of a noise, then Nothing. The charge is gone. Five years, and nothing but

brt

(KRAB 5th Year Anniversary Program Guide)

STATEMENT BEFORE THE COMMITTEE

"...They are even calling into question the matter of our right to profits. They are saying that our policy of raising price to meet costly demands is unconscionable, that we shouldn't make so much money...

"Well, all I can say to our critics is this: if you can fill the public need as well as we do, then *you* go into the business. You take the responsibility for the delicate pickups along the inhospitable and blazing African coasts. You pay the seamen, and the insurance for transshipment across the violent Atlantic. You pay all the extreme costs of discharge on the American shores, to fill the voracious American needs. You do all that: and then, and only then—tell me if we are not performing a public service of the highest order, no matter the price."

—Testimony before the Parliamentary Anti-Slavery Hearings by Lord Wilbur Acton, Owner, nine slavers. (June, 1809)

(From the Rand-McFeelie Roadtrap Roadmap)

EXPLOITING ANOTHER NATURAL RESOURCE

Dirty old men. One of the most important unused resources of this country. Why can't we follow the lead of England during the victorian period. Transship all our sexual deviants to some other country—so that we can (at home) come to a new period of prudery.

I could see a project under the CARE program—where everyone who had been accused of taking terrible liberties with the young of the land could be given the choice of a life in jail, or a life feeding and caring for the youth of another land, in aboriginal Australia, or Central Africa, or Calcutta.

Instead of morosely squandering the rest of his days in guilt-ridden agony in some bleak citadel of penal reform, our Dirty Old Man could be the Santa Claus of the cold, and anguished, and ragged youth of another, less prosperous country. This man, a pariah in his own United States, would become involved in the distribution of food and clothing and (of course) affection for a thousand small, pitiful, unloved waifs, distorted by hunger and cold.

Inside each molester must beat the heart of one who cares for children *too much:* and this surfeit of devotion can be turned to a program of fighting for and caring for the desperate, unloved, and desolate of the whole world.

By distributing love and warmth and nourishment, our suffering hero, castigated and shadowed in his own home, will abroad become the symbol for the generosity and kindness and affection of The American. And, certainly, no one will fight harder for the survival of his charges—be they six, or ten, or thirteen years of age. No one will have more heart for fighting the bureaucracy and the administrative coldness of private charities and the government distribution agencies—than this once-besmirched and derided Saint of Child Care and Love.

(KTAO Program Guide #27)

THE PEOPLE WANT BLOOD...
And We'll Give It to Them...

For those of you outside the Passion Belt—California voters recently opted, by a margin of 2-to-1, to reinstitute the Death Penalty. Which all goes to prove that vengeance and the pleasures of Death to Society's Wrong-Doers shall never be denied, even by the firm hand of the Supreme Court.

For that reason, we have asked one of our state representatives to introduce legislation in the next session of the California State Assembly. That is, that all executions which are to be carried out by our penal authorities be available for film, television, and radio coverage.

Although radio and television coverage might be difficult to find sponsors for (Kentucky Fried Chicken? The California Gas Institute? Northern California Power & Light?) we figure that the people deserve what they voted, and what they'll be paying for.

Give us two or three fully televised executions. Give us some close-up shots of a man's face in grimace as life is leeched from him. Show us the twist of a man's arms tied down as he is poisoned through the official fiat of the state: show it to us in full, livid color, around dinner-time: an intimate shot of death coming into every home in California—and I would guess that the passion of this state to wreak vengeance by death would be somewhat curtailed. I would say that after a couple of full-dress public deaths, we might see a revision in the attitudes of a state so eager to reinstate brutal measures under the guise of law...
(KTAO Guide #115)

To the Ute Indians—the wind was not the function of some outside disturbance coming from 'heaven.' Nor was it a disturbance of 'low' or 'high' pressure areas.

"No, the wind came from within the trees themselves. From inside the bark and limbs and leaves, there was a ghost in turmoil—a raging, rolling, moving ghost which tormented the structure of the trees in an angry effort to get out. And this spirit—tearing at the limbs, making such a racket—was easily communicable to the near-by trees: so that when one started, sooner or later, the whole forest would, with branches and leaves whipping and turning and rattling and moaning—the spirit trying desperately to free itself from the prisons of bark.

"Thus, in their innocence, they saw the tree and the wind as one force. This is how the spirit of earth and life affected their minds."

—E. Jing
The Sexual Practices
of the Savages

(KTAO Guide #111)

A BRIEF HISTORY OF RADIO

Radio was discovered some 50 years ago by a dog named RCA Victor. RCA Victor discovered radio accidently by looking into a horn, and discerning the voice of his master. Ever since then, RCA Victor has been a tradition, and many have capitalized on his cocked ear and puzzled face.

In the early days of radio, there were many exciting inventions. The Father of The Tube was Lee deForest. He evacuated a bulb left by the Gardener (coincidentally, a friend of RCA Victor) and stuck his in his thumb and pulled out some mysterious little bugs called electrons. When he put the whole thing in a wall-socket, he said "Yreka." And he heard the voice of London Calling. The voice said, "This is London Calling!"

Radio grew apace after that. There were modifications of DeForest's evacuated tube. One of them was put together with some verve by Maj.-Gen. Edw. Armstrong. He called it the Heartstrong receiver. He was able to hear Trenton on his receiver. He also said 'Yreka!' which was a favorite quote of radio inventors.

Television also grew apace. The first signal was a picture of Howdy-Doody sent from Seacaucus N.J. to Weehawken, N.J. The effect was electrifying. Howdy-Doody was seen from as far away as Bayonne. CBS then was invented to steal patents from RCA Victor and his friends. There were many suits.

The transmission of radio signals is amazingly simple. A voice makes the diaphram (later called the 'IUD') tremble because of a basic flow of electrons. Electrons are also fondly called 'Little Boogers' by inventors who couldn't find them too easily.

This amplified signal flows through a series of coils and feeders (The Islets of Langerhans) in the first stage of amplification. The first stage leads to the second stage, which in turn leads to the third, and so forth. Finally the last stage is reached, and everyone goes out for tea.

Radio developed apace with the coming of singing commercials. RCA Victor and CBS bought up everyone and their grandmother, including Saul and Roweena Triode who helped to found the Heaviside Layer, the Aether, and the tube which ultimately became their namesake: The Pentode.

An unsung hero of these days was Senator Wheatstone, builder of the Wheatstone Bridge connecting Biloxi and W. Biloxi. He stated on the floor of the U.S. Senate that he would die content if he had his rye, Don Ameche, and The Breakfast Club. He was buried with honors in Athens, O.

After the war, radio went into its infancy. The continent was leaped in a single span, and a mother in Regina could hear the same Drano commercials as a truckdriver in Omaha.

Familiar to broadcasters is "The First Time on the Air" also known as "Beginning Stomach". This quickly changes with experience to "The Oriental Clam".

With the advent of Television (also called "The Third Eye"), radio came to be transformed into something else again. No longer would listeners depend upon the laughter and songs of G.J. Told of WOOD. No, soon the eyeball had replaced the ear; the cathode tube had put a single white dot on the sentence called radio. Instead of being an instrument for information and commercials, with brief sieges of entertainment, or top pops. New engineering techniques made possible the arousal of HiFi, which in turn led to Quadraportographic Sounds and Stereomagick Musics. The new horizon of radio is cloudy but bright.

And so it is with a friendly wave that we say 'Goodbye' and 'Godspeed' and 'Godamercy' to our old furry friend, Radio. From RCA Victor, through Roweena Triode and The Joy Boys, it has been a fun-filled adventure into the electronic tingling of a whole continent. The Future of Radio is no larger nor smaller than we can imagine. Long may she wave!

(KTAO Guide #62)

Part Seven
appendix

> *The attempt [to abolish the Commission] would not be a wise thing to undertake...The attempt would not be likely to succeed; if it did not succeed, and were made on the ground of inefficiency and uselessness of the Commission the result would very probably be giving it power it now lacks. The Commission, as its functions have been made limited by the courts, is, or can be made, of great use...It satisfies the popular clamor for a government supervision...at the same time that that supervision is almost entirely nominal. Further, the older a commission gets to be, the more inclined it will be found to take the business view...of things. It thus becomes a sort of barrier between the...corporations and the people and a sort of protection against a hasty and crude legislation hostile to [our] interests...The part of wisdom is not to destroy the Commission, but to utilize it..."*
> —Richard Olney
> Commentary on the ICC, 1892
> as quoted in *The Superlawyers*
> by Joseph C. Goulden.

References

Last week we reported that the Ford Foundation had announced new grants totalling $1,326,000 to two Washington-based "public-interest" law firms that had represented litigants against broadcast operations—sometimes called strike applicants. McGeorge Bundy, president of the Ford Foundation, never friendly to broadcasters, heaped encomiums on the groups in announcing the grants.

We are constrained to wonder how Henry Ford, the automotive genius, the hardnosed industrialist who was not without his hangups about minorities during his lifetime, would have welcomed the distribution of funds that sprang from his acumen to underwrite the ultraliberal, antiestablishment, socialistic concepts of pressure groups.

—editorial, *Broadcasting Magazine*,
May 13, 1974

The worst and the best of the trade magazines is *Broadcasting*. The weekly issue contains at the back all the small print decisions of the FCC—grants, application, hearings, station sales (along with prices), and major FCC policy decisions.

However, you have to wade through all manner of churlish, 19th century Robber Baron editorials, catatonic attacks on Nicholas Johnson and Al Kramer, and the most appalling sort of name-calling on anyone who is working to better American garbage radio.

If you can handle all that, you should subscribe: most especially, for your purposes, for the used equipment-for-sale section, and occasional honest and well-written 'special reports' (which are slipped past Sol Taishoff when he's out to lunch, which is most of the time, we suspect).

Subscriptions are $14 a year, and should be addressed to 1735 DeSales St., NW, Washington D.C. 20036. Their subscription department is about as archaic as their economic view of American radio and television—so you can expect to wait up to 6 weeks for your subscription to be processed.

Broadcasting also issues a Yearbook, which is supposed to come out in January, and manages to make it around April or so of each year. It costs $17.50—and is a goldmine of information about every radio and television station in the U.S., when it went on the air, its power, its ownership, its management, its programming. As well, there is a summary of FCC rules, lists of agencies, group owners, and a general plethora of necessary information.

Be cautious, though. By the time the Yearbook gets to you, the information is six months to a year out of date. Furthermore, it is one of the most poorly proofread information manuals around—because facts about stations (power, antenna heights, directional patterns, hours of operation) are often at variance with the official FCC figures.

Those who are aficionados of FCC day-to-day subscribe to the airmail edition of The Daily Releases. What it consists of is 15-20 Xerox sheets a day—exact duplicates of the news releases that come out of 1919 "M" Street. You have to be a total printword fool to take this gusher out of the FCC: the most inconsequential decisions; the most far-reaching and significant; the most arcane. All blurred together in a rush of paper which (as of this writing) is supplied by ABS Duplicators, 1732 "Eye" Street Northwest, Washington D.C. 20006. After Spring of 1975, contact the FCC for the name of the new contractor.

It will cost you from $20-$35 per month—depending on the quantity of information. But you will be the most well-informed FCC nut on your block if you subscribe. Every goddamn diddly decision on Safety & Special Services, or Docket Case #19988, or proposed rule-making proceedings, or fixed-land-mobile decisions, or satellite rate cases, or $25,000,000 television transfer application announcements.

It may be gibberish to you when you start plowing through it—but soon enough, the whole will begin to make some sort of addled sense, at least a Catch-22 sense. There is *something* going on over there; maybe they really know what they are doing—maybe not. You will learn which papers you can wastebasket at once; you will start to save out the notices of interest to you: the applications in your area; the changes in specific rules for broadcast licensees; the decisions on docket-hearing cases. You will come to learn the peculiarities of the different parts of the FCC—the contradictions within one supposedly monolithic government body.

You will learn the *truly weird* criteria of hearing cases: in which $100,000 or $1,000,000 worth of broadcast license is handed out after exhaustive mountains of bullshit for the most bizarre reasons. Not, as we would hope, because of proposed superior broadcast service—but because one would-be broadcaster shows greater financial resources, or promises six, not seven salesmen on his paid staff.

You get some feeling for the scoundrel-nature of most applicants—and wonder that the rest of the FCC pays no attention at all to the Kafkaesque standards set up in the hearing process.

The Daily Releases are worth what you pay for them. After crawling over jungles of paper, making you a Cortez of legalese—one can begin to get a glimmer of the direction of this Hydra we have created in Washington. You will read through 7500 pages of words words words to find an occasional nugget: a piece of information on licensing procedure that will save you untold trouble down the way; or some information about a way you can save a few bucks on 'type accepted' equipment. Or maybe you will be one of the few—among the 2000 hard core FCC crazies, lawyers, and engineers who subscribe to this stack-full—to jump when you read the ominous words "call letters deleted" or "revocation proceedings scheduled." That might make you a broadcaster—without much investment at all.

The same duplicating service puts out reproductions of the official FCC lists of proposed FM stations, those holding construction permits, and those on the air. It lists station call letters, owning entity, frequency, power (into and out of antenna), height, weight, bodily dimensions, scars, deformities, birth signs, and coordinates for the transmitter. Some of these figures will be essential for the early part of your engineering application to see, for instance, if you will have to use a directional antenna—something that is perfectly legal in educational and commercial FM broadcasting.

There are two magazines which you can get for free, which contain useful information and some interesting ads—*BM/E,* 820 2nd Avenue, New York City 10017; and *Broadcast Engineering,* 1014 Wyandotte Street, Kansas City, Mo., 64105. Both are available for free if you print up some fancy stationery and convince them that you are an honorable member of the broadcast industry.

For information on legal and political aspects of the future of FCC or telecommunications policy—you might subscribe to *The Network Project* out of Columbia University. Despite a terrible layout and occasional obscurantist articles in subjects such as public or private use of international satellites, the Notebook is a fascinating compilation of information on the future, and communications, and You. As they say it, their job is to "inform people about the profound ways electronic media affect their lives and their communities." It is available by subscription for $10 a year (four notebooks) or $2 each from *The Network Project,* 101 Earl Hall, Columbia University, New York, NY 10027.

Middletown California is the home of *The Alternative Radio Exchange*—a magazine which tries to hold together the various community stations and their devotees. It is written, laid-out by, glued up, inked, printed, mailed, and blessed by David Lane Josephson D/B/A Middlefield Fred Oyster. His address is Box 191, Middletown, California 95461. His mag costs $10 a year, and is worth it.

For the obscurantists in the crowd, there are two peculiar and high-specialized publications which come up with specialized information. One of them is the Federal Communications Bar *Journal*—issued four times a year from 1225 Connecticut Avenue, Northwest, Washington D.C. 20036. It costs $2 per copy and, up to recently, its political and social sensibilities have been slightly to the right of Genghis Khan, as befits the Communications legal profession in Washington. However, a new younger editor, elected in the Spring of 1974, promised some change and perhaps awareness of the 20th century. [It never happened.]

The other magazine which has to do with the formalities of radio is *The Journal of Broadcasting,* issued by the School of Communications, Temple University, Philadelphia, Pennsylvania 19122. Subscriptions for students (that's you: we have a lot to learn) are $5 annually. Some of the writing is that fine-weave clinical computer readout crap about the viewing habits in Media, Mass., of 63-65 year old spastic widows on long summer Thursday evenings. But often there will be some insight into questions of bias in television news, or some discussion of broadcast traditions in other countries, or a peek into the history of radio in this country. These reports make your subscription well worth it, and, if you really want to suffer, you can order all 75 or so back issues, dating from the founding of the *Journal* in 1956. Volumes 1-17 will cost you $90, unless you are out of school. In that case, they will cost you $302.50. I would suggest you dig up some pot-smoking friend with a student card and have him do the ordering for you.

There seems to be a paucity of worthwhile books on the theory and practice and history of radio and broadcasting in this country; and, outside of Pike and Fischer, almost nothing on the legal aspects of it. Most books available to you are descriptions of stuff-dull production and managerial techniques, and some high-class comic books on sales written by the clowns at the Columbia School of Broadcasting or Elkins Institute.

There is one history book—a pompously titled but carefully researched work by Eric Barnouw of Columbia University. There are three volumes: *A Tower in Babel*, *The Golden Web* and *The Image Empire*. I prefer the name *The History of Broadcasting in the United States* (Oxford University Press $29.25). Some of his writing has the grace of a turtle walk—especially when he gets into the 1950s and 1960s. But he is excellent when he is describing the ghosts of the unbelievable Nora Blatch, and Edna Purdy Walsh, and Michael Pupin, and our own beloved Reginald Aubrey Fessenden. The story of how the Fessendens and Armstrongs and The People (our parents; you, me) were screwed, ripped off, driven looney, cashed out, sold down the river by the David Sarnoffs and the George Washington Hills and the Albert Laskers and the Orestes Caldwells (Orestes Caldwell!)—why it's enough to make us all retch, in close harmony.

There are two adulating biographies of the man who dug FM out of the darkened dungeons of the aether—the good mad talented taunted troubled tinkery Maj. Edw. W. Armstrong: *Man of High Fidelity* by Lawrence Lessing (Lippincott, 315 pp.) and *Armstrong's Fight for FM Broadcasting: One Man vs. Big Business and Bureaucracy* by D.H.V. Erickson (University of Alabama Press, $7.75).

Lessing's book is the classic study of Maj. Armstrong—and came out in the early 60s when FM and Armstrong's name weren't too important to too many people. However, since the man has passed on to the Great Heaviside Layer in the Sky—the Concept and Fight of Armstrong have become important enough for some hokey New York Types to set up an "Armstrong Foundation" to perpetuate his name. You can be sure that their grants are made to the safe-and-sure broadcasters (not you and me) and that if, by some electronic feedback flash of superhetrodyne wisdom, Edw. W. came back to the electronic world of the living, the Foundation and its stodgy Board would, no doubt, drum him out of their rarefied and precious lodge.

Both Lessing's and Erickson's books are worth reading—if only to get the feeling for the early history of radio and FM. Lessing's book is the better—with the fascinating attic history of the early Armstrong, with some passages of true poetry; read the chapter on the discovery of the principal of feedback: when the voices started shouting into the earphones, Armstrong *knew* he had tapped the rush of the gods. Erickson seems to have more of a hard-on about RCA and Sarnoff and the Sarnoff-Armstrong feud which was, really, a silly battle between two naughty head-strong babies, two pretend grown-up men. (They carried their 20-year Civil War into the bedroom, as befits the anguish of two essential brothers.)

Even more seriously, both books ignore the karma wheel concept of life: that Maj. Edw. was a soul so haunted by the fear that everyone was going to steal his marbles that, indeed, he helped it to happen. He never learned the wretched lesson of benign genius; namely: the gods of invention have so many seed secrets in the craw that they can hand out, give away, throw away, award, set loose their magic on the whole world for free. Because there are a thousand more where they came from. And a thousand after that. Or, saying it another way: if I fear the loss to you of my flashes of insight—then I obviously don't have very many; and should suspect the roots of my insight.

Both books give you a tortured history of the growth of FM—and both ignore the fact that you and I wouldn't even be here if it weren't for the controversial move of FM in 1945, by the FCC, from 42-50 mHz to the present 88-108 mHz. If that fortuitous set of circumstances had not occurred, FM would have been successful (probably) from the start—from the first early boom between 1947 and 1950. Then you and I and KPFA and WGBH and WFMT would never have had the leisure and lack of competition of frequencies during the 50s—making it possible, especially, for us johnny-come-latelys to get set up before the new and now ugly 'renaissance' of FM. The medium had to grow up around us: we never would have had the resources to grow up around the medium.

Furthermore, if the FM band had stayed at the lower frequencies—there would not be the richness and plethora of stations available today: only 42-43 mHz were set aside for non-commercial stations, or 12.5% of the band, as opposed to 20% available today. And instead of getting 20 stations into one market, we would have been limited to 10 or less.

All in all—Major Armstrong had a bitch—and they handled him cruelly during his tortured life (he finally killed himself in the mid-50s). But the propaganda of his adherants ignores the limitations of his dream.

There is a PhD thesis by Dr. Eugene V. Stebbins which you should buy, even though it costs $15.00. It is called *Listener Supported Radio: The Pacifica Stations* and is available from him at: 2075 Maplewood Rd., Stow, Ohio 44224.

The reason it is important for you to read and memorize is because it delineates in careful detail the rip-and-tear of internal politics that almost killed off KPFA 20 times between its founding in 1949 and its present affluence and flatulence. It describes the early structure of the corporate board of Pacifica—and gives you some idea of why your community station would be well advised to have an outside board, not a board of directors of staff...who, if they were to operate like the early Pacifica people (they will) would spend most of their time tearing each other apart as an adjunct to their beautiful sweet spot-of-dew-on-the-rose radio station. The death of the dam of all of all of us—Lewis Hill—is described, and in the whole, there are some heady lessons for all of us.

The Pacifica experience has produced a couple of books. One, *The Exacting Ear* by Eleanor McKinney can be ordered from the offices of the Foundation at 2217 Shattuck Avenue, Berkeley for $3. I found it a bit tame—a bit defensive, I thought, as the Pacifica people were themselves when the book was put together 15 years ago. McKinney missed the opportunity to tell those of us in the country-of-the-mind how to put together our own Pacificas—nor did she give us more than a peep and snip of the rich inner turmoil of KPFA in its bearded constructs for us all in that madhouse. There are, however, some exquisite transcriptions of KPFA's most popular programs from the 50s—which will give you a good feel for what the station stood for when the rest of American radio was such a poop-pile.

Another Pacifica book is *Playing in the FM Band: A Personal Account of Free Radio* by Steve Post. It came out this year, from Viking ($10). ($10!) It is a fine book—it gives you (it gave me) a feeling for the delight of using the palette of a radio station not only to paint the aether horizon, but to grow inwards, within the broadcasted self. It shows you how Post blossomed in the hot-box of 99.5 megahurtz, and it is, as well, a close description of the one station that seems to be a jewel in the diadem of Pacifica—that being WBAI.

If you have an urge to be an obscurantist and historian, then you should volunteer for KPFA and spend your spare time reading over the program guides that came out every two weeks between 1949 and 1957. Lewis Hill wrote and opening essay for most of them. Although he fancied himself a poet, and tended to write with the same general clarity of Ludwig Wittgenstein, or Emanuel Swedenborg—the excitement of what he and the others were doing helps to enlighten his words, if not simplify his thoughts.

Ben Dawson tells me that the FCC has put out a good brochure on starting ten watt FM stations, although I haven't gotten it yet since I only asked for it six months ago. And the National Citizens Committee for Broadcasting (Nick Johnson's group) has put out a folder called *Demystifying Broadcasting* which has all the things in it that I am too lazy to write about here. You can get it by writing them at 1914 Sunderland Place, NW, Washington 20036.

Canada seems to be the site of some of the most exciting community broadcasting in the world—outside of Micronesia. The Master is a man by the name of Douglas Ward (I have included further on his essay about some specific experiments with turning radio over to the people). You can order the following booklets from Marilyn Mercer at the CBC:

Community Radio in Canada—a reproduction of the applications filed with the CRTC, including such communities as Tuktoyaktuk, NWT, and Montreal, Quebec.

Local Broadcasting in Remote Communities—the report I have reprinted further on.

Evaluation of Audio Components for Community Radio—an excellent summary of all the cheap, good, hardy, trustworthy, loyal and useful equipment needed for your station. Although most of the equipment ordering addresses are Canadian, the reviews of the equipment are understandable for the layman, and perceptive.

All these can be ordered free from the CBC at
>Box 500, Station A
>Toronto M5W 1E6
>Ontario, Canada

If and when you go on the air, you should have a copy of the *Communications Act of 1934*. This and the next 4 items can be ordered from the Government Printing Office, Washington D.C. 20402. But something has gummed up their usual inefficient bureaucracy—I ordered a set of the Rules two years ago (with check enclosed) and have still to receive them. You might have more persuasive methods than my own.

Anyway, the Communications Act of 1934 costs $1.25. *Volumes I and III of the FCC rules* cost $11—and these are the ones you are required by law as a licensee to have in your possession. For suffering would-be lawyers, there are the *FCC Reports,* weekly pamphlets with the decisions, reports, public notices, and other mind-killers. $14 a year.

For $25 a year, you can receive the daily *Federal Register.* FCC legal stuff, including rules amendments, proposed rules, and miscellaneous notices are cheek-by-jowl with FPC requests for bids for power stations in Bagloosh, Minn., FTC show-cause orders on varieties of dental floss, and AEC decisions on burning you (and me) into easily disposable cinders. Hardly worth it.

Tables, charts, and see-what-we've-done are in the FCC Annual Reports at $2.85. Like *Broadcasting Yearbook*, they'll always be 6-9 months late: the *Report for Fiscal 1973* is not a hand yet, and here it is damn near summer-melting time.

Health, Education & Welfare grants are described in intimate detail in the Federal Register Vol. 34, #18 (January 28, 1969). Another Federal Register Reprint is called "The Public & Broadcasting, Procedural Manual"—and it embodies many of the Petition Devices I list later on. You can get the former reprint from HEW, Room 406, Reporters Bldg., 300-7th St. SW, Washington D.C. 20004. The latter is available from the FCC, 1919 "M" St., NW, Washington D.C. 20554.

For the non-profit aspects of your corporation, you should order "Tax Information for Private Foundations and Foundation Managers," from the IRS (Publication #578: it's free.) It'll tell you the difference between a "Private" and "Public" foundation, in elaborate, disinteresting, and confusing detail. However, since it might improve your fund-raising chances by having a "Public" foundation—you would do well to order it and maybe even read it.

Some broadcasting people seem to think that it is vital to subscribe to *Billboard* and *Variety*. Bless me if I can figure out why: both seem dedicated to the furthering of the shuck aspect of communications, and the style in which they are written is strictly flapper stuff. If you are really into trade 'hype' magazines—about the best is one that is called *Radio* (formerly *Bob Hamilton's Radio Report*). It has occasional good articles and interviews with people who are doing something in commercial radio, but you will be paying for all the top 100 pop reports, so that for your 52 issues a year, you will be paying something staggering like $130. If you can survive that tab, their address is Box 14869, Memphis, Tennessee 38114.

THE REFERENCE ROOM

There is—in broadcasting and FCC Common Truth—only one Reference Room. It is at the FCC offices 1919 M Street Northwest in Washington D.C. The public (you) have as much right to visit and reference there as do all those $100/hour hot-shot Connecticut Avenue double-breasted communications attorneys in their knit cordovans and blue-stripe shirts. The only difference is that you are reading that stuff for you and for good radio; they are doing it for you-know-who and getting paid megabucks for their time there.

The room number is 239. It's back in the back of the building on the second floor. As you are looking for it, wander around and see what our $45,000,000 a year is buying for us. If a hearing is going on (open to the public) sit in for awhile and see if you can figure out what that nonsense is all about: watch the lawyers playing up to their clients (if they happen to be present) and shouting at each other and doing a little hearing-examiner ass-kissing. Un-bee-lee-vable. Stick you head in some of the offices and maybe even talk to one or two of the people who work there: in certain branches, if they are not rushed, they will show you around. They won't answer any controversial questions—but they will tell you what they are about.

The last time I was in the public reference room two of the clerks behind the counter were doing the boogaloo and acting like *humans*. I think the FCC has changed a bit since I wrote the first chapter of this book some 3 years ago. You even hear occasional snatches of laughter in those right-angle walls and halls, indicating that, yes, humanity may well have come to the great pristine father-of-us-all. As a matter of fact, Ed Hackman who—believe me—did more to insure the standards of FM in this country in the last 40 years (and has consequently had more to do with FM transmission than 1000 broadcasters) told me recently that he was looking forward to retiring in 1975. "This place is turning into Disneyland," he said, a bit sadly, I thought. Most of his friends have died off or retired—the old timers at the FCC—and he is very suspicious of some of the characters who are beginning to come up the ladder. I watched him chase two engineers from the TV broadcast bureau out of his office: they wanted those records he has so carefully built up over his time there: but like any character out of Gogol—Ed knows that the information he has is right (it is) and he isn't about to turn it over to some kid who was in diapers when he first started to work. In some countries it is called a 'fiefdom.'

Back to the reference room: Sam Buffone calls it "the last egalitarian institution in the United States"—but Sam is prone to such sweeping remarks. In any event, it is a fascinating mixture of people poring over the business secrets which are a part of the public filing required of each and every AM, FM, Short Wave, and Common Carrier in the country.

The Reference Room is divided into two parts: the station file section, and the dockets division. In the file section lies every scrap of material filed by any broadcaster in the last ten years (the rest are over in some nightmare paper warehouse in Arlington—order time: 5 days) every technical change, most communications from the FCC, every document of a non-confidential nature filed by every station from Boo-hoo, Ore. to New York City.

It is here that you can look into the greasy heart of your home-town station, and see what they have bundled together to justify their miserable existence to the FCC. It's called Promise vs. Performance.

On the other half of the counter lies the Dreaded Dockets Division. I say Dreaded because it is the repository of those $100,000 drain-'em-dry hearings that some broadcasters have had the temerity and gall and tax write-offs needed to put themselves through. All the transcripts of hearings are smushed together between blue school-boy notebook folders marked in black-run printers' ink *Do Not Remove. Property of the FCC.* You can flip through some of those folders (I have seen one that stands about as tall as meself) and learn The Rule of FCC hearings. That is, there are too many words. And they are not really that important. But they are trapped forever: like some random fly, buzzing endlessly and maddeningly against the clouded windowpane of last summer's hottest afternoon. These dockets are the buzzes of tragic misspent American broadcasting—caught forever.

Broadcasters are notorious childish tattletales. And they send their lawyers over here to spy on their schoolmates. That's what those guys are paid for. But over the past few years, the visitors to the Reference Room have broadened out—become more universal looking. There are always one or two beards from one of the citizens' groups who are preparing stuff for license renewal time. You should spend a few days here if you are fascinated by American broadcasting. You will learn more than you ever wanted to know about FCC forms and filings.

Across the hall from the Reference Room is the Broadcast Bureau (Room 242). It looks like *sanctum sanctorum* of the FCC—but any of us can wander in there to use the circular files. This is the heart and soul and secret of the paper world we have come into: circular and horizontal and vertical files that have every radio and television station in the United States *that exists or ever existed* carded, along with every filing made for or against that station, along with the date.

It is an amazing compilation, kept up-to-date, the center of it all. If you want to find out the relations of any station with the FCC—if you want to find out if there has been anything filed, *ever*: this is where that information is kept. And the people there are nice and helpful. I tested the system by asking to see the information card on a radio station I had heard of, but never read about: it was one of the few cooperative community stations set up in the history of the country. I knew it was FM, served the Washington D.C. area, that it had gone on shortly after WWII, and had expired sadly in the mid-50s. Suzy—who happened to be there that day—directed me to the Stations Deleted File, I looked under District of Columbia—and there it was: WCFM, the Potomac Broadcasting Cooperative, granted a construction permit January 5, 1948, with a license deleted during The Dark Days of FM—January 20, 1955. All the material that anyone could ever want for an MA thesis—packed into that sad yellowed frayed card in the Deleted Forever file.

Elsewhere, I have told you how to subscribe to the FCC releases. If you live in the D.C. area—or if you have a friend who will go by 1919 "M" St. every day—then you can get the Public Notices for free. The Public Information Office is in Room 207. The rush times are 11 AM for the morning releases, and 3 PM for the afternoon stuff. The FCC—you might have guessed—is kind enough to underwrite the lawyers and engineers by printing up 1550 copies of each of the summaries of basic decisions and pending cases—and 200 copies of documents like the Initial Decisions of the Hearing Examiners and the full FCC decisions.

The FCC library is on the 6th Floor—Room 639. It is open 5 days a week, 8 AM-4:20 PM. Anyone may use the facilities—although only FCC staff and schools and colleges may borrow books. It has complete copies of Pike & Fischer, the basic legal cross-reference on FCC decisions which you could subscribe to if you were going into communications law or going crazy—but otherwise, is all there for free for reference in the quiet and pleasant library. You'll find hundreds of other technical and scholarly magazines there—all the back issues of *Broadcasting* and *The Journal of Broadcasting,* for instance. There are also more books than you would ever care to read on all aspects of American radio and television, the good and the bad, the sordid and the sweet, the self-serving and the highly critical. You could become a wise grey-beard by hanging out here for a few weeks. Their fiction shelf is a must.

As you wander around the FCC, you will notice a funny phenomena—not funny tee-hee, but funny queer. That is, that the FCC staff and attorneys seem to be very familiar, maybe too familiar with the visiting private lawyers. The high-paid Cohn & Marks or Smith & Pepper types move in and out of the various offices like fish in the dark caves at the bottom of the ocean. All doors are open to them except when (supposedly) there is a contested case. That is where the lines get a bit fuzzy. What is a contested case? When does it become illegal for a lawyer to call up some guy over at the FCC working on a file, or to come over to see him, or even take him out to lunch? Freedom of access means that all of us—you included—can drop over to the FCC and talk. But it also means that those tweeds in private practice can take someone from the FCC to lunch, and drop a not-too-subtle hint about employment opportunities available, on the other side of the fence, at double or triple the FCC salary—a nice job, awaiting them, when they get through with that particularly tough case which would benefit everyone concerned with a positive decision.

I'd say that a goodly portion of the lawyers I have met sense that subtle dividing line between persuasion and bribery: but not all; god knows, not all. Payola doesn't exist only in the broadcasting end of the radio-tv biz.

LAWYERS AND ENGINEERS

You can't afford legal payola—but even so, there are real and good reasons to have an FCC attorney, even for a simple application. Their intimacy with the processes of the FCC can help inexperienced you a hundred way—even though the bloody FCC was set up to *do away with lawyers* through administrative law.

Broadcasting Yearbook lists all the attorneys who are members of the Federal Communications Bar Association, and who are thus more or less competent to work on your application. Unfortunately, *Broadcasting* has no rating system: thus, if you pick at random, you might find the many who will harm you more than help you. You would be wise to call the people at *Citizens Communications Center*—1914 Sunderland Place, Northwest in Washington—and ask for a reference. CCC itself is in the business of bonking on existing broadcasters, not helping out would-bes like you: but they are familiar with the Washington jurisprudence jungle, and can name a couple who will help rather than harm you. Before you take on *anyone,* talk fees with him. You might end up with a $150 a month retainer (common in this industry, so awash in money, and looking for the usual business expenses)—which you can ill afford. Don't be shy: most of these characters adore bucks, and aren't ashamed of telling you of their needs. However, the ones recommended by CCC will more than likely be sympathetic to your poverty.

Broadcasting Yearbook also lists the engineers who practice before the FCC, do all them spaghetti maps, plot contours, and charge $100 an hour to give expert testimony in hearings for that $1,000,000 channel. Most of the FCC engineers I have met who practice out of Washington D.C. are true-to-life meat-head Eichmann types who will work for any aether-waster who comes down the pike. But if you get in a pickle with the Section V-B of your form—you will need a consulting licensed engineer; and in any case where you might have an overlap with other stations, or are in need of the perfectly legitimate directional antenna for educational FM stations—you are going to need all the help you can get. I know a couple of good and honest and inexpensives who can help you: some who practice with a conscience, and who deplore the waste of radio as much as we do. They will charge you for the complexity that is contour prediction, radial elevation (necessary for over-10 watt applications), and equipment—but they won't overcharge you.

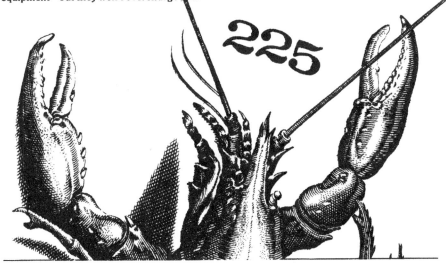

EQUIPMENT (NEW AND USED)

They don't advertise it too much, but each of the major broadcast equipment makers carry a large stock of reliable, used, and reconditioned (and often good) transmitters, antennas, consoles, turntables, microphones, and the like. Further, they will regale you with countless free catalogues if they think you might buy any new stuff. The major ones are

Andrew Corporation
10500 West 153rd,
Orland Park, Ill. 60462

CCA Electronics
716 Jersey Avenue
Glouchester City, N.J.

Gates Radio Corporation
123 Hampshire Street
Quincy, Ill. 62301

Jampro Antennas
6939 Power Inn Rd.
Sacramento, Calif. 95828

RCA, Broadcast Systems,
Building 15-5
Camden, N.J. 08102

Collins Radio Corp.
Dallas, Texas 75207

Sparta
5851 Florin-Perkins
Sacramento, Calif. 95828

Andrew and Jampro are about equal when it comes to new antennas; CCA is rumored to have superb FM transmitters—but they are sort of downtown pushy types. Gates are the Dodge Rebellion Boys of the broadcast world—but they might give you a free lunch; RCA is all fucked up: their equipment can be the best or the worst, but because they are such a monster corporation, they'll always get at least *one thing wrong* with your order, or the billing (sometimes this can be to your advantage);* but the best of all to deal with, and with quite adequate equipment, is Collins. Of all these engineering types, their people seem a bit human, anyway. Whoever you buy the equipment from, if you are buying new equipment, try to get them to give you a payout. It's the American Way. But don't let them fool you with their "add-on" vs. "straight" interest gibberish. The former ends up being 12-18% —even though they mutter something about "6% interest." Remember—they are only trying to sell you something in the time-honored debt mode.

*When I sent a copy of the first edition to RCA with this description, Don Forbes of RCA Broadcasting Equipment Sales in Austin Texas wrote back: "May a pregnant hippo in heat hump your mother for your lies about RCA..."

There is a list of manufacturers in *Broadcasting Yearbook*—but the most comprehensive one I have seen is the Electronics Industry Telephone Directory which is issued free from the Harris Publishing Co., 33140 Aurora Rd., Cleveland, Ohio 44138, even though the cover sez $10.

For used equipment, try Maze Corporation, 1900-1st Avenue, Birmingham, Alabama—or some of the others I listed in the first chapter of this book. But you should remember that old equipment can drive you crackers—even if you (or your chief engineer) pride yourself on your creative engineering ability. If you can possibly roust up the money, or get an extended payout from the manufacturer—you will definitely be better off with the new stuff. If you can afford it.

"Extended Pay-Out"—what we have come to call E-Z Squeezy—means 1/4 to 1/3 down, and the balance over 3 or 4 years, at some awful interest rate. The equipment itself is used as security. You may fear getting into debt—especially a $15,000-$30,000 debt that you will need for a 5-10 kw station. However, I remind you that you cannot change and grow without a gamble here or there. And the gamble of insolvency is no more than every broadcaster has taken since beginning: your contributions, if your station is worth a shit, will come to $1000-$2500 a month after the first four months (unless you are broadcasting to Pee-Pie, Mont.)—and the need to pay off the equipment is always a high incentive to gamble. Lew Hill, I am told, always felt it was best to plunge ahead and go into debt—rather than contract and shrink on the vine. And it always worked: the need for what you will be doing is great enough to allow you to gamble huge debts for even greater rewards. So sez Mr. Aether.

TOWERS

If someone tries to sell you an existing tower, forget it. It costs more to dismantle an existing tower and cart it to your location, than it does to build one anew. The only existing tower you might be interested in would be one near your proposed location: that can be a gold-mine. Ecological considerations are making it harder and harder for us to build towers, so it is worth your while to own your own—and you can rent out extra space to two-way services, or land mobile, or the FBI, and the CIA. These last two seem to have an insatiable need for tower space all over the world—and they are good tenants: quiet, don't eat much, very protective and always prompt on their rental payments. WYSO has been sharing space with some secret US-NKVD mystery bunch, and hasn't had a peep from them in the more than three years of comfortable neighborhood. If for any reason, your tower gets blown up, you know that you will have instant rebuild from the very best that The Boss can provide.

The towers that I prefer were made by Buckminster Fuller and were called Tensegrity Towers. I tried to get the Fuller combine to build me one in Seattle, and they treated my request with elaborate disregard. At present, my friend Rusty is designing one for Loma Prieta that will be an exact, 250 foot replica of the Eiffel Tower. Barring all this, the most common are put out by Rohn Manufacturing (Box 2000, Peoria, Ill.) or the sturdy 4-leg expensive Martian upright put out by *Upright* (right!)—and they are located at 1013 Pardee Street, Berkeley, California.

If you have to build you own, try to get enough land so that you can install guys: self-supporting towers cost almost twice as the guyed variety. For your guys, you will need enough land so that you can reach out 60% of the height of the tower from the base. Thus if your tower is 100 feet tall, you will need guys to stretch out 60 feet in three directions *from the*

base of the tower, or about 2,500 square feet of land.

Towers are very good for climbing up on on rainy or windy days: if there is a radiating element working while you are up there, you will find that your ears will tingle as the VHF-RF dissipates from your body, which may make you feel a bit like Frankenstein getting charged up.

CIDER-MINDER

Holt and I got into a hectic argument about apple cider. We were talking about the deplorable taste of the run-of-the-mill pasturized see-thru apple juice, and comparing it unfavorably with the non-pasturized, raw, and cloudy variety. He pointed out that it cost $2.75 to buy the real, unprocessed cider from a local health food store, and I stated that it was available on a back country road just outside Sequim, Washington for $1. Holt said that we could make a fortune in the apple juice business—importing it in bulk from Washington to California. I said that he should consider approaching Standard Oil of California with using their giant tankers for the shipment of raw apple cider rather than bunker oil. "Not only would this lower the price of the juice for its California fans, but should serve to allay the fears of militant conservationists about tanker collisions in the foggy reaches of the San Francisco Bay." Neither of us, of course, were willing to hazard a guess as to the ecological consequences of a spill of 500,000 gallons of unpasturized apple cider on the marine life of the Bay, but we guessed that—at least—it would not gum up the feathers of all those damn grebes.

(*KTAO Guide #61*)

SYMPATHETIC BROADCASTERS

There are a few honest broadcasters in the U.S.—and the smaller of these will do everything they can to help you get on the air with your own station, *if* they are convinced that you will be doing a community type of operation. I list all of them below in alphabetical order. You should try to get to at least one of them to see how it operates, and what it is like to be in their peculiar and particular atmosphere. A week at one of these places—if they want you—can teach you more than a dozen fatter, thicker copies of this and similar books.

ON-THE-AIR

1) KAOS, Evergreen State College, Olympia, Washington
2) KBBR, Bi-lingual Broadcasting Foundation, 4010 Finley Avenue, Santa Rosa, California
3) KBDY, Montgomery-Hyde Park Neighborhood Advisory Council, 2505 St. Louis Avenue, St. Louis, Missouri
4) KBOO, The Jack Straw Memorial Foundation, 3129 SE Belmont, Portland, Oregon
5) KOPN, 915 East Broadway, Columbia, Missouri
6) KPBX, Spokane Public Broadcasting, 24 West 25th Street, Spokane, Washington
7) KPFA-KPFB, The Pacifica Foundation, 2207 Shattuck Avenue, Berkeley, California
8) KPFK, The Pacifica Foundation, 3729 Cahuenga West, North Hollywood, California
9) KPFT, The Pacifica Foundation, 618 Prairie, Houston, Texas
10) KPOO, Poor Peoples Radio, 532 Natoma, San Francisco, California
11) KRAB, The Jack Straw Memorial Foundation, 1406 Harvard, Seattle, Washington
12) KTOO, Box 1487, Juneau, Alaska
13) KUSP, The 'Pataphysical Broadcasting Foundation, Box 423, Santa Cruz, California
14) WAFR, Box 1166, Durham, North Carolina
15) WBAI, The Pacifica Foundation, 359 East 62nd Street, New York City
16) WRFG, Radio Free Georgia, 1091 Euclid Avenue, Atlanta, Georgia
17) WYEP, Pittsburgh Community Radio, 4 Cable Place, Pittsburgh, Pennsylvania
18) WYSO, Antioch College, Yellow Springs, Ohio

QUASI-ALTERNATIVE ON-THE-AIR

[*Stations which have open access, but because of commercial or university ties or naked fear, are not exactly as free as the ones listed above.*]
1) KDIC, Grinnell College, Grinnell, Iowa
2) KERS, Sacramento State College, Sacramento, California
3) KDKB-Am, KDKB-FM, Box 4427, Mesa, Arizona
4) KKUP, Pasadena and Granada, Cupertino, California
5) KSAN, 211 Sutter, San Fancisco, California
6) KUMN, University of New Mexico, Albuquerque, New Mexico
7) WAMH, Amherst College, Amherst, Massachusetts
8) WBCN, 312 Stuart Street, Boston, Massachusetts
9) WGTB, Georgetown University, Washington, DC
10) WREK, Box 32743, Atlanta Georgia
11) WTBS, MIT, 3 Ames Street, Cambridge, Massachusetts

OPEN-ACCESS STATIONS WITH CONSTRUCTION PERMITS
AS OF THE DATE OF THIS WRITING

1) KCHU, Agape Broadcasting Foundation, 2516 Maple, Dallas, Texas 75201
2) Waif, Stepchild Radio, Box 6251, Cincinnatti, Ohio
3) WDNA, Bascomb Memorial Radio, Box 330069, Miami, Florida, 33133.
4) WFAC, Columbus Community Educational Broadcasting, 490 Oakland Park, Columbus, Ohio
5) WORT, Back Porch Radio, RR 2, Midtown Rd., Madison, Wisconsin 53711
6) ---, Pacifica Foundation, National Press Bldg., Washington, D.C.

GROUPS WITH APPLICATIONS BEFORE THE FCC, OR IN THE PREPARATION PROCESS

1) All Indian Pueblo Council, Albuquerque, New Mexico
2) Austin Community Radio, Jester Center A 231-A, University Station, Austin, Texas 78712
3) Birmingham Public Radio, N. Dormitory, Birmingham Southern College, Box 125, Birmingham, Alabama
4) Community Broadcasting, San Luis Obispo, California
5) Double Helix Corporation, Box 8187, Pierre LaClede Station, St. Louis, Missouri
6) Fresh Air, Inc., Minneapolis, Minnesota
7) GranFalloon/Denver Educational Radio, 222 Logan, Denver, Colorado
   ?!B£$(;)⁂⁒
8) Kauikeaouli Institute, Honolulu, Hawaii
9) Makah Communications, Neah Bay, Washington
10) Northern Community Radio, Grand Rapids, Minnesota
11) San Antonio Community Radio, 225 Castano, San Antonio, Texas
12) Southern Communications Volunteers, Memphis, Tennessee
13) Sound & Print United, Route 1, Box 177F, Warrenton, North Carolina
14) Sunrise Communications, 2418 "Y" Street, Lincoln, Nebraska
15) Sunrise Community Radio, Telluride, Colorado

In all, the best people to tell you about starting from scratch, with no scratch, are those who have done it: David Freedman at KUSP, John Ross at KBOO, Tom Thomas at Double Helix, and the folks at KBDY in St. Louis. If you are not after the usual make-a-buck duck, they will give you time and information—lots of it if you show interest in them and their divinely appropriately cheap and scratched up transmitting apparatuses.

PROGRAMMING SOURCES

I told you it is ridiculous to think about all the great programs you are going to do until you get your official construction permit from the FCC.

However, during that long boring time between the filing of your material, and the grant of the CP—you might begin to make contacts with some services which can provide you with a strong talk and music section of your library—and which can be a source of inspiration for your local live origination programs.

All of the radio stations I have listed above have some sort of tape source-tape exchange program. The Pacifica tape library at 2217 Shattuck Avenue in Berkeley is the most comprensive, and the most carefully put together—but their prices are murder, especially for a small, poverty-stricken operation like yours. The KRAB Nebulae stations used to send good tapes in apparently random order to each other. The secret of course is for you to get to know someone at one of these stations and have them pirate tapes for you when no-one is looking. Really: this is the best way for you to assure yourself a continual flow of quality talk programs.

Radio Free People at 133 Mercer St., New York 10012 has a small, growing catalogue of radical material—some of which is a little heavy. But they do strive to make this available to all those who want to use them for any purpose (no copyright) at a minimum charge. Another weirdo group is something called Zero Bull Shit (honest: *ZBS Media*) at R.D. #1, Fort Edward, N.Y. 12828. I think (but am not quite sure) that they are to contemporary recorded material what the Ashcan School was to American art in the 20s. I may be wrong.

For a dollar or so, record companies will sell to you any records in their current catalogue. You have to have a permit to broadcast from the FCC, or a construction permit before they will do so, and you have to make your order on appropriate stationery, and you have to sent them a company check. The price varies—some are charging almost $3 per record now, but in any event, you will get records for your station for a price cheaper than even the local distributors get them. This is a great chance to get a meaningful rich collection of ethnic, classical, folk, blues, and jazz music for your proposed outlet.

You will need an address for each of the record companies, and by far the most comprehensive listing is Billboard's "International Buyer's Guide" which is issued each September. The address is 9000 Sunset Blvd., Los Angles 90069. You won't (I won't, none of us can) believe the rich variety of record companies who are producing discs—anywhere from RCA down to some dibbly little things with a total catalogue of two records. You might do well to print up a general letter, and send it to all the record companies listed there, asking for free new releases, and for their price for older selections in their list. I would personally like to suggest these manufacturers as a must for *outre*, rich, subtle, sometimes rare and unknown material:

Ansonia	Folk Lyric	Folkways
Telefunken	Argo	World Pacific
Monitor	Arhoolie	Seraphim
Historical	UNESCO	Biograph
County	Maype	Oiseau-Lyre
BAM	DGG-Archive	Regal
Request	Odeon	
Topic	Dogmouth	
Orfeon	Ocora	

S E X & B R O A D C A S T I N G

BAM, Regal, Topic—and some others—are handled by Peters International of 619 West 54th St., New York City. They have some of the finest discs in the world, and deserve your especial attention.

Some other important American companies are Vanguard, Decca, Takoma, Nonesuch Elektra, Angel, Vox, RCA-Vintage, and Colmbia—but you know about them already.

There are some international broadcasters who will be superbly generous, and who will send you records and tapes at no cost. A few of these are outright boring propaganda programs— you will be able to recognize them easily enough, and at least you can erase the tapes and use them for your own purposes, like we always did with the bullshit religious rant-and-rave tapes that we got into KTAO. The international broadcasters with the best quality programs are:

The Canadian Broadcasting Corporation, Box 500, Toronto
NHK (Radio Japan) Uchisaiwai-cho, Chiyoda-ku, Tokyo
Radio Moscow, Moscow
The New Zealand Broadcasting Corporation, Bowen Street, Box 98, Wellington
East German Radio, East Berlin, DDR
The Australian Broadcasting Commission, 145 Elizabeth Street, Sydney NSW 2001
Radiofuziunea Televiziunea Romina, Str. Moliere 2, Bucharest
The South African Broadcasting Corporation, Durban
The British Broadcasting Corporation, Broadcast House, Wood Lane, London W12
Radiotelevisione Italiana, Vitale Mazzine 14, 00195 Rome
The Netherlands Broadcasting Service, Hilversum, Holland

I give you their home addresses here—although you can also contact their US office (see Augie Blum's article further on in this chapter.) By direct contact with the home office, sometimes you can get programs that the representatives don't even know exist.

The British Broadcasting Corporation put out some superb discs of music and drama from 1948-1964. Which they sold to a few, a very few experimental American broadcasters of the time. Most of these are rare and unobtainable, but if you ever dig them up, let me know. Their plays of Beckett, their music of India or North Africa, their Reith lectures on contemporary thought—set a standard for broadcasting that most of us tried to emulate at the time, and could never quite succeed in doing. The BBC—since its founding—had been the touchstone for those of us who believe in the art of radio: and no-one this side of the Alantic has ever been able to reach their artistry and care. Someday, somehow, someone will get the permission of the Queen (or whoever's in charge over there) to re-duplicate these rich, leisurely, immortal programs—and broadcast them on this side. You and I could create a world of wonder and emotion by merely playing and re-playing the august diamond-work of the BBC Third, and news, and public affairs, and drama, and music.

This was the Holy Grail that Lew Hill was seeking when he put KPFA on the air in 1949—and sometimes managed to touch.

There are two services which specialize in distributing international programs. One, the North American Broadcasting Corp., at 8900 Bohemian Highway, Monte Rio, California, serves as free distribution for some international broadcasters. Then there is the Broadcasting Foundation of America (52 Vanderbilt Ave., New York City 10017) but they tend to be

unspeakably dull in their talk programs—the most exciting things they do are the giant music festivals of Europe which are recorded in super fidelity, with classy announcers speaking four languages.

While we are on the subject of unspeakable dullness—avoid association with the National Association of Educational Broadcasters like the plague. These nurds are paid to produce programs which inflict sleeping sickness on the audience, and as far as I can tell, have yet to issue one interesting, controversial, meaningful program in their entire sordid (and expensive) history. Pap for the intellectual masses is what it is.

Somewhat less dull—at least when it comes to their coverage of live congressional hearings—is National Public Radio. If you fulfill certain qualifications, you can easily become a member of NPR, with operating grants from CPB. However their public affairs programming should never replace your own strong locally oriented material. In addition, in their need to grow, NPR seems willing to operate in association with 3 or 4 broadcasters in the exact same service areas. This means that we are treated to the same program being heard at the same time at a variety of places on the dial. This is a hoaxing waste of valuable frequency, and should be yet another reason for you to avoid duplicating the creepo local educational stations in your area. Remember, what I tell you again and again: The people who run these stations are *scared*. You don't have to be. They are *imitative*. You don't have to be. They were educated and think in terms of the Dark Ages. You should be enlightened and alive and different.

For you see love: you do not *have* to be dull to instruct and inform and enlighten people. Your best, most exciting source of programs will be the thousand thousand people in your area who have never never had a broadcast station come to them and offer them time *for free* to talk, to be interviewed, to sing. American broadcasters—both commercial and 'educational'—have built in a mystique: a mystique that says you must have magic to have access to the microphone. The magic is sometimes called 'money,' sometimes 'a degree from the Grundge School of Broadcasting,' sometimes 'proper training.'

This is, of course, the big sin in the whole system. For America—unlike most any other country in the world—has more broadcast outlets, and less official (government) restraint and requirements on *who* can reach the microphone. So that you and I are being falsely discouraged; being discouraged from thinking that the aether is our own.

This is so evil. A radio station should be the place in the community for concerned and talented and plain-home-folk individuals to have a chance to express themselves. In the place you live right now, there are hundreds of secret talents: there is someone who collects (and loves) old jazz; there are politically aware people—who can speak to reality, and raise so many consciousnesses in the process. There are readers—who can do fine 30 or 40 minute readings from novels or plays or poetry or children's books.

There are individuals, walking down the street right now, right there: live, loving people who can play the guitar or the kazoo or the harp—people who would be delighted to know that there is one door to the aether which is open and free to them: a door to all the hungry minds and souls of so many people who will, at last, know (through your station) that they are not alone.

You can end that mystique. The forbidding wall that American commercial broadcasters—and their in-school counterparts—have built against the rest of the world. You have the chance to destroy the cruel walls built by the media barons to keep out the dispossessed, the thoughtful, the wondering. Your frequency can be the one place in your community of men where the angry, and the frustrated, and the knowing and the creative and the perceptive and the hopeless and the lost: The one place that they can know that they are free to speak their piece—without censorship, without fear, without cost.

You will then give them the secret—which is really no secret at all: that is: the radio station owners and the schools and colleges are not the priests and magicians who control the voices of 10,000 American transmitters; but rather, just frail ghosts who we are programmed to think have the right and the duty to keep us out.

And you and I know, thus, that the door can and should be opened to all who care and who want to be heard. This is the secret: the one that I am trying so hard to share with you. Please, let us share it with all the others.

OTHER PROGRAM SOURCES

Augie Blume put out the following list of foreign broadcasters with available programming. It is valuable because it is the first in-depth listing of international broadcasters who will send you their material, most of it for free. (Here, I say you: meaning, of course, existing radio stations, or permittees for radio stations; the international broadcasters deem that this material is good for their image in the United States.)

FOREIGN PROGRAM SOURCES FOR INTERNATIONAL RADIO

Broadcasting Foundation of America. 52 Vanderbilt Avenue, New York, NY 10017. Both music and spoken word.

German Information Center. 410 Park Avenue, New York City 10022. A weekly German view of the news, also a 15 minute review of the German press. (For German language radio plays and special music programs, contact Mr. David Berger, U.S. Representative of the Association of German Broadcasters, 635 Madison Avenue, New York City 10022.)

The Swedish Broadcasting Corporation. 1290-6th Avenue, New York 10019, offers a monthly taped program called *From a Stockholder's Diary*, as well as others on music.

The Danish Information Office, 280 Park Avenue, New York City 10017, offers a monthly program entitled *Copenhagen Report*. News & music.

Norway provides a free series called *Spirit of the Vikings*, which mainly accents their cultural aspects. Norway House, 290 Madison Avenue, New York City 10017.

The Netherlands Information Service at 711-3rd Ave., New York City 10017, has free programs available on music.

Music and life in Austria is available from the Austrian Information Service, 31 E. 69th Street, New York City 10021. German language records of drama are also available.

Free tape loans of operas (many rare ones), classical and baroque music, and some outstanding contemporary works are available on loan from RAI, the Italian Radio-TV System, 717-5th Ave., New York City 10022.

Masterworks from France is one of 5 weekly series currently available from the French Broadcasting System, 1290-6th Avenue, New York City 10019.

Vistas of Israel is the title of a continuing series of programs available from the Israel Information Service, 11 East 70th St., New York City 10021.

Interviews, music and talk is the focus of *A Look at Australia*. Write the Australian News and Information Bureau at 636-5th Avenue, New York City.

South African cultural materials are available from the Information Service of South Africa, 655 Madison Avenue, New York City 10021.

It is practically impossible to say the phrase "New York Unique" rapidly three times in succession.

For English or French programs from Canada write the English Language Transcription Service, CBC, Box 6000, Montreal, Canada.

Another overseas source of programming comes from the Belgium Broadcasting Services, Director des Emisions Mondiales, Radiodiffusion Television Belge, 8 Place Flagey, Bruxelles 5 Belgium.

NHK is the Japanese organization for recorded information on the culture and news of Japan. Overseas Broadcasting Dept., Nippon Hoso Kyokai, Uchisaiwai-Cho, Chiuoda-Ku, Tokyo, Japan.

Free Indian programs on classical music, talk and features in Hindi, French, English, Arabic, Indonesian, or Tibetan are available from The Director of External Services, All India Radio, Broadcasting House, New Delhi, India.

Malaysia offers music and features—Radio Malasia, Peti Surat, Pos 1074, Bangunan Persekutuan, Kuala Lampur, Malaysia.

Folk music and talk programs are free from the Director General, New Zealand Broadcasting Corp., Box 98, Wellington C.1, New Zealand.

Programs are available from Finland, Spain, Korea, and many other countries. Write to the local consulate or embassy. The information lady can get addresses for you at no charge.

Lastly, the British Broadcasting Corporation offers programs in the categories of Features & Drama (including Alice in Wonderland), Variety, Serious Music, Light Music, Talks, Schools, Children's Programs, and others. Write to the BBC Transcription Service, Kensington House, Richmond Way, Shepherds Bush, London, W. 14, England.

> *(Under the section NO SYMPATHY FOR BROAD-CASTERS, Blum quotes from Friedman, the "conservative" economist, and Nicholas Johnson:)*

We guess that Milton Friedman, the conservative economist, was right after all when he said, "a government system cannot be devised which will not be taken over by vested economic interests, and exploited for the preservation and enhancement of their own wealth." Industry won't regulate itself: it's better to be rich and guilty because you can always give to the church. Looks like government won't regulate it either. Even if it is hopeless, Nick Johnson came up with the following good ideas for the FCC to pursue:

...we could require that a given proportion of gross income be invested in programming.

...we could require certain size news staffs and budgets as a proportion of gross income.

...we could require an appointment of someone to argue the public interest side in cases before the FCC, sort of a consumers' rep.

...we could require more public participation in TV. In Holland, any group that gets 15,000 people to support its proposed programming gets free time on the air.

237

...we could require that the three networks provide one hour of prime time for public affairs.

...we could require $500,000,000 for Public Broadcasting instead of Nixon's paltry $5,000,000.

...we could require informing the public at license renewal time by requiring full page ads instead of a line in the obituaries.

(Finally, Blum explores the depth of common ownership of radio and tv stations by conglomerates, with exhaustive facts and figures:)

The FCC allows a single party to hold broadcast licenses for up to seven television stations, no more than five of which may be VHF (the AM of TV), seven FM stations and seven AM. The eleven largest cities do not even have one VHF television station that is not in the hands of a network, a newspaper, a newspaper chain, an owner of a group of stations, or an industrial or financial conglomerate.

There are 7,350 radio and television stations. There are 1,547 cities with daily newspapers. An FCC study of the ownership of media revealed that in spite of the numbers of newspapers and broadcast outlets, the control of the largest share of audience, profit and political power lies in the hands of a very few – usually a newspaper.

Today there are 1,483 cities with monopoly newspaper ownerships, compared with 64 cities who have competing ownerships. Des Moines, Minneapolis, San Diego, Louisville, and Portland (Ore.) are important cities with newspaper monopolies. Whether a city has a single or multiple newspaper ownership, one or all of the proprietors may be a large chain like Hearst, Copeley, Newhouse, or Scripps-Howard.

(*Radio Times #118*)

WOWIE! ZOWIE!

THE 100: KTAO'S GREATEST HITS!

We asked all the KTAO people to give us the name of their favorite record in the KTAO collection. In addition, those of us who work here on a day-to-day basis picked out ours. We came up with about 300 records, so we tried to prune it down to what are the true, great, soulful, meaningful, stunning, rich, magnificent sounds in the record room of KTAO.

It's a silly project, of course. I found myself deluged with the diversity of the musics we have available to us: and by the universality of the tastes of most of us who work here. There are some records with great cuts—but the whole wasn't what we call 'great.' There are some records which have slipped like slugs out of our gentle hands: tapes, too; whatever happened to that fine recording of *The Dramatic Music of Hunan* that we recorded from Garfias' collection? And all those tapes of the music of Atahualpa Yupanqui. We can't possibly list them all—although we should; then this paper would be as big as *The New York Times,* only twice as fun.

So we picked and chose, and chose and picked. I have listed the record numbers—and from time to time, there are comments we have appended. I dare say, though: if you bought all the records we list herein, you would have a rich and fine collection of all that the world record companies have been able to produce; all that we have heard, anyway.

(CLASSICAL)

Peasant, Dance, and Street Songs in Germany. The Early Music Quartet of Munich (Das Alte Werk SAWT 9486). Contrary to most recordings, this medieval music is fun and jouncey—and like most Telefunken recordings, excellent in fidelity.

"Come Ye Songs of Art" by Henry Purcell. (L'Oiseau Lyre 50166) Alfred Deller turned into a turnip after his early successes—but this record was made before his vegetation. The famous "Sound the Trumpets" with John Whitworth is a thrill-chiller.

Cantate Domino by Jean-Joseph Cassanea de Mondonville (Music-Guild MG-119) One of Billy Baroque's favorites. With Louis Martini and the Jean-Francois Paillard Orchestra.

36 Fantasias for Harpsichord (Complete) by Telemann. (Dover, Box E) A real sleeper, with sumptuous performances by Helma Elsner.

Igor Kipnis. *English Harpsichord Music* (Epic BC 1298) and *The Harmonious Blacksmith.*

The Pope Marcellus Mass by Palestrina. (Angesl S-36022) The Roger Wagner group.

Motets for Double Chorus, Brass & Organ. (Westminster 14090)

SEX & BROADCASTING

Celestial Music and *Now Does the Glorious Day Appear* by J. Henry Purcell. (Vanguard Cardinal VCS 10053)

Spanish Songs of the Renaissance. Victoria de los Angeles (Angel 35888)

The Virtuoso Recorder, with Frans Brüggen. (Decca DL 710049) Music of Telemann, de Fesch, Veracini, and Loeillet.

Baroque Organ Works performed by Helmut Rilling and Douglas Haas (Turnabout TV 34135S)

The Siena Pianoforte, Scarlatti & Mozart. Charles Rosen (Counterpoint-Esoteric 3000)

L'Arte del Violino, Vols I & II of Pietro Locatelli. Suzi Lautenbacher (VOX VBX 40). There is a lot of dull crap of Vox, but Suzi makes up for all of it, playing Locatelli (it must be) as he would have done, had he had our time and equipment and leisure.

Six Double Concertos for Two Organs. Padre Antonio Soler (Columbia MS 7174) Another of Billy Baroque's favorites.

Glenn Gould Plays Bach. The 6 Partitas, the Two and Three Part Inventions. Was, has been, and always will be the great interpretation of Bach, even though the Partitas were never meant to be heard on the piano. (COLUMBIA D3s 754)

Gamba Sonatas: Sonata in G Major. J. S. Bach. (Telefunken SAWT 9536)

The Wedding Cantata [#202] of J. S. Bach. Agnes Giebel with Jaap Schroder and the Concerto Amsterdam. Telefunken SAWT 9513). You wouldn't believe the words are
> Oh Maytime's the gay time for cooing and wooing
> far better than flowers' so fleeting delight.
> The clover's soon over, but never will sever,
> the bonds of the devotion that true love unite.

because it don't make any difference (it never does with Bach).

The Six Solo Concerti after Vivaldi by J. S. Bach. Janos Sebestyen (Turnabout TV 34287)

The Complete String Quartets of Shostakovitch with The Borodin Quartet. (Melodia/Seraphim SIC-6034)

The String Quartets Nos. 1 & 2 of Leos Janacek. The Janacek Quartet (Crossroads 22 16 0014)

Kindertotenlieder of Gustav Mahler. Kathleen Ferrier with Bruno Walter & Orchestra (Columbia)

Bartok, *Mikrokosmos* (Complete). Gyorgy Sandor, pianist (VOX VBX 5425)

The Complete String Quartets, Franz Schubert. The Endres Quartet (Vox Box VBX 5004, 5005, 5006). The richest and most sensitive performance of Schubert. In mono (you should avoid the echo tunnel rechanneled for Stereo SVBX).

(ETHNIC)
(English)

The Lark in the Morning. Dave and Toni Arthur (TOPIC 12T190)

Leviathan! A. L. Lloyd, Ballads & Songs of the Whaling Trade. (TOPIC 12T174)

Chorus from the Gallows. Ewan MacColl with Peggy Seeger. (TOPIC 12T16)

Waterloo-Peterloo. The Critics Group. (Argo ZDA 86) Stupendous—just stupendous.

Frost and Fire. The Watersons. A Calendar of Ritual and Magical Songs. (TOPIC 12T136)

(Africa & the Near East)

Messe des Desherites. From Mali, dir. Jose Gourgois. (Barclay 920177) There are a great number of 'folk masses' floating around, most of them pure horseshit—like the dimwitted 'Missa Luba' and another wretched one from Spain. The *Messe des Desherites* is distinct, beautiful, and honest.

Escale en Guinee. (Pathe CPTX 240.746) Dig especially the "Homage a l'O.U.A." at the end of side 2.

The African M'Bira. Music of the Shona People of Rhodesia. Nonesuch Explorer is the most honest of the U.S. folk-ethnic companies, and this record, collected by Robert Garfias, is one of their best. (Nonesuch H72043)

The Iranian Dastgah. Philips International has put out the UNESCO collection and this, a collection of music for voice, tar, and zarb, is one of the finest. (Philips 6586 005)

Abdallah Chanine. The Oriental Bouquet. Piano music if you believe it tuned in the Arabic (rast) mode. Changes your ideas about what a piano can and should do. (Parlophone LPVDX 134)

Music for the Classical Oud. Although Hamza El Din is good, he gets overplayed—and this one with Khamis El Fino is as good. (Folkways FW 8761)

Musique Populaire Marocaine.
  "I go through life without a care
  And sing 'May Allah deliver me
  From material care...' "
(Boite a Musique LD 435)

Flutes Orientales Sacrees des Derviches Tourneurs. Islamic flute and vocal from Turkey. One of Cese's favorites. When she can find it. (Disque Vogue CLVLX 542)

(CENTRAL EUROPE AND RUSSIA)

Gheorghe Zamfir. Pan pipe music from Roumania. Pan pipe music is beautiful if it's done right: here it is. (Electrecord ST-EPE 0432)

The Pennywhistlers. Seven women singing songs from all over Central Europe. (Nonesuch H72007)

Music from Rumania. One of the first great ethnic records released by London, and collected by Deben Bhattacharaya. (London SW 99456)

Folk Music of Kosmet. A real sleeper. One of the Monitor records—which range from superb to terrible. This one is superb (Kosmet is an obscure region of Serbia with terrific music). (Monitor MF 386)

Yugoslav Folk Music. Lyrichord is another of those record companies which has music ranging from the super-duper to miserable; usually the recording quality is such that you can barely hear the music. Not so with this one, which features (ready?) Sopila music from Krk. (Lyrichord LL 189)

Songs and Dances from Bulgaria. Argo—like most English recording companies—uses extraordinary care in the recording and presentation of their ethnic series. This one, complete with map and Deben Bhattacharya, is so pretty you would want to cry, or at least I would, when I first heard it. (ARGO ZRG 562)

Music of Bulgaria. This was one of Nonesuch's first records, and although you have to suffer through Stereo rechanneling, it's worth it for the ensemble conducted by Phillipe Koutev. (Nonesuch H-72011)

Folk Music of Albania, recorded by A.L. Lloyd. A.L. Lloyd. Jesus, I want to meet him. Not only does he sing all gravel and grunch great (see earlier listing) but he collects and masters discs like this one in which our KTAO commentator wrote next to the selections things like 'wow' 'nice' 'wow' and 'etc.' (Topic 12T107)

The Piatnitsky Song & Dance Ensemble of Russia. Back in the dreary old dark ages, when all trade (or mention) of Russia was considered to be disloyal, crusty old BRUNO records continued to grind out a huge catalogue, drawn from Russian masters. Most are unlistenable, because of wretched fidelity, but if you are willing to defy that, then their series of Piatnitsky Ensemble are well worth it. (Bruno 50170)

Folk Music of the U.S.S.R. Sometimes I think that Folkways is best at collections. They have in their catalogues some 1000 or 1500 recordings (they never delete) of the most boring music from the Yourbi country to endless wastelands of Pete Seeger and the Country Gentlemen. But when they put together a collection, they sift through the mass and come up with something like this one which is brilliant as a collection. Henry Cowell did the compilation. (Ethnic Folkways FC 4535 A through D)

Skaggmanslaget "Pjal, Gnall & Ammel." Lennart Wretlind brought us a whole series of Swedish fiddle records, and this one is the most interesting because not only are they all hippies carrying on the tradition of the style, but Side 1 Cut 7 is a stunning rock rendition of 'Pekkos Pers Storpolska.' (Sonet SLP2510)

Music from the Far North. Music from Sweden, Finland, and Lapland. (ARGO ZRG 533)

   SOUTHERN EUROPE

La Toscana di Caterina. Vernon Buck found this one for us. An antidote to the miserable Italian Neopolitan drinking shit—this one is an old craggy lady (Caterina Bueno) who knows how to pull the strings in *real* fashion. (Tank MTG 8010) You might also get their record of Jewish Harp music (with no Jew's Harp on it but one) (Tank MDG 2001)

Fados from Coimbra. You'd never suspect United Artists could put out anything but bad mush, but this one is another sleeper; such a sleeper that it's now deleted. The Coimbra Quartet of Portugal. (UA UNS 15539)

Songs of the Spanish Civil War. This should be under American Folk—since it was recorded by Pete Seeger in the early 1940s—but it gives the feel and misery of that miserable war. (Folkways FH 5436)

Fiesta Andaluza. The Spanish Regal company under the 'Series Azul' label has a series of Flamenco performers most honest and all of them superior to the dimwit records which pass for Flamenco in this country (such as "Manitas de Plata" which knowing Spaniards refer to as "Manitas de Plomo" [or "Little Hands of Lead"]). This Regal record features Los Giraldillos and Los Gitanos de la Cuevas. (REGAL-EMI LREG 8.006)

Pepe Marchena [Nino de Marchena]: Sus Estilos Flamencos. Pepe Marchena has two records out on Odeon of Spain. By listening to him, you can learn what Flamenco was supposed to be all about. These are all 78s 'reconstructed' (in the Spanish phrase) unfortunately. But great, nevertheless. (Spanish ODEON LCLP 168 and 251)

Archivo del Cante Flamenco. Vergara, the Spanish recording company, has compiled an exhaustive six record set of Flamenco which includes extremely rare 78s, and equally rare recordings of Pedro del Lunar. (Vergara 13.001-13.006) Nothing phony here.

(CENTRAL AND SOUTH AMERICA)

Musique Populaire des Llanos Venezueliens. Maracas, harp, cuatro, and voice. By Los Quirpa. (Boite a Musique C 441)

L'Ame Musicale du Perou. Harp, guitar, charango, and quena music of Peru. (Boite a Musique LD 5442)

The Pinata Party Presents Music of Peru. One of the first records available in this country of harp and religious music. (Folkways FW 8749)

Musique des Andes by L'Ensemble Achalay. (BAM LD 5422)

Atahualpa Yupanqui. Unfortunately, the music of this great and good and masterful Argentine artist is practically impossible to get in this country. It is distributed on Argentine Victor—but the only one that Peters International (the fine import distributor in New York) carries in their catalogue is imported through Spain: 'Y el dolor, quien se lo paga?" (Spanish RCA-LPM 10 383)

Ramito: El Cantor de la Montana. This is just one of nine Ramito records put out by Ansonia—the best company for real Puerto Rican mountain music. There is a lot of lesser stuff in their catalogue, but if you look for the records with the pictures of the obvious hicks on the covers, you should be getting the 50 or 60 best in their collection. Ramito has recorded 9 discs alone—but you would be wise to stick to the first 4 or 5. (Ansonia ALP 1277)

Navidad con Priscilla Flores. La Alondra de San Lorenzo is one of the great Puerto Rican artists—and this Christmas music is like none you have ever heard before: rough and tough and good. (Ansonia ALP 1371)

Caribbean Island Music. Songs & Dances of Haiti, the Dominican Republic, and Jamaica. Another of the great (and cheap) Nonesuch Explorer series put out by Teresa Sterne. (Nonesuch H72047)

Music of the Bahamas: Bahaman Folk Guitar—Joseph Spence. Samuel B. Charters found Spence and recorded him for Folkways and that is how the KTAO Spence Top 100 Cult started: he sings "Coming in on a Wing and a Prayer" which I mean has little to do with the 1940s song that we all knew and loved as young jingoists. (Folkways FS 3844)

Baile al Estilo Oriental. This is a recording of the funkiest, realest, most-ethnic version of mechanical-human music we have in the collection, from the Oriente province of Cuba. (Discuba LPD 107)

Recordando a Guty Cardenas. Orfeon has a series of Saul Martinez records which some people call sentimental but which I call pure Mexico Yucatan *realismo*. When most people think of Mexican music, they think of that drippy 'Ranchero' stuff which has little of the soul and none of the artistry of this homage to the great artist "Guty." (Orfeon LP 1270)

Mexican Panorama: 200 years of folk songs. Vanguard has collected some of the finest funkiest artists in Mexico and put them together on this disc. (Vanguard VRS 9014)

Nara Pede Passagem. This is Cese's favorite recording of Brazilian bossa nova. (Philips P 632.787L)

The Real Bahamas in Music and Song. Nonesuch does it again: of all the records in our collection, this seems to touch the most who are in touch with what we call 'ethnic.' The harmonies are so bizarre that you want to get on a boat and go there just to see how they can possibly *ever* sing so sweet. (Nonesuch H-72013)

(India and Far East)

Shakuhachi Duets. Imported records from Japan—and some of the best of Japanese flute music. (Japanese KING LKD 24; Japanese Polydor LPJ 36)

The Music of Japan. Barenreiter-Musicaphon puts out the greatest ethnic-real-honest-quality-true-blue-recordings ever. If you ever see their UNESCO series in a used record shop, *buy it*—no matter what the price. For instance, their six-record series of Japanese music is the best general survey of classical and folk Japanese music *ever*. One of my favorites is the Noh play "Hagoromo"—but the others are as careful and true. (Barenreiter UNESCO 30 L 2012 through 2017)

Anthology of the World's Music. The Music of China. Somehow also related to UNESCO—this lo-fi record was one of the few of Chinese music to be released to the Western world by the China Recording Company. The Mainland Chinese, of course, are doing wretched re-writings of classical Chinese music—but this is pure and, despite the fidelity, good. (Anthology AST-4000)

The Music of Indonesia. If they stick me away on an island, this is the one I want to take with me. Hell—they can send me to Java and I won't need it. For years, we used *Water Music* from this album as the sign-on theme for KRAB. And it still gets me when I hear it. If you buy nothing else in this compilation, get this one: and learn (as we all have) that the subtlety and harmony of Indonesian music is as magnificent as an Beethoven or Tchaikovsky or even, dare we say, yes, a Bach. (Folkways FE 4537 A, B, C, & D) Edited by Henry Cowell.

Java: The Historic Gamelans. Somehow Philips has gotten into the UNESCO business—and this record is one of the most stunning. There are birds in the Martapura Slendro gamelan selection—and sometime you should get Garfias to tell you about them. (Philips 6586 004. So far there are 7 records in the Philips UNESCO collection 6586 001 through 6586 008. They are all worth however much you pay for them.)

The Gauguin Years. Songs and Dances of Tahiti. (Nonesuch Explorer H-72017)

Musique de la Route Interdite du Nouristan "L'Afghanistan." Barclay put this one out: flute, tulla, tablas, voice, and tambourines. Great. (Barclay 920 086)

Afghanistan and Iran. Somewhere, somehow, there is a great warm luscious rich heaven for people who put together a record like this one: it is so haunting that I found myself wondering about the divinity of the man who can intone the love song of Uzbek (Side 1, band 1). I suppose you can say that disdain for the whole American musical-artistic culture grows out of the fact that there is no other radio station that I know of which has played this, ever. Ever. Ever. For that reason alone, they deserve all the bad that Al Kramer and Nick Johnson and Tracy Westen can put on them. Forever. This recording has made some of us want to get to Afghanistan, and not for the usual hip reasons. (Disque Vogue LVLX 191).

Call of the Valley. Santoor, flute and guitar from India. Despite the corny title, a great work of Indian classical music. (ODEON SMOCE 1149) By the way, if you want to build a true and honest collection of Indian music—don't go out and buy all the World Pacific records of Ravi Shankar. Although he is an honest enough musician, the true great Indian music is recorded by Odeon of Dum Dum, India. The best way to get a collection is as follows: get hold of the Peters International catalogue (600-8th Street, New York City 10018). Buy *all* of the Odeon India records *except* for the Movie Music. Zingo—you're an Indian music collector—and a good one too.

Vishne Sahasranamam. Anyone who gets into the KTAO collection of Indian music finds, at last, this, the apex-be-all-and-end-all of the collection. The great Indian singer Smt. M.S. Subbulakshmi singing the one thousand sacred names of Lord Vishnu. Don't ask why: go out and get it. Your karma and you need it. (Odeon SMOAE 5011)

The Sitar Quintet. Another sleeper. Abdul Halim Jaffar Khan. (Odeon SMOCE 115)

Santoor and Guitar Duet. This is one of Cese's favorites. Shivkumar on Santoor and Brijbhushan on guitar. (Odeon MOCE 1063)

The Bauls of Bengal. I almost didn't put this one on the top 100—because like "Finbar and Eddie Furey," Sandy Bull, Leo Kotke, Robbie Basho, Chatur Lal ("The Drums of India"), Dueling Banjos—these records get played so much that we lose concept of the greatness of them. But the Bauls of Bengal deserves to stay—and be heard. (Elecktra EKS7325)

Folk Music of the Punjab. We include this one because Indi says its so bawdy—but she may be prejudiced, because she's from the Punjabi region. (Odeon MOCE 1174)

(American)

The Anthology of American Folk Music. This series of three Folkways albums (each, with two records) is divided arbitrarily into "Ballads," "Social Music," and "Songs." Ignore the classifications. They are probably the most honest and true compendium of American folk, bluegrass, gospel, and Black music. All the recordings are from the 78 era—and this is the cheapest and best way to hear great artists, in their greatest songs: people like Buell Kazee, Bascom Lamar Lunsford, Blind Lemon, John Hurt, Henry Thomas, Blind Willie Johnson, and the great Rabbit Brown. (Folkways FA 2951, 2952, 2953). Collected by Harry Smith. Praise be to Harry Smith, whoever he might be.

Smoky Mountain Ballads. Some of us think that Bascom Lamar Lunsford was the greatest and best of the Mountain music collectors. Although the best record is his Library of Congress recording, the Folkways 10" is possibly more accessible. (Folkways FA 2040) Can you imagine: Folkways is still manufacturing ten-inch lps?

Goofing Off Suite. I mentioned earlier that Folkways has a long and boring mass of indifferent discs by Pete Seeger. He is an artist who has not, we think, aged too well: and the best he ever did for Folkways was the aforementioned *Songs of the Spanish Civil War* and this one, a series of experiments for banjo. Dig, especially, the duet from Beethoven's 7th Symphony on *Banjo* Another ten-inch weirdy. (Folkways FA 2045)

Gospel Songs. All *a capella*, with the Missionary Quartet. One of Cese's favorites. (Folkways FW6824)

Negro Religious Music: Sanctified Singers, Vol. I & II. For honest true and good recordings of American Black musicians—especially repressings from 78s, you can be totally sure with *Arhoolie*, and *Biograph*, and *Historical*. In addition, for more modern music of this genre, Arhoolie again and *Takoma*. Chris Strachwitz, the *Arhoolie* man—brought out these two albums under the *American Music Series* label. They include the great "Little Boy" of Rev. Kelsey and the stupendous Washington Phillips records. (American Music Series BC 17, BC 18)

Southern Prison Blues. From Tradition, including the great Hogman Maxey (honest!). (Tradition-Everest 2066)

Moving Star·Hall Singers. For honest Black country musicians, Folkways is the most consistent: sometimes funky (dig their Country Brass Bands album—I'd put it in here if I could find it—and "Negro" singers like Horace Sprott)—they have done the true work on what used to be called 'field' recordings. This record, from a festival at Johns Island, South Carolina, is Moving, and a Star. (Folkways FS 3841)

The Six and Seven-Eight String Band of New Orleans. We never knew how to classify this one: blues, blue-grass, Cajun. Dr. Edmond Souchon. (Folkways FA 2671).

Berkeley Farms. Although Folkways isn't always the best for contemporary modern records (their sound, for one thing, is a bit weak—and they were the last to go to stereo of the big record companies) this one is good, and recent. (Folkways FA 2436)

More Clawhammer Banjo Songs and Tunes. For authentic folk and bluegrass, the record companies who are most true-blue and *Rounder, County,* and *Folkways.* This is P. Sheehan's choice, and is recently recorded. (County 717)

Joan Baez. Well, they've tried to use her all up, but this one will never fail or die. For those of us who were folkniks in the 50s, it was this record, and the *Weavers at Carnegie Hall* (Vanguard) that got *us* started on modern type folk music. Aficianados of later Joan Baez are often puzzled by the purity of this particular record—which is why we like it. (Vanguard VSD 2077)

Mingus Ah Um. It's too bad that we couldn't get the KTAO jazzniks to put together their list of the true great jazz records. So we have to depend on our own (rather arcane) taste, which includes this, the next record, and a scattering here and there of early MJQ and Sidny Bechet and Django. Thass all. That doesn't obscure the durability of this one: C. Mingus at his silliest and best. (Columbia CS 8171)

Patty Waters Sings. The ESP catalogue was always the most stupendous exploration of the perimeters of jazz. And this one remains our favorite—largely because of "Black is the Color of My True Love's Hair." (ESP 1025)

The Doukhobors of British Columbia. This one gets our obscurantist-excellence of the year award: the Doukhobors have achieved a certain notoriety because every time they think they are getting screwed by the tax-laws of BC, they strip off their clothes and burn their farms, sort of pee-pee Mau-Mau action, as it were. The sensationalism of this obscures the fact that they are a great anarchistic culture, with their own music and customs. The music—they are original *emigres* from Russia—has a strange blend of East and West, haunting harmonies which do more to humanize them than a thousand silly stories of their nudity-in-the-snow. (Folkways FR 8972)

248

(Speciality Records)

Phono-Cylinders Vol I & II. For those of us who never get a chance to play cylinders, this is one of the best collections of that genre—when the voice transcribed on wax was such a novelty that they announced each selection as if they were a radio station—long before radio stations existed. Except, of course, for our own dear Mr. Fessenden's. (Folkways FS 3886-3887)

Oh, What a Lovely War. The title is for real. There was a record from WWI by Courtland & Jefferies which sang of the trenches as if it were a great & fun lark. Where is that innocent time now, Jack? The whole album bespeaks the fun of war which makes some of us glad that we didn't participate in the shock of actually getting gassed at Ypres. (WRC SH 130)

The Golden Age of Mechanical Music. The Saydisc Company puts out records of jazz and 'actuality' which are all well chosen and well presented. This is one of music boxes and automatic pianos. Would you believe it is distributed by "Ahura Mazda" of New Orleans? (Saydisc SD 218)

Gilbert and Sullivan: the Pirates of Penzance. Those who really know D'Oyly Carte claim that these 1931 recordings have never been matched, never beaten for wit & life. For the time, great fidelity. Malcolm Sargent conducts (EMI MFP 2143)

(KTAO VOLUNTEERS PICK THEIR TOP RECORDS)

Karyn Alice Baker: "The Irish Pipes of Finbar Furie." Because it makes me dance every time.

Joe Morrow: "An old 78 RCA Victor recording 'When It's Nighttime in Italy It's Wednesday Over Here.' Lou Holtz is the artist. It's a comedian with orchestra. The title sez it all..."

P.P. McFeelie: "*Kathy & Carol*—a real sleeper from Nonesuch Elektra (EKL-289). Some of the sweetest close harmony ever recorded. And they're from San Jose, my home town."

Merrill Kelly: "My favorite record is 55.6 seconds. Because when I was in the Sixth Grade at good old Will School we lost everything. We lost in football. We lost in basketball. We lost in baseball, and then they wouldn't even let us play in the volleyball tournament. We lost everything except at the City Track Meet we won the relay race and made a new record which is still good today."

Geoff Leskovsky: "My favorite record is *The Girl in the Chair* by The Roots of Madness (Dogmouth BALH-1760). The Roots of Madness is a San Jose group with enormous potential. They play trumpet, guitar, ocarina, short-wave radio, and with themselves. I would say they are the new Santana, or at least The Penguins, and I think they will look back at the historic year of 1972 and say 'This was when *The Girl in the Chair* got pressed.'"

Doug Cruickshank: "Merle Haggard & The Strangers 'Live in Philadelphia.' Absolutely one of the finest live albums ever to be recorded. Listen to 'I take a lotta pride in what I am,' also 'Today I started loving you.' He also does his old favorites 'Okie,' and 'Fightin' side of me.' He's great. If you wake up some day feeling down in the dumps just put on Merle. He'll bring you right up. 'White lightnin's still the biggest thrill around.'"

David McNeil: "My favorite is Damian Luca's *L'Eblouissante Flute de Pan Roumaine* (Pathe-Marconi C 054-11362) It's very sweet stuff and brings in subscribers."

THE E VERNON BUCK SELECTIONS

When E. Vernon Buck came to KRAB at the age of 15, he could hardly spell "Huapangos Huastecos," much less play it. We stuffed him on the air because we couldn't find anyone else who would give their Friday nights to engineer for us.

Now, ten years later, he are one, and he is also our Mr. Ethnic. He learned through the excellent collection of ethnic music we had built at KRAB, and studied under the spiritual music-pie of KRAB [Robert Garfias, of the University of Washington Department of Ethnomusicology), and now has one of the most subtle collections in the Western World, or at least in North Oakland where he lives. I asked him to write up a list of Buck's Great Ethnic Top Poppers, and this is the one he came up with. You could do worse than to order all of them.

Latin America

Bolivia
"Folklore, Vol. 2"
by Dominguez, Favre, & Cavour
Campo LPS-005. Distributed by Columbians Potosi, 1286 Casilla 3051, La Paz, Bolivia

Chile
"Las Ultimas Composiciones"
Violeta Parra
RCA Victor, Chile, CML-2456

Argentina
" Soy libre!   Soy bueno!"
"Preguntitas Sobre Dios"
Atahualpa Yupanqui
Le Chant du Monde (French), LDX-S-74371 and LDX-7-4415

Ecuador
"La inolvidable"
Carlota Jaramillo
Remo LPR 1553, 664 10th Ave., New York, N.Y. 10036

Peru
"Canta el Peru"
Los Morochucos
Sono Radio, LPL 1117. Distributed by Remo (see Ecuador)

Mexico
"Huapangos Huastecos"
Los Cantores de Panuco
Cisne C1-1041. Carracci 64-Mixcoac. D.F. de Mexico

Cuba
"La Vida es Una Carcajada"
El Jilguero de Cienfuegos
Rumba LPR-55535. Distributed by Met-Richmond Record Sales, Inc. 1637 Utica Ave., Brooklyn, N.Y. 11234

Venezuela
"Musique Populaire des Llanos Venezueliens"
Los Quirpa
BAM (French) C-441

Brazil
"Batucada No. 2"
Jadir de Castro
Companhia Brasileira de Discos, P-632.180-L

Asia

Japan
"Gagaku Taikei"
Nivico SJ-3002 (3 record set), and Nivico SJ-3003 (3 record set)

China
"Ehr-hu and Yang Ching"
Art-Tune Company COL-3132 (Hong Kong)
"Folk Music"
Art-Tune Company ATC-146 to 150 (5 record set)

U.S.S.R.
"Ensemble of Hornists"
Mezhdunarodnaya Kniga (MK) 14795 (7")
"Tatar Folk Songs"
MK 16839-16841 (7")
"Armenian Music"
MK 13149-13153, 13185

Indonesia
"Java Pays Sounda"
BAM LD 110

Africa

Music of Africa Series
33 LP set
International Library of African Music
  P.O. Box 138 Roodepoort
  Republic of South Africa

Madagascar
"Valiha"
Ocora 18

NOTE BY DON CAMPU, ONETIME MUSIC DIRECTOR OF KTAO:

These twenty records are the best we've run into,
should be a part of anyone's basic jazz collection.

Cecil Taylor/Buell Neidliner: *New York City R&B* (Barnaby KZ 310 35)
Patty Waters' *College Tour* (ESP 1055)
Greatest Jazz Concert Ever (Jazz at Massey Hall with Charlie Parker, Dizzy Gillespie, Bud Powell, Charlie Mingus, and Max Roach) (Prestige 24024)
Jazz Meets India (with Irene Schweizer Trio and Indian Musicians) (SABA SB 142ST)
Don Cherry: *Eternal Rhythm* (MPS 15 204ST)
Don Cherry: *Relativity Suite* (JCOA 1006)
Ornette Coleman: *Free Jazz* (Atlantic 1364)
John Coltrane: *A Love Supreme* (Impulse A-77)
John Coltrane: *Kulu Se Mama* (Impulse A 9106)
Miles Davis: *Filles de Kilimanjaro* (Columbia CS 9750)
Duke Ellington/Charlie Mingus/Max Roach: *Money Jungle* (UAS 5632)
Bill Evans/Jim Hall: *Undercurrent* (UAS 5640)
Jimmy Giuffre: *Western Suite* (Atlantic 1330)
George Gruntz: *Noon in Tunisia* (SABA 15 132 ST)
Friedrich Gulda: *Air from Other Planets* (MPS 15225ST)
Pedro Iturralde: *Jazz Flamenco Vol. 1* (HispaVox HHS 11-128)
Thelonious Monk: *Greatest Hits* (Columbia CS 9775)
Charlie Parker: *The Immortal* (Savoy SB 15148ST)
Archie Shepp: *Life at Donaueschengen* (SABA SB 15148ST)
Eric Dolphy: *Out There* (Prestige 7652)

Finally, we asked all our friends and fans and ex-on-the-air people to pick the best records they could find in the KTAO collection, or their collections at home. This list is a knock-out—especially when you realize that the people who made up the selection couldn't five years ago, name one "ethnic" record.

Music from the Morning of the World (BALI) (Nonesuch Explorer H 72015)
Dances of the Renaissance, Tieleman Susato & Pierre Phalese (L'Oiseau-Lyre SOL R330)
Guitare Argentine, Raul Maldonado (Arion ARN 30 U 154)
Columbia World Library of Folk & Primitive Music: France (91A 02003)
Kphth Moy Lebento-Manna (Greece) (Odeon OMCG 28)
Folk Music of Norway (Folkways FM 4008)
Spelmanslatau Fran Uppland (Sonet SLP 2025)
Music from the Far North (Argo ZRG 533)
Folk Music from Albania (Topic 12T107)
Music from Rumania (Argo ZRG 531)
Bulgaria (Argo ZFB 47)
Yugoslav Folk Music (Lyrichord LL189)
Greek Folk Music (Lyrichord CL 188)
Greek in Music & Song (Argo ZFB 70)
Dances & Instruments at Turkey (Request RLP 10074)

Turkish Music (UNESCO Barenreiter BM 30 L 2019)
Pechers de Perles et Musiciens du Golfe Perisque (Ocora OCR 42)
Iran II (UNESCO Barenreiter 30 L 2005)
Andre Hossier—Rhapsodie Persane (Pathe 2C 064 11000)
Maria Escudero (Montilla FM 57)
Federico Garcia-Lorca—Homage (Philips 836-985)
Musical Treasures of Portugal (Philips PHI 414)
Anthology of Portuguese Music (Folkways FE 4538)
Columbia World Library: Southern Italy and Islands (91A 02025)
Sicily: Music and Song (Argo ZFB 71)
Easter Week and After—Dominic Behan (Topic 12T44)
Gaelic Music from Scotland (Ocora OCX 45)
A.L. Lloyd—The Bird and The Bush (Topic 12T135)
A.L. Lloyd—Leviathan! (Topic 12T174)
Subhulakshmi: U.N. Tour (Odeon: EMI 9870)
Les Flutes Roumaines (Arion 30T073)
Les Flutes Roumaines Vol. II (Arion 30 T 095)
The Living Tradition: Music from Rumania (Argo ZRF 531)
Ethnic Folkways Library: Rumanian Songs & Dances (FE 4387)
Muzica Romaneasca (Request RLP 8114)
The Penny Whistlers (Nonesuch H 72024)
The Penny Whistlers (Nonesuch H 72007)
Chants et Danses de Yougoslavie (Mode Vogue CMDINT 9800)
Madagascar: Trio ny Antsaly (Philips 849.481)
Programme Music of the Baroque Era (Telefunken SAWT 9549)
Musiques du Vietnam (BAM LD 434)
Cambodia (Production LD 112)
Laos (Barenreiter BM 30 L 2001)
A Bell Ringing in the Empty Sky (Nonesuch H 72025)
Cacjemire Vaiies Himalayennes (BAM LD 400)
Na Fili (Ireland) *Farewell to Connacht* (Outlet SOLP 1010)
La Calandria (Vol II: Puerto Rico) (Ansonia ALP 1399)
Le Steel Band de la Trinidad (Trinidad) (Arion ARN 33 167)
Caribbean Island Music (Nonesuch Explorer H 72047)

SEX & BROADCASTING

Early Cante Flamenco (Folk Lyric 9001)
Janacek: Quartets #1, 2 (Crossroads 22 16 0014)
Penderecki: St. Luke Passion (Philips PHS2 901)
Kurt Weill: Mahagonny (Columbia K3L 243)
Ligeti: Adventures (Wergo 2549 003)
The New Music (RCA Victrola VICS 1239)
The Varese Album (Columbia MG 31078)
Assassination Songs & Ballads (Library of Congress AFS L29)
Negro Blues & Hollers (Library of Congress AFS 659)
Singing Preachers & Their Congregations (BCs American Music Series BC 19)
An Evening with Rev. Louis Overstreet (Arhoolie F 1041)
Benny Goodman 1938 Carnegie Hall Jazz Concert (Columbia OSL 160)
Jelly Roll Morton (RCA LPV 559)
James P. Johnson (Folkways FJ 2842)
William Burroughs (ESP 1050)
Folklore of the U.S.: Jack Tales (AFS L 47: Library of Congress)
Masterpieces for the Chin & Pipa (Lyrichord LCST 7142)
Korean Epic Vocal (Nonesuch H 27049)
Music for the Teusung — Sulu (Ethnosound 8000/1)
Priscilla Flores (Ansonia ALP 1371)
Haitian Dances (Folkways FW 6822)
Mountain Music of Peru (Folkways FE 4539)
Musique de Tchad (Ocora OCR 36/37/38)
Musique du Burundi (Ocora OCR 40)
Musique Populaire Marocaine (BAM LD 435)
La Serva Padrona (Telefunken SLT 43 126)
Frans Bruggen: Blockflöten (Telefunken SMA 25073 T/1-3)
Arp Schnitge Organ (Telefunken TK 11521/1-2)
Bartok String Quartets (Columbia M31196)
The Sergio Mendes Trio: Brasil '65 (Capitol)

Helpful Hints from Radio Experts

(Some solid—and free—engineering advice on filling out the dangerous, difficult, duopolous, ding-bat Form 340 from the FCC, subtitled "Application for Authority to Construct or Make Changes in a Noncommercial Educational TV, FM, or Standard Broadcast Station." The information necessary for parts I, II, III and IV were given in the earlier part of this book. This essay, on Part V—was written by Benjamin Franklin Dawson, III—who is a professional consulting engineer, and a member of the Board of Trustees of the Jack Straw Memorial Foundation.)

The portion of FCC Form 340 that deals with the technical aspects of an application is section V-B. (Section V-A is for educational AM stations—damn few of them and no new ones since World War II—and section V-C is for educational TV stations, which are for high rollers only). Section V-B consists of two pages of questions, and all of the second page can be ignored in 10 watt applications, save for the signature required at the bottom of page 2. I say, can be ignored, but not always, since if you are applying for a frequency which presents a serious possibility of overlap with another authorized station, then you had better prepare a showing that the *possible* overlap won't in fact occur. To do that, you will need to understand what the statement in Note 1 to Sec. 1.573 of the Commission's rules means, and you will need copies of the 50:50 curves and perhaps the 50:10 curves. Good luck on the 50:10 ones, as they are no longer in the Rules and are out of print. The 50:50 curves are about the same as the 50:10 ones for distances less than 10 miles, however, so you can probably use them. In order to estimate distances to contours, you will need to be able to figure the average elevation of a radial, using the method described in the FM Technical Standards section of the *commercial* FM rules, Section 73.313, and to use the curves, together with the antenna height above the elevation of the pertinent radial, to figure out the distances to relevant contours. If this isn't enough explanation, after you have looked up the rules I have just mentioned, then find someone else to whom it makes sense, or find a friendly engineer to prepare the exhibits for you. I hate to see Ed Hackman having to process applications prepared by people who didn't understand what they were doing, as then he has to do all the work and he has enough to do already. If he gets an application that is really ineptly done, he will just send it back.

If you just can't figure out the basics of the questions section V-B asks, then ask your RCA salesman to get for you a copy of the booklet RCA has called "Guide to Applications for Radio Stations" or something like that. It has a green cover, anyway, and will be helpful in figuring out things like the answer to paragraph 6 of Section V-B, wherein you are supposed to show the efficiency in percent of the transmission line you propose to use. (Get it right on the application—length and all—see below.)

I want to say something here about antennae. When you fill out Section V-B you will have to specify an antenna by manufacturer's name and model number, and you will have to give some technical information about it. Unless you want to fill out another Form 340 all over again, you will have to use that model of antenna when you get around to building the station. You can build your own. For a 10 watt station you can use a simple omni-directional antenna like the superturnstile receiving antenna that Olson Radio sells for 10 or 15 bucks. You can also use a directional antenna, if everyone lives on one side of your transmitter, or if it is on the side of a hill. A very nice rugged one is sold by McMartin for use with background music multiplex receivers—I think they want about $18 for it—it is a 5 element Yagi. Jerrold and Channel Master also make 5 element Yagi's for around 10-12 dollars. If you are wise, however, and the population distribution is all around your transmitter site, then I recommend that you use a 2 or 3 or 4 element omni-directional antenna made especially for low power FM broadcast use. The best one I have used, and the cheapest I have any experience with, is the one made by Phelps-Dodge Communications Systems. They make a lot of 2 way radio antennas, and their low power FM antenna is very well made, out of stainless steel, heliarc welded, and although the super low power one is rated at 250 watts or so, I put one in a commercial station with 1 kw. in a windy and icy part of the world, and it is still working fine, 2 years later. It costs $200 per bay. Others, made by Gates, CCA, and other equipment suppliers, are fine, but somewhat more expensive. It is sometimes possible to obtain for free or cheap a used FM antenna from somebody who has discarded it, usually when they put up a circularly polarized one, but retuning them, and mounting the heavier ones, can be a real double A stinker of a problem unless you have a high frequency RF bridge and accessories and know what you are doing.

I recommend that 10 watt stations, and even moderate power ones, use horizontal polarization. With a 10 watt license, you want to get *out* as far as possible, and you hopefully can educate folks to run the line cord of their receiver horizontally for a couple feet in order to get your station. Using circular polarization means twice as much antenna for the same ERP, and besides the antenna will be much more expensive.

If you have to hire someone to do the engineering portion of the application for you, make sure you get someone who has done *educational* FM applications before and knows what he is doing. Standards of allocation for educational FM stations are on a demand basis, and, especially in metropolitan areas, require that the person doing the work be familiar with the method of allocation. There are some things the FCC will accept that *aren't* specified in the rules, and someone who isn't familiar with the allocations procedure and with past precedent can make large magnitude mistakes in some critical area. Above all, don't ask your local FCC field office what to do, as they don't deal very much in this particular aspect of the Commission's work, and therefore don't have the time or the information to be able to do much more than give you 10 copies of Form 340.

In line with the above I would like to point out several situations which will be a problem requiring some study for anyone proposing a new station application that is affected by them: 1) presence of a channel 6 TV station within 40 or 50 miles, 2) sometimes, if they are miserable jerks (and some are—even stations owned by bleeding heart liberals), presence of a channel 5 TV station within 40 miles or so, 3) location within 50 or 60 (and sometimes more) miles of the Canadian border, and especially within that distance of a large Canadian city, 4) location in Texas, New Mexico, Arizona, or California within 50 or 60 miles of the Mexican border (see the *very* recently added sections of the FCC rules about this), 5) location in an area such that your station's second harmonic falls within the passband of some channel 7 or 8 TV station that is significantly viewed in your area but has a crummy signal, 6) proposing the use of a transmitter site within a few hundred yards of some extensive two-way radio site or of the head-end (receiving site) of a cable TV system, (this may be no problem at all, however, depending on the competence and sensibility of the owner or technical supervisor of the two-way or CATV outfit) and 7) proposing a site or location on a tower very-very-very close (a few feet) to the transmitter of some high-power (50 or 100 kw) FM station. (Down 50 or 60 feet on the same tower is ok—across the tower on another leg may not be.)

WHAT TO DO WHEN THE CONSTRUCTION PERMIT ARRIVES

In your application for the construction permit, you had to demonstrate financing adequate to construct the station and to operate it for one year. The purpose of this chapter is to tell you how to construct the station as cheaply as possible, so that any capital that you have can be reserved for emergencies, the maximum possible program and production facilities, and for unforeseen expenditures. An example of this sort of unanticipated expenditure is the city tax on radio station of $200 per year that you don't find out about until you go downtown to get a building permit so that the electric company will go ahead and install a power drop. There are doubtless other pitfalls I haven't encountered yet, so be prepared for one or two of them.

I will assume that your application and CP are for 10 watts. Later on I will point out some specifics which apply to higher power stations only. Most of the material about 10 watt stations applies to higher power ones.

You will have specified in your application the type of transmitter you intend to use, as well as the type and length of transmission line, and the type and make of antenna. If you are going for more than 10 watts you will also have specified a type of modulation monitor. You can change the type and model of transmitter, and of modulation monitor to any other type-approved unit up until the time you apply for license. But you *cannot* change the type and length of transmission line, or the type and model of antenna unless you reapply on Form 340 for a modification of your construction permit. You cannot change the height of the antenna on the tower or pole you are going to use, either without such a modification. If you are going to use a homemade transmitter (and it is done occasionally, even by commercial stations) or a transmitter that is made up of sections of various different ages or manufacturers, that wasn't type approved as a unit, then you must specify that in your application for construction permit, and provide details as to what you are doing. The part you answer is the question which says: "If the above transmitter has not been accepted for licensing by the FCC attach as Exhibit No. _____ a complete showing of transmitter details. Showing should include schematic diagram and full details of frequency control, & etc." The Commission has always been willing to let a permittee use a handbuilt or much modified (the official jargon is "composite") transmitter, so long as it can be demonstrated to be in compliance with the "standards of good engineering practice"—meaning that its performance as to frequency stability, audio performance, and lack of spurious emissions is up to snuff. A couple of paragraphs from now I will talk about what to do about old 10 watt exciters that aren't specifically type approved as transmitters.

Please heed this advice about what you can and cannot change from what it says on your CP. If you file your license application with different information it will be a source of needless delay while you could be on the air, and even if you have a smooth-talking Washington lawyer, the Broadcast Bureau of the Commission won't budge until you get it all straight. You may, however, change the studio location and the remote control point (if you are remote controlled) at any time up to the application for license, on Form 341.

Now some specifics about equipment. You can buy a brand new transmitter from a new equipment manufacturer for something like $2000. Better, find a used one. What you want is an "exciter" section from a higher powered transmitter. Among the FM stations within a hundred or so miles from you is likely to be one that put in a new exciter when they went stereo, or replaced a tube type stereo exciter with a newer, solid-state one. You don't need stereo for a 10 watt station—really don't want it at all—since it will cut down your available audience by increasing 1) multipath distortion and 2) front end nonlinearity with low signal levels in a great many sets. What you want is a nice old tube type mono exciter (or the mono section of a stereo one) made by Harvey, ITA, RCA, Gates, Westinghouse, Western Electric, REL, GEL, Raytheon, or Standard Electronics. You do not want one made by Collins, GE, or Federal, since they depend for operation on a funny tube called a Phasetron (type 2H21) made by GE. They work wonderfully but unfortunately the tube is no longer available. I have a beautiful GE exciter in my office (I keep it around as a personal albatross) which works flawlessly, but I unfortunately have no Phasetron tube for it. There are two considerations about getting one of these exciters licensed and working as a 10 watt transmitter that you should keep in mind. Some of them are not specifically type approved as 10 watt transmitters, and you will have to specify them on your Form 340 as "X-brand type 1234 exciter, modified for 10 watt operation with external metering of plate voltage and current, and composite power supply," or something roughly similar. It will be necessary in many of them to use an external power supply and to supply metering of the plate voltage and current of the last RF amplifier or multiplier stage. At least one of the exciters, the RCA MI7016 (known to engineers as the 'iron fireman,' for its similarity to a well known brand of furnace burner), has its output at half frequency. If you are fortunate enough to come across one of these really superb exciters (they are used in old RCA TV aural transmitters, too) you will have to convert the output stage to a doubler to get an output in the FM band. (The MI7016 is my favorite old piece of FM gear—they are so good that they are type approved for stereo, despite the fact that the design is from the late 1930's, and for the old 44 mHz FM band.) You can obtain the correct crystals for exciters from a couple of different folks. I have had very good luck with Eidson Electronics, Temple, Texas.

If you come up with a GE or Federal or Collins exciter, you can make a transistor and IC substitute for a phasetron tube, but you might as well put your resident electronic genius to work building a whole new transmitter from scratch. If you cannot find a local source for a transmitter, then you may resort to one of the used equipment suppliers, like Boynton, Morris, New York; or Maze Corp., Birmingham; or Broadcast Equipment Sales, Bristol, W. Va. Boynton is very heavy on audio gear. Maze has a mixture, and BES deals with parts for ancient stuff, like a filament transformer for a 1933 Western Electric AM transmitter. Boynton gives you a salvation message with each month's mailing that is best avoided if you have a weak stomach. If you search diligently, you can usually come up with a useable exciter for $300 or less, sooner or later. I have bought them for as little as $100. If you have nothing but money, then by all means buy a new one, solid state and reliable and beautiful.

Of the new ones I personally like the Collins best (I have had good luck with the one I regularly work on). The RCA, actually made by Moseley, is good but kind of complicated—full of $8 IC's that zap once in a while. I still haven't seen a Wilkinson, although it has super specs and I have talked to one guy who is in love with his. After the first edition of the book came out I got a heat letter from CCA, because I badmouthed their AM transmitters, although I did say, and still do say, that everyone I have talked to who has one of their FM rigs likes it. I think the guy who runs CCA ego-identifies with his products too much. Anyway, all of them new jobbies are awful goddamn expensive. I think that Collins has the best customer service.

Your transmitter needs an antenna with which to work. Circular polarization is neat if you have transmitter power to burn. You don't. You want a horizontally polarized antenna. Make sure you specify the one you plan to use in your original Form 340 application (see above). I tend to deplore high gain antennas. About 4 bays is enough. You will need at least 50 feet of tower space for this—the bottom bay should be at least 20' off the deck. If you are going to have less than 30 or 40 feet of line from the transmitter to the antenna then you can use good old cheap and dirty RG-8 coax. If you are going to be more than 50' from the antenna, then you should use (and specify on your original Form 340 application) some 1/2 or 7/8 inch air dielectric Heliax, made by Andrew. A good source for this, fairly cheap, is Sierra Western Electric Co., in Oakland. It runs about a buck a foot, plus the connectors at each end, which are about 18 dollars apiece.

If you have a combined studio/transmitter space, skip this paragraph. The rest of you are going to have to extract the most possible out of AT&T. If your studio is at a location within a block or two of the transmitter, do your damndest to run a couple of sets of shielded twisted pair cable of your own from one to the other. Do a neat, workmanlike job and you can even (usually illegally) run pairs over streets and down alleys. If you are located in the same telephone office at both locations then order an unequalized program pair from Telco. It will be run in unloaded 22 gauge cable, and if less than 2 or 3 miles long, can be equalized nicely by terminating it in 50 or 150 ohms at each end. To control your transmitter you merely simplex a DC control voltage between the pair and ground, and use a sensitive relay at the transmitter end. Check out a book on telephone and telegraph systems from the library if this sounds too confusing. If the line is longer than 3 miles, up to about 6 or 8 miles, depending on the cable gauge, you can get it out flat to 15 kc with the aid of a simple series resonant equalizer at the receiving end and low impedance terminations. Check out a book on equalizers and filters from the library. You may have to feed the line at fairly high level (like +12 dbm) to get signal-to-noise ratio high enough, and this may get you into trouble, but play it cool, as most local telephone repairmen will never catch you. The reason that I have gone to all this explanation is that the unequalized line will cost you $10 a month, and will transmit both your program and your control holding voltage. If your transmitter is in a different *office* but the same *exchange* then you will have to shell out $30 a month for a 15 kHz program circuit, because interoffice cable is almost always loaded unless you pay for it not to be. You will also be stuck with having to order a separate circuit for transmitter control. Try to find out from your Telco special services office the rates for 1) intercom circuits, 2) 30 baud telegraph circuits, 3) unequalized program circuits (usually $10).

One of the first two may be under $10, in which case order it. If it is DC continuous, great. Use a simple DC control circuit. If it isn't, build a small audio oscillator and a companion detector and install them so that the presence of a 750 Hz tone on the line will hold on your transmitter. All of this presumes 10 watt operation, where all you have to do is be able to turn on (and off) the transmitter from the remote control point and monitor it off the air with a receiver. (If you are at a higher power, you will have to build or buy a remote control unit which will transmit certain transmitter meter readings back to the operator at the studio. There are some simple DC and tone telemetry methods for doing this, and I recommend that you borrow the instruction book for a commercially made unit and copy it more or less closely, with allowances for the state of the surplus electronic parts market in your locale.)

If your studio and transmitter are located in different telephone exchange areas, you are one of Job's descendants, no doubt, for you may have to pay as much as $60 a month for a 15 kHz line, and you will almost assuredly be unable to get a DC continuous circuit for remote control. At this point you should think seriously about an S T L.

An STL (for Studio Transmitter Link) is a transmitter and receiver combination for point to point operation in a tiny band from 947 to 951 mHz, with highly directional transmitting and receiving antennae. There are ways to do this fairly cheaply, but it takes a lot of UHF savvy and test gear, and a comprehensive knowledge of the surplus communications equipment market. For remote control you will probably have to use a voice grade private line telephone circuit, which costs from about $20 a month or so, and requires telemetry gear that works on frequency-shift keyed tones. It is possible to use 30 baud telegraph circuits for 10 watts, since all you have to do is turn it on and off, and they can be used for higher power installations, but someone with a good knowledge of digital control logic will have to build the remote control, and there are no commercially manufactured units for this service.

All of this by now should have clued you to try to keep your transmitter and studio in the same telephone company exchange area. Generally, rates are about the same all over the country for the kinds of services we have been talking about, but there are local anomalies, especially when independent small telephone companies are involved. If you are dealing with a small 'phone co., and you can get into their public relations guy, it is perfectly legal (now that NET and the Corp. for Public Broadcasting are in the act for big money) for telephone companies to charge you less or nothing since you are a bonafide educational broadcaster. You will also need a standard business phone. Get one, pay the $5 external equipment fee monthly (a rape job for sure) and hook up your own extensions, telephone taps, etc. *Don't* get lighted push-button key system phones. They charge too much for them.

One of the hazards of having a transmitter on some beautiful hilltop, which doesn't have power or telephone service, is the fact that both of these fine public spirited utilities will try to charge you the cost of construction for your power and telephone circuits. Unfortunately, they have sandbagged the public regulatory agencies so badly that they can usually get away with it—so check it out in advance.

We will now talk about audio equipment. You will need two good, low rumble turntables. Good ones are Russco, QRK, RekOKut (some), and Presto (some). If someone has donated you a couple of old RCA monsters of the 76B variety, they will do *if* their center bearings are in good shape. Buy two Shure M232 or M236 (12" or 16") tone arms. Get 16" ones *only* if you need to: for instance, if your turntable is 16 inch. Install Pickering PAC or PAT pickups in them. They are the cheapest and are rugged as hell, and sound better to me than any Shure. Don't use Hi Fi type turntables like the AR or Garrard. They are nice for Hi Fi's, and are lower rumble than some commercial types, but they will not last under the strain of radio station use and misuse. You will need at least two tape machines. There are a lot of tape machines in the world, and most of them won't last a week in a radio station. Even expensive ones like the Revox A77 won't hack it in a radio station. I know of some commercial stations using Revoxes in automation systems, and also sometimes for production, but I think that volunteer personnel are so damn hard on equipment that they are still risky.

If someone gives you an old Berlant, throw it away. Trying to keep it running will give someone of your volunteer technical staff so much grief that he may commit suicide. Best are commercial service Ampexes, of the 600, 601, 602, AG600 series. Usually one of these can be bought used for from 150 to 300 dollars, depending on condition and how much its owner knows about its worth. Sometimes you can get a nice unit used by some preacher or teacher or musician that has never seen commercial service, for a hundred dollars. They are rugged as the devil, and can be rebuilt an infinite number of times when they quit. They resemble the Model T in other ways: they are mechanically rather unorthodox. You must have at least one of them, for they will keep you afloat when all else has failed. Try to have two. If not, then buy the cheapest Sony deck, like a Sony 350 and when it wears out fling it out and buy another. If you can afford an even more expensive machine, like an Ampex 350 or 351, please do so. Love it and it will love you. Also nice but rarely in good shape by now are Ampex 400's. Rare and nice are Ampex 400A's. PR10's are ok but don't pay more than $200 for one as their motors are prone to fail and new motors are $150 last time I heard. Scully 280's and Ampex AG440's cost more than $2,800 and are out of our area of discussion. Forget about all those goddamn Ampex home models, and Panasonics, Teacs, Sony 777's, Robert's Webcors, Akai's; all of those. They are typical consumer market *junk*. You might want a Sony $100 cassette machine for out-of-the-studio interviews.

When it comes to a control board, things aren't so easy. It is nearly impossible to buy a good mixer used, as there is a big market for them. Every radio station can always used another board for recording, and the like, and so ones that are worthwhile for FM use are very scarce. One solution is to build your own. Find the schematic for a Sparta 5 channel jobbie and copy it, generally or exactly, depending on your talent for that sort of thing. If you are careful, parts for it shouldn't run you more than $150. Shure has a series of mixers and things two of which can be assembled into a passable control board for a total cost of about $300. It is the cheapest commercially made item that I know of that will suffice. Don't buy a big old tube type console because you may spend a lot to get it working correctly (noise down 60DB? Response out to 15 kHz? Flat, really flat? Distortion down under 2%?—those are tough to make with crummy old tube type equipment that has probably never been properly maintained). Even sleazy solid state gear will give you a lot less grief. For mikes buy Electro Voice 664's and maybe a 635A or two. Good sounding, and durable enough to drive nails—no kidding; cheap too, especially if you buy them used from a local pawnshop. If you are going to build any of your own gear you will need an audio oscillator and an harmonic distortion meter. Buy the kits from Heathkit and build them. My Heathkit distortion meter is just as good as any but the very latest 1500 dollar Hewlett Packards, and I have been using it for years.

Remember that you are running a non-profit non-commercial radio station, and that it is licensed to a non-profit (hopefully) tax exempt corporation. This means that if you can con anyone who has something you want into giving it to you free he is entitled to use it as a tax deduction at its fair market value. Sometimes it is tough to interest anyone in this, but every once in a while somebody will have a thing you can use that he is willing to give up on this basis. Don't be afraid to beg. Don't be afraid to poor mouth. Don't forget that the fair rental value of your transmitter or studio space can be tax deductible to your landlord if he gives it to you free. I don't know for sure how little it is possible to build a radio station for. I know a character who built a commercial station for under $10,000. I thought that was great, but I supervised the construction of KBOO, and we did that for under $4000. Both of those examples were for stations with 1000 watt transmitters. I think that a 10 watt station could be built for less than $2000 if all the contingencies went the right way. Give it a try.

> *Benjamin Franklin Dawson III is author of the above, and loves talking about obscure facts of FCC engineering lore. You can reach him [and his ear] through KRAB, Seattle.*

TURNING ON YOUR NEW TRANSMITTER

The following incantation is to be spoken the first day that your engineer has gone up to the mountain, and pushed the magic red button marked *"On."* At that moment, the walls of your life are fallen away, and the burnished tongue of communication is finally your own.

This is what you should pray:
O Wheatstone Bridge, I tote you
O great son of de Forest,
Great master of Armstrong man
Down the burlap, egg-carton halls
Of our days: I salute you.

O Fat Tongue of Freedom,
Words come bubbling across the
Rectified spaces between grid and grind—
Fat Fates of Radio, the Three of you
[Fessenden, Findereski, and Fass words],
We raise your promise to the heavens of sporadic E.
You, great gods of Aetherville,
We spark-gap into tantric nevers,
The Always Word of modulated mothers,
Stretching bands of wonder
Forever cut fore skin the circular/polarized/parts/
Of self.

Sweet god of transmission, we raise
Nights Nectar sound to you and your thousand progeny
In transmission heaven.
         Amen

(Health, Education & Welfare grants, and the Corporation for Public Broadcasting are eternal mysteries to those of us who grew up in community radio before there were any Approved Funding programs from the gummint. However, Tom Thomas, author of the following, and head cheese of The Double Helix Corporation, is a print nut, and has spent some of his happiest hours filling out forms for CPB and HEW. It works, sometimes, when Congress gives a minute thought to the desperate needs of radio and television transmission in this country. Thomas will answer further questions you have about these two organizations, or about the FCC forms, or layout, or the reality of setting up a radio station which is not run as an autocracy.

"HEW AND CPB" BY TOM THOMAS, DOUBLE HELIX CORPORATION

There is pie in the sky. In fact, there are two of them. And before you apply for your license you should decide if you can cut yourself a slice or two. Allow me to introduce you, friend, to the Corporation for Public Broadcasting and the Educational Broadcasting Facilities Program of the Department of Health, Education, and Welfare.

Several years back, the U.S. Congress passed the Public Broadcasting Act to encourage "the growth and development of noncommercial radio and television broadcasting." Out of this Act was born the Corporation for Public Broadcasting and its adjuncts, National Public Radio and the Public Broadcasting Service—and the Educational Broadcasting Facilities Program.

The Corporation for Public Broadcasting is very straightforward. They have some basic qualifications; if you meet them, you get the money. Simple as that. The details of CPB will be unfolded here, however, only as we track the miasmatic path of the Educational Broadcasting Facilities Program, a far more complicated proposition.

The Broadcast Facilities Program allocates money to buy equipment for public broadcast stations. No programs, no operating costs, no staff salaries—just equipment. The idea is to give public broadcasters physical capability on a par with the commercial broadcaster. For some strange reason, EBFP was buried deep in the bureaucratic heart of HEW.

As one might expect of an HEW enterprise, the program was first exploited by large state universities and state "educational networks" that, in general, built lavish, showcase production centers far beyond their real needs. But of late, the program has begun to open up to independent local groups that hold the prospect of becoming full service broadcasters.

Does This Mean You?

The HEW folks aren't interested in religious groups, labor unions or campus minded students; they aren't interested in the school stations with dustbin instruction programs. They want to fund noncommercial broadcasters who will sing and dance for everybody, without special interest syncopation, to a tune that all can hear.

There are three fundamental requirements to qualify for a grant: 1) You must meet FCC requirements as a noncommercial broadcaster; 2) Upon completion of the grant project, you must meet the minimum qualifications of the Corporation for Public Broadcasting, outlined below; and 3) You must have some money of your own. Allocations are based on a matching grant formula: up to 3 federal dollars for every 1 local dollar.

While the Corporation for Public Broadcasting grants are independent of the EBFP grants, the "minimum service criteria" of CPB provide a good index of the kind of station, or at least basic level of service, EBFP is willing to consider. CPB starts by asking for at least three full time staff (at minimum wage), 14 hours of programming, seven days a week, with a minimum power of 250 watts at 500 feet (or the equivalent). Within three years, you are expected to increase staff to five employees and program service to 18 hours. In return, of course, you receive CPB funds to help pay the bills, $10,500 at the minimum level, $17,000 for full qualification. Yet this is clearly no easy obligation to meet.

Some of you might well stop reading at this point. If you have no hope of ever taking your power above 10 watts, if you are planning a campus radio station, or if you expect to broadcast only part of the day for years to come, the HEW and CPB programs are not for you. But if you dream the dream, have the time and will, and think you can find some money, please read on.

How to Start

The first thing you must do is write away for the application and information forms. When doing this, specifically ask to be put on the continuing mailing list, so you will be kept up to date on changes in the program. The address is:

    John Cameron, Director
    Educational Broadcasting Facilities Program
    U.S. Office of Education
    400 Maryland Avenue, S.W.
    Washington, D.C. 20202

What you will get back is a complicated set of forms that seem to have little relationship, if any, to your cosmic radio dreams. This is because it is a standard HEW form and HEW has little, if any, relationship to your radio dreams. Except for their money.

Don't let the form scare you—it was designed for people who want HEW money to build schools and hospitals rather than radio stations. The folks at the EBFP mostly want to know what you are going to do with your radio station, but they also have to file categorical reports with their superiors, reports that match the categories of the form. I can guarantee that they have no more affection for this paper labyrinth than you.

The staff of this program is, for the most part, made up of individuals who have worked in public broadcasting themselves. They share many of your hopes and dreams. They also recognize broadcast bullshit at least as quickly as you.

They have had to fight and struggle for their allocation, just as you will have to fight and struggle for your piece of it. They have limited funds and, like you, they want to spend their money wisely.

They are also stuck in the middle of the second largest federal bureaucracy. They are responsible to (though often at odds with) their own superiors, who, needless to say, know little about broadcasting and care even less about you. And they are also subject to the pressures and influence of countless pea-brained Congressmen, committees, and educational agencies. Keep this in mind as you work through the process. Don't make it hard for them to give you money.

But if you are truly a part of your community, if you really want to serve that community with a full broadcast operation, they will be your most helpful friends in negotiating this long and arduous path. They will go out of their way to assist you in answering your many questions. Don't hesitate to ask. But don't make yourself a pest, either.

Just keep in mind that EBFP has a very small staff, possibly smaller than your own organization, and that they are working with hundreds of potential applicants each year.

You can call them collect: they will refuse your call and then call back on the federal WATS line, but only if you manage to leave a number. Sometimes you will have to call person-to-person collect for this to work. It is usually advisable to confirm any phone conversation with a follow-up letter, though.

Having mentioned letters, I'll also note that it is a good idea to save all your correspondence: both their letters to you and carbon copies of your letters to them. Go buy some carbon paper tomorrow. And take notes on your phone calls; save the notes with your corresponsdence.

Shuffling the Papers

Read through everything they send you, carefully and thoroughly. Underline anything that seems important. Read the original Act of Congress; read the regulations of the program, which come in a reprint from the Congressional Record; read the program memos. Then read it all again. The key to success is figuring out what kind of projects the program will fund—and then telling them why you are that kind of station. It's all a process of shoehorning your reality into this paper reality; figuring this out now will save you much time and agony later.

Then read over every page of the application. This will be your basic vehicle of expression. Think about the person on the other end. With all the subtleties of regulations, priorities, criteria, and common sense tenuously in mind, that person will read through your answers.

The questions are the opportunity to tell your story. Master the Questions: don't let them use you.

Much of what Lorenzo has written about Form 340 holds equally well here. The HEW application is, as much as anything, a test of will. You must simply grit your teeth and work your way through.

If you do a half decent job with the FCC application, much of your material will be already written. It may have to be shuffled and juggled a bit, but most of it should be there. I have the impression that what EBFP wants is really just an improved and more detailed broadcast application plus some basic HEW information.

But there are also many subleties at work. Unless you face a competitive situation, the FCC mostly applies a formula of money and technical criteria. Not so with HEW. To be sure, you must have the money for the match; you must technically be able to complete the project. But every HEW application is competitive—with all the other applicants who want a slice of the pie.

Slicing the Pie

There are a score of items on which you win or lose comparative "points." The first and most obvious of these items is the "priority system." All project proposals are assigned a priority according to the nature of the project. As of this writing, improvement of existing, low-service stations in areas without other public broadcast outlets are at the top of the list. Starting a new station in an area that already has a public outlet is at the bottom. The priorities may change from year to year.

In any given year, funds are divided among the various priority categories. The most funds are allocated for I-A priority projects; less for I-B projects, etc.

Projects then become competitive within each priority category. The standards employed are the so-called "criteria," which are to be found in the annual regulations of the program. Many of these criteria are objective; some are subjective. Some of the objective items include: the size of your potential audience, the number of hours you plan to broadcast, your connections with other public broadcasters, the number of hours of local programming in your proposed schedule. Some of the subjective items include: minority representation in your organization, efficient use of your frequency, the recommendation of your state educational broadcasting agency, the relationship of your proposed programming to the needs of your community, the broadcasting needs of the nation.

And finally, there is the total *gestalt* of your project. The gestalt requirements are to be read between the lines. Not everything can be answered true or false, or be given a number value. Somewhere along the line it will make a difference whether they like you or not. This is a living program; the staff wants to use its money to make a difference in the quality and force of public broadcasting in this country. As you answer all the questions, fill in the blanks and slowly tell your story, you must show that you will make a difference for your listeners, for your community.

Appetizer of Just Desserts

Facilities grants can be obtained for a variety of purposes: everything from building a whole new station to increasing your power to adding a satellite studio. But some projects are considered more important than others (the priority system) and, hence, are more likely to be funded. Projects are also limited by the amount of money you have; total costs cannot exceed four times the funds you have to invest (the 3:1 match formula).

There is a great deal of strategizing that goes into this grant-getting business. Let's consider two different approaches.

You may want to seek funding from EBFP for your initial construction, known in the parlance as an "activation." This will allow you to start off with a fully equipped station, use every dollar you have (almost) for matching and, using the proposed equipment, put in a stronger application to the FCC. Unfortunately, activations have recently been relegated to relatively low priority, even lower in areas that are already covered by at least one public broadcaster. This means you might have to wait for a while at the back of the line; it also means a delay in FCC approval; HEW and FCC grants are made almost simultaneously.

An activation grant has the advantage of more efficient use of your probably meager funds. And if you anticipate long delays between starting to organize your station and actually filing for a license, the anticipated grant can be an incentive in raising local funds.

Activations have the disadvantage of keeping you off the air until not one, but two (FCC and HEW) bureaucracies bless you. And this can run to many, many months indeed. Enough months to make you and your friends bone weary.

On the other hand, and most likely a better idea for your struggling group, is a facilities improvement grant. With this strategy, you get on the air as quickly and cheaply as possible, then apply for a grant to raise your power, build a real station, add a production room and do all the things you are now dreaming about. And while you are waiting for your grant, you can be on the air, flexing your acoustic muscles, bringing in those folks who said they would help you once you're on the air, giving the community a taste of what's down the road.

Facilities improvement grants have been moved to the top of the priority list for grant giving. And you can start the application process as soon as you receive your construction permit fom the FCC.

The only disadvantage of going the improvement route is that you will use up some of your precious capital going on the air—money that cannot be used later for the match with HEW. Every dollar you spend outside the HEW project is really costing you four dollars: the one of your own and the three that you might get with a matching grant. Equipment already owned cannot be used for the match.

There are, of course, other factors as well. For example, a project with a high priority that does not do well at meeting the other criteria may be less likely to be funded than a low priority project that otherwise excellently meets the published standards.

This is a time when it would be wise to consult with the EBFP staff. Spell out several alternative scenarios. Ask them for an indication of which would appear to be more successful.

The important thing is to decide your strategy early on. Regardless of whether you will seek an activation grant or an improvement grant or whatever, you should shape a financial plan before you spend your first dollar or sign your first contract. As we shall see below, even the timing of gifts can affect your prospects.

SEX & BROADCASTING

Filling Out the Form

There are two important steps in filling out the application: 1) telling them what they want to hear and 2) telling them what you want to say.

Telling them what they want to hear is somewhat of a decoding process. Each question, indeed, almost every word, has a very special meaning. Some of the definitions are in the form's instructions, some are in the regulations, some are in the Act, some are only in someone's mind. This is where it becomes extremely important that you have familiarized yourself with the totality of the program, its purposes and objectives.

I can tell you some of the answers. They want to know that you are truly going to be a public broadcaster. They want to know that you have secured the participation of a variety of community interests. They want to know that you have thought about local problems and issues and plan to deal with them. They want to know that you have thought through this entire project, that you know how you want to spend your money, that you know how you will construct your facilities, that you know how to maintain them in a professional way. They want to know that you are going to be around two and five and ten years from now, still plugging away on the airwaves.

I cannot tell you other answers. You must solve a local equation: what are the needs of your community, what is being done to meet those needs, and where will you fit in. Only you can answer that question—there is no right answer. You, in fact, must provide at least part of the question.

Here, again, it is wise to consult the EBFP staff. If you don't understand a question, ask them what it means. Ask them what they want to hear; they will tell you.

Then there is telling them what you want to say. As you read through all the material, you should come upon various points that seem to perfectly fit your proposed operation. Form a checklist of those criteria that you meet best; remember that these are the comparative standards. You must come up with a good list of reasons why you should receive a grant. A successful application will clearly present these reasons in the course of answering the different questions.

Don't omit any favorable point just because it wasn't specifically asked for. Find a way to work it in. KPOO sent along pictures of their shoddy equipment to make clear their needs; KRAB sent in practically every letter they had ever received from their listeners. Granfalloon in Denver included notices of public meetings.

As you work through the application go back to the published criteria. Ask yourself if you have made the best possible showing on each and every point. Talk through the application with your friends; see if it all makes sense, if it all holds together.

How It Works

The Education Broadcasting Facilities Program is, albeit loosely, tied to the federal fiscal year. A cut-off date for applications is set, generally, sometime in late fall. Applications are evaluated over the winter and about half the money is given away in the spring. After further consideration, remaining funds are handed out in late summer.

Applications are not always funded in the year they are submitted, since there is always more asked for than there are funds to be given. Low priority applications that nevertheless qualify are held over to the next year. The waiting process is often aggravating. If you are ready to apply in April, for example, you face a wait of at least a year. This is a major incentive for taking the improvement route rather than the activation grant, as outlined above.

And that's if all goes well. Sometimes Congress fails to appropriate the money at the right time or someone in the higher echelons decrees the need of a new form or something else goes wrong. And then it takes another month, or two, or six.

When You Get Your Money

You will not be handed a blank check. On the other hand, you don't have to spend your money exactly as you wrote in your application so many months ago. On the third hand, you will have to carefully account for every thing you do.

HEW requires that all expenditures be preceded by open bidding. This is to give everyone a chance at the federal bucks. You can do this in a variety of ways. Announcing required items in a local paper will do; so, too, will simply sending a list to at least three equipment suppliers. Then keep account of every penny. There will be a final reckoning at the end, before you receive final payment.

You are allowed to make "reasonable substitutions" in your project. As long as expenditures follow the general outline that was originally approved. Thus, if you get a great price break on your $4,000 audio console, you can use the extra money for several cassette recorders or an extra mike or two. All changes, though, have to follow the "approved equipment list" guidelines.

HEW money cannot be used to buy land for your tower, though it will pay for the tower itself. It cannot be used to buy a building for your studio or remodel it, though it will pay for equipment installation. It cannot be used for records or tape or pencils or paper or your lunch.

Some Further Notes on Strategy

It is important that your board of directors genuinely reflects the diversity of your proposed coverage area. One of the advantages an independent group has over the institutional broadcaster is its theoretically closer ties to a wider spectrum of the community. This advantage is sacrificed if your board is just a bunch of friends.

It is similarly helpful if your sources of funding reflect wide support. While it is commonly accepted that the majority of capital funds are raised from a few wealthy contributors and the remainder comes from everyone else, the more contributors, the better, even if they give only small amounts.

Documentation of your community support can help. Call on lots of community groups, art groups, political groups, anybody with a letterhead. Tell them about your station and invite them to write you a letter of support. Send the letters with your application.

Donations of equipment can help build your local matching share. Each item of equipment, appraised at fair market value, counts toward your local total. Even a $100 piece of junk will get you a free $300. So contact your local broadcasters for their basement supplies. Bait them with the fact that their gift will be matched. Tell them you'll be happy to let the FCC know of their cooperation in your worthwhile project.

But be careful with these donations. Equipment that you already own, whether purchased or donated, does not count. So what you really need is a *pledge* of a donation, contingent upon your receipt of the grant. Tell HEW about the pledge, pick up the equipment after your application is filed. Read the regulations carefully on this point.

Back to CPB

Since you must eventually meet CPB funding criteria to get the HEW grant, you might as well get CPB money to help you do it. So write to:

                    Corporation for Public Broadcasting
                    888-16th Street Northwest
                    Washington, D.C. 20006

Again you will receive some forms and information, only this time much simpler. In fact, they explain what you have to do so clearly that I won't bother here.

CPB is not competitive. If you meet the criteria, you get the money: simple as that. In fact, you may want to seek CPB funding independently of the HEW program. It is important to be thoroughly familiar with the CPB requirements before filling out your HEW application, but your CPB application should not be filed until you are actually ready for funding.

Take note of the cut-off date, however, usually in late spring. All qualification decisions are made once a year. If you miss the date, you wait twelve months. This, too, must become a part of your overall plan.

The Anticipated Summary

There you have it, at least a start. Some readers will perhaps wonder about taking money from the government. WBAI in New York, for example, refuses all government funds. But neither HEW nor CPB place restrictions on programming other than those imposed by the FCC—with the single exception of requiring that broadcasting is presented to the general public, not to a classroom or religious congregation.

The whole process is time-consuming, frustrating and often boring. But these two programs hold out perhaps the major source of funds for many community stations—and can make the difference between friends talking to each other and a whole community speaking its mind.

PBS GO BOOM

(The following is a telegram sent to Tom Warnock, Washington Director for the Corporation for Public Broadcasting. It comes from John Ross, the manager of KRAB Nebula station KBOO in Portland, Oregon. It is a reaction to The Public Radio Plan—*put out by CPB which encompasses, among other things, a desiderata that member stations finance themselves with CPB grants of almost $100,000 to construct the operation, and another $100,000 to run it each year.)*

"The Public Radio Plan" calls for a national noncommercial radio service with a "distinctive sensibility which reflects its values while developing a style that is highly listenable and stands out for its fresh, natural sound." The plan points out that the service should be "actively engaged in developing new talent, new forms of expression, and new audiences."

Unfortunately, the Public Radio Plan gives no consideration whatever to that form of noncommercial radio that comes closest to fulfilling that ideal—the true community station with no institutional ties whose sole source of funding is direct listener contribution. Since 1949, these stations have produced some of the most vital, exciting programming ever heard in the United States, and when they're blessed with a decent quality signal, their audiences are usually large and always tremendously loyal.

Unlike the institutional station, listener-sponsored radio can and does provide free access to all points of view and all stripe of opinion. It serves as a medium for experimentation and innovation. Most important perhaps, the stations all have a "style" with which their listeners can and do identify.

As of January 1973, there are eight such stations on the air: four licensed to Pacifica Foundation, two to the Jack Straw Memorial Foundation, and two others. Three other groups have received construction permits—in Atlanta, San Francisco, and Columbia, Missouri—and at least six others are in the organizing or application stage. Of the eight stations now operating, five are qualified for CPB support.

The Public Radio Plan describes a station operating on an average current budget of more than $120,000. KBOO operated last year on less than $500 a month. As the plan is presented, a listener-sponsored station in a small city has little or no prospect of reaching a level of outside support large enough to become eligible for a community service grant from CPB. How, for example, will KOPN (Columbia, Missouri) "make new efforts to raise existing minimal levels of local support and performance?"

A radio station in Columbia serves a market of maybe 150,000 people. There is already an institutional station on the air—KBIA—but it is operated by the university and not accessible to students or other community groups, except as a potential audience. Given this situation, enter the New Wave Broadcasting Foundation, which holds a construction permit for KOPN. If the people running the station are halfway competent...they will gather a large enough base of listener-support to just about pay for utilities each month. But even if they meet the target set by Lew Hill (founder of Pacifica) of subscriptions from 2% of the listening audience, at $20 a head, that's only $3,000 a year. No listener-sponsored station has ever come close to that 2% goal.

Where then, does the cost of a power increase, three (or is it really five) full time salaries, and

additional studio facilities come from? As the Public Radio Plan is proposed, the answer might be development grants, but these too are geared for the station with sources of funding larger than the ones we are discussing.

The Radio Plan quotes the Public Broadcasting Act in its goals for radio in the United States, "...to encourage programming...which will constitute an expression of diversity and excellence." It repeatedly talks about the value and need of worthwhile local programming. But those stations who have the best chance of meeting [these values and needs] are practically ignored.

If we accept the premise that small, listener-sponsored stations have the potential for fulfilling the goals outlined in the plan, the question becomes, "How can the Corporation best encourage such stations?..." As presented, the Public Radio Plan implies that only large, costly radio stations can provide the kind of service it hopes for. On the basis of past performance, it would appear that an independent community group, running a station on $30,000 or less can do as much. It is the writer's hope—not that the corporation give up on institutional radio—but that it also consider the alternative of true community radio for a portion of its activities.

Because of the small scale on which these stations operate, a little bit of money spent in the right places can go a very long way. A radio station has been put on the air for $500 [That station being KUSP, in Santa Cruz—operated by The 'Pataphysical Broadcasting Foundation —*Ed*]. If advice and coordination were available, as it is now through CPB for 'Institutional' radio, the number of community stations, both high and low power, could be increased to provide a valuable service to audiences throughout the country.

In Canada, the CBC is experimenting with several forms of free-access and community programming on their low power AM stations. Much of what they are investigating has equal validity on this side of the border. CPB should watch [these] very closely, and seriously consider initiating some of its own. There are a few minority-operated stations on the verge of signing on, such as those in St. Louis and Santa Rosa, California, but these groups accomplished their objectives without CPB support. The potential impact of this kind of 'third level' low power radio service is tremendous. Based on the experience of stations now on the air, it's possible to put something like 32 of these low-power operations on the air for the $96,000 that the plan proposes as the cost of starting a *single station*. One wonders which would have the greater effect on the life of the listeners—one more large institutional station, or thirty-two small stations, each serving the interests of a particular community and open to participation by all who would come through the door or pick up a telephone.

On the other hand, offering tremendous amounts of financial support to a station like KBPS (which had been primarily instructional) does not so much encourage the licensee "to critically review the purpose of the station," as to give them an impetus to expand their schedule while maintaining the same deadly philosophy of broadcasting. The result is most likely to be either hours on end of "serious" music badly presented, or one syndicated program after another. Or perhaps they could follow the lead of WSIE [Edwardsville, Ill.] and install an automation system.

Requiring stations to expand their schedules is a fine way to insure national coverage for the national program service, but so long as that NPS is available, it does nothing to encourage local production. Combined with the stated goal of two or three qualified stations in a large portion of the country, all of whom are able to utilize the same NPS, the result is most likely to be an inexcusable duplication of programming. Rather than an individual, distinctively local style, the requirement for increased hours on the air combined with a freely available national program service can create a sameness in the sound of noncommercial radio as bad as that of commercial radio. As an example of the problem, more than 90% of National Public Radio affiliates now carry "All Things Considered." In several cities, it's available on more than one station, sometimes simultaneously.

The plan's recommendations for financial support of qualified stations seem to be geared to allow the already large station to grow larger while the small station has to stay where it is. As a general rule, the large stations are the ones who can count on institutional support, while the smaller ones must go directly to their listeners for support from direct contribution.

I would hope that the Corporation thinks about basing its support—not on a station's success in raising other money—but upon the station's fundraising prospects. For example, a listener-supported station in a medium sized city can expect to receive less money from non-federal sources than a station operated by a branch of the state university in the same city. Some sort of graduated system for CPB support might prove more equitable than the proposed blanket ceiling on grants. The kind of program I'm suggesting might call for 100% matching on all other income up to $25,000, 75% up to $50,000, and so forth. As income from other sources increases, the eligibility for CPB grant decreases, as it does in the plan's proposed system. The only stations affected would be those on the bottom end of the scale.

The Plan's proposals for the station facilities and support also seem to be a case of the rich getting richer while the poor stay poor. Rather than directing funds for equipment replacement to the stations with the least or oldest facilities, the plan would give the most money to the stations who already have the largests amount invested in equipment. Somehow there ought to be a better way. If it remains as it has been under Health, Education, and Welfare, the facilities program for new stations and expansion is no help.

If a station qualifies for Corporation for Public Broadcasting support with hand-me-down equipment of marginal usefulness, and it's in a state in which large areas are not yet served by public broadcasting, it's not likely to be very high on the priority list for a facilities grant.

The alternative, while a bit cumbersome to administer, would spread more of the total available for replacement to stations who need it in a given year. A relatively small portion of the total might go to well-equipped stations for maintenance and replacement, while the bulk of the money would go to the stations whose equipment most needs replacement. Another possibility might be to provide those stations whose equipment is not up to an established standard a grant to replace the junk for new gear across the system, with a proviso the same for other stations as they qualify for support. I'm unable to project the cost of such a program, but it would have to appear as an additional item on the corporation's budget.

All the foregoing should not be taken as a total repudiation of the Public Radio Plan. But the writer's special perspective—as station manager of the CPB-qualified station with the smallest operating budget—makes the stated "ideal" of a station running on half a million dollars a year, with 33 people on salary, totally ridiculous. It is my feeling that the writers of the plan somehow became so taken with the Great Dream of Hundreds of Large Stations Blanketing the Country with "Public Radio"—that they lost sight of the real problems of making ends meet in a small station, and of the tremendous value of small, rough-around-the-edges, community radio.

I sincerely hope the Corporation will take these comments in the spirit in which they are intended.

(Radio Times #122)

John M. Ross
Manager, KBOO
Portland, Oregon
January 28, 1973

THE TAX MAN WILL GLADLY FUND YOU

The American Tax System is designed to hurt the middle class and the wage earner. It is designed to protect the very rich and keep the very poor very poor. You will probably be of the latter school—but you should use the benefits that accure to the very rich in your organization and capitalization efforts.

I claim that you get only a few benefits from applying for a radio station as a non-profit, tax-exempt organization. The use of tax exemption as a fund raising device is highly overrated. It will help you—I believe—in three ways only: (1) There will be a few, a very few large contributions or grants, mostly from government institutions, which require such status; (2) You will be able to eliminate the usurous, rip-off, cruel Social Security withholdings from all the employees of the corporation if they so choose; (3) the FCC requires that your corporation have tax-exempt status so that you do not have to make filing, grant, or annual licensing fees.

Outside of that, being a non-profit corporation (and then applying to the Internal Revenue Service for designation as a tax-exempt organization) is not such a big deal. Further, the application to the IRS is a shit-load of trouble, almost as boring as your FCC application, and can take anywhere from 4 months to three years to be approved.

What is little known in the radio biz is that many of the community radio stations either started as commercial outlets, or operated under the aegis of so-called "profit-making" corporations. KPFA went on the air in 1949 as a commercial station—and only went non-commercial listener-sponsored after they couldn't find and commercial businesses that would support their zany programming. Both KDNA and KTAO were owned by "Subchapter 'S'" corporations, and KTAO especially would be commercial some days and non-commercial the next.

The big distinction, I guess, is between non-commercial and non-profit. WEVD*, a cruddy commercial radio station in New York City, was actually set up and run by a non-profit corportation which tested (and won) the right to retain its tax-exempt status even though the station turned highly commercial. The corruption of the innocents was complete when WCFL—licensed to the Chicago Federation of Labor—turned to the crappy programming it does today, and O Irony! had to be filed on by the Citizens Communications Center for lying to the FCC about its commercial practices and public service programming.

You can use the techniques of commercial radio and commercial tax-law without being absolutely corrupted by them. You can run an enlightened community public-service shit-kicking radio station and still use the techniques of big business to get perfectly legal tax advantages. Or to put it another way—most commercial radio stations use the tax laws to legally avoid taxes (eg., get the government to subsidize them)—and you would be foolish not to use the same techniques. The big one is in equipment leasing. And that's a honey.

Most school, colleges, and community radio and television station buy their equipment in the

*Named for Eugene V. Debs if you can believe it.

name of their tax-exempt corporation, which is a big waste of depreciation and tax-credits. If they knew what they were doning they would do what I am now going to suggest you do:

SET UP A LEASING COMPANY.

Who me? Yes you. Shit: even if you are reading this book for pleasure not radio and would no more operate a broadcast station than eat a toad—set up a leasing company. If you pay more than $1,000 a year in Federal Taxes, you are dumping money into the Federal War Machine and when you don't have to. I mean, you can refuse to pay your $93 a year telephone taxes and get your name in the paper—but why be small time: all the Texas business people can afford to be conservative and support Life Line and bitch about the Federal Government because they get subsidies from the Federal Government *via* the IRS, and you might as well join the crowd. Here's how it works:

There is a system in the IRS codes which is described as "depreciation benefits and tax credits on tangible personal property." Personal property does not necessarily mean your cat or dog or hair pie: a transmitter is "tangible personal property." So is a tape recorder. Turntables. Antennas. Audio consoles. Monitoring equipment. Anything that you need to start a radio station *that can be easily installed and fairly easily uninstalled.* A building is not considered such an item; nor is the tower to support an antenna. All your portable equipment, including desks, chairs, and typewriters are clearly TPP.

As far as I can figure it, there are about four different tax benefits from your setting up a leasing company, and either investing in the equipment yourself, or doing it with the support of some local businessmen who like your radio idea, but don't want to give away money. Everyone, I mean everyone looks for a way to keep from paying taxes—and if the money that would normally go to Washington will come to you instead, there are enlightened businessmen everywhere who will do just that.

I will now give you an example of how the purchase of a piece of heavy equipment, tangible personal property, can benefit you and your friends with the assistance of a necessary CPA, you can extrapolate from this and figure out countless ways for the Feds to support you. But remember—your leasing company which buys all the equipment, and then rents it to your non-profit corporation, must be legitimate, and do what it says it is going to to, and must get monthly payments from your radio station. The entire IRS code is a joke, but we have to pretend that it is real.

You set up the Heavy Dooty Equipment Leasing Company. It can be an individual proprietership (you); or a partnership (you and a couple of friends who want to stop paying taxes legally); or a Subchapter "S" Corporation (a corporate entity which has all the protections against liability of a corporation, but is taxed by the IRS as an individual partnership: and which you deduct your losses personally each year according to the amount you put in it).

For simplicity's sake, let's say that you have set up Heavy Dooty as a partnership with you

putting in 10% and Sam Insull Carnegie—the young hip doper down the street who inherited some $1,000,000—as your 90% partner. Young Carnegie, when he isn't stoned out of his fucking gourd, worries about the fact that he pays $75,000 a year in taxes to support the wars he is supposed to hate. But you can show him through your partnership how to evade, unh, avoid a goodly portion of those.

You talk Carnegie into putting up $9000 and you scrape up $1000 to make a downpayment on $30,000 worth of transmitting and studio and monitoring equipment from Collins, including a delicious 5 kilowatt brand new hum-thump transmitter with the sweetest exciter you ever saw, makes you all excited it's so new and perky—and you *know* your station will work without a flaw once you get this package installed.

The Heavy Dooty Equipment Company with you 10% and Stoned-Out Carnegie 90% can be a "General" Partnership or a "Limited" Partnership. Get a lawyer to tell you the difference— but by all means, get a lawyer to set it up for you, because as you are the normal run-of-the-mill paranoid who is attracted to radio, you will need a firm leasing contract so that you won't worry about your partner coming in the night with Sheriff John Sloan and taking away all that shiny equipment.

Now about your taxes. For your combined $10,000 downpayment, assuming you buy the equipment at the beginning of a calendar year, you will get three advantages: deductible interest payments, depreciation, and tax credit.

1) *Interest.* Any payments that you make on your Collins note in the form of interest can be deducted from the amount of income on which you are taxed. This law was enacted so that we could all go into deepest most miserable debt, and thus stimulate the economy. The $20,000 indebtedness during the first year will cost at least $2000 in interest; your 10% will net you a reduction of taxable income of $200.

2) *Depreciation.* Depreciation is actually an existential concept: and also a bit anthropomorphic. In the eyes of the IRS and consequently most businessmen, machinery has a life span, and you are allowed to reduce your taxable income *each year* by that amount of reduction in life of your lifeless piece of equipment. Transmitters have a real life of infinity if you have a good engineer—or at least, they will last as long as he (the engineer) will. The IRS says that your transmitter is a dead duck after ten years—so each year, you can knock 10% of its cost off your taxable income. Assuming roughly the same life for all the equipment you have bought, you two partners will be able to deduct a total of $3000 annually off your taxable income through regular depreciation.

Then, in order to encourage new equipment buying—and, of course, to help us borrow from the future—the Congress decided to allow us a "bonus" depreciation the first year only. If you so choose, you can write off an additional 20% of the cost of the equipment from your income during the first year. In this case, with $30,000 worth of equipment (even though you are buying it on time), you and your partner can deduct a further $6,000. You lose in subsequent years: after Year One—you only have $30,000 minus the regular depreciation ($3000) minus

the bonus depreciation ($6,000) for a total of $21,000 to deduct over each of the remaining nine years, $2,333.33 per year: $233.33 for poor 10% you, $2,100.00 for piggy dope-fiend friend Carnegie.

3) Finally, and the most delicious apple in the whole fruit-filled basket—is an actual *tax credit*. In the first year of purchase (whether it's January 1 or December 31)—you and your partner are allowed to knock off 7% of the total purchase price *directly off your taxes*. For your $30,000 purchase—the two of you can subtract $2,100 from your taxes: $210 for you, $1,890 for him. That's after everything else has been figured out—at the Bottom Line as Haldeman would say: you bop $210 from your taxes. Wow! This is the equivalent of another $7,000 in equipment purchase.

* * *

All these figures might be confusing—but what I am telling you is that if you are in the standard middle class tax world, where you get $12,000 a year, and end up paying $3,000 in taxes, you get the following benefits if you put $1,000 plus $200 interest into Heavy Dooty Equipment Leasing Company:

REDUCTION OF TAXABLE INCOME:

$200 paid in interest to Collins
$300 standard depreciation of the equipment
$600 bonus depreciation

Reducing your taxable income from $12,000 to $10,900 or your taxes from $3,000 to $2,750 plus the $210 deducted from the tax for the 7% credit or $2,540. Or, to phrase it another way, for your $1,200 investment, you have saved out $460 the first year alone (e.g., the federal government has paid for about 1/3 of your cost of the equipment) and this does not include the subsequent tax advantages each year, and, as well, the fact that you will be getting leased income for the next ten years from your station. When it can afford to pay.

The figures I present above are complicated—but well worth studying. For someone even in the lower tax brackets—the advantages of such a partnership are extremely great. If you can figure out what I have told you here, you can go to a variety of businessmen and friends and have them all help you on the station without *giving* away money; rather, without giving away *their* money; but (doing as American's all love doing so much) giving away the *government's* money to support your project. $460 doesn't sound like too much, but for your friend Carnegie who *could* be in the 50% tax bracket, you are offering him a reduction in taxable income of $10,800 the first year, and and additional reduction in actual-taxes-paid of $1,890. For him, the *entire* investment comes from Washington, DC. He gets the good will of helping you out—and instead of being a lifeless "contributor," you are helping him to be a businessman. Which, even your local stoned-out hippy millionaire freak likes—because the old man keeps bugging him about his useless ways, anyway. Think of the self-respect you are giving him: why, you are doing him a public service!

Footnote: you should check out these facts with your CPA. At present, there is some dispute as to whether you can get full tax credits when you are leasing to tax-exempt organizations.

LOCAL BROADCASTING IN REMOTE COMMUNITIES

By Douglas Ward,
Radio Producer,
The Canadian Broadcasting Corporation

(I mentioned at the beginning of this book that if American broadcasters needed a model for great and good radio and television, all they had to do was to look over the border at what has been going on in Canada for the past 40 years. The CBC is an exquisite combination of real radio, and good commercial broadcasting, and idealism, and authoritative culture and vision. They have spawned some good men—and one of these is Douglas Ward. His essay which I have included here will give you an idea of exactly how far the CBC has gone to keep the doors open to the world out there. Or to phrase it another way, as you read this—think of KERA or WSIE or KETC or any of those rich booby state educational broadcast organizations going out to the boonies, setting up a tiny station for the people, and then having the 'broadcaster-animator' work himself out of a job! We here in the U.S. have not developed the wisdom of people's participation in transmission—not yet. You could learn a lot from the Canadians; the Educational Broadcasters in this country could learn lots more.)

The CBC has recently undergone an intensive study of the future of its radio broadcasting services. Among the recommendations now being implemented, one emphasizes the development of local information programming. (This includes news, comment, consumer reporting and other information which helps equip listeners for personal and community decision-making.)

This emphasis is not to be at the expense of our national and regional programming responsibilities. Rather, since urban and community politics are becoming increasingly important, we have a responsibility to the listener (who listens to a local transmitter) to provide him with the full range of information, including local. We also believe our national information programming will become enriched by the contributions of broadcasters with deep roots in many local settings.

This is a big enough task in the twenty-five cities in which we have radio stations with program staffs. But what can be done for citizens in remote, mainly northern communities, who presently receive CBC network radio programs from an unmanned, low-power, repeater transmitter (LPRT)? These are smaller communities (500 to 10,000 population) often without alternative local mass media. Many of them are undergoing serious social changes at present, as resource industries move in and out, or as native (Indian and Eskimo) and white populations come to see the need for self-expression.

In the spring of 1971, we decided to mount an experiment in which we would provide studio equipment and a broadcaster-animator to such a community for half a year, to see if the local people were interested in preparing, presenting, and operating their own local programming. Any experiment had to respect the fact that very little money would be available to provide local broadcasting for each of two hundred isolated communities. If large amounts of money were forthcoming for hinterland broadcasting, it should be concentrated on the development of sub-regional information programming production centers, rather than on local centers.

In September, a small studio facility was set up in a hotel room in Espanola, Ontario, a pulp and paper mill town of six thousand people, whose civic leaders had been requesting local programming for some years. A young CBC broadcaster with a good record for encouraging citizen access to one of our regular stations was chosen, and he and his family moved into the community for the next five months.

One of our fears was that the local programming operation could fall into the hands of a small elite or other unrepresentative local group, and render the operation useless to the majority of the citizens. It was the people with more education and more economic status who came forward first to offer their services to the station, but our broadcaster did all the preliminary programming and community organizing until he felt assured that there was a group of people around the station who reflected the fact that Espanola was a working class town, half women, with a large teen-age population, and a school center for young Indians from surrounding reserves.

Over the months, the local programming developed, and Pat Reilly, our broadcaster-animator, worked himself out of one function after another. By the end of December, he was not needed for technical operations, announcing or paper-work. He continued to help in the areas of program content development and the building of a permanent community radio organization.

By the end of January, Pat and the Committee decided that he had to leave the community in order for the Citizens' Committee to be recognized as the local broadcasting authority and assume full responsibility for its work. A letter of agreement was drawn up and signed by CBC and by each member of the local radio committee, in which CBC agreed to provide studio facilities and air time on the transmitter in return for local programming and adherence to basic broadcasting regulations. CBC, as the licensee, remains fully responsible for the station and its program content. The Canadian Radio-Television Commission, which regulates broadcasting in Canada, has been kept fully informed of this project and, in fact, approved it at the outset.

At this time, three months after Reilly's departure, the Espanola Community Radio Citizens' Committee produces twenty hours a week of local programming, and they receive an additional one hundred and ten hours a week from the network. From 0645-0900 hours, a team of young shift workers from the mill present a wake-up program featuring community announcements, local sports results, weather, an occasional consumer report on local business, a story for the children, popular music.

In the late afternoon, some thirteen-year-olds present a half-hour of stories and records for pre-teenagers, and in the next half hour, older teenagers produce a hard rock music program. Late Friday evening, a furniture store employee hosts Party Line, a phone-in show which encourages people to call and talk about a specific community issue (health facilities, summer jobs for young people, etc.). Saturday afternoon, a crew of young Indian students presents an hour of native music, history and current events.

No one is paid for any of this work. The Citizens' Committee has over one hundred members, of whom perhaps forty are active participants in the work of the station. Twenty of these—ranging in age from thirteen to forty—have learned to operate the studio console and to announce. A group of women have recently come forward requesting time for a program for, by and about women, The Committee is also encouraging senior citizens to develop their own program to include interviews about the early days of this region. Meanwhile, Committee members have fund-raised and banked $1600 and have proceeded towards legal incorporation, as steps to assure that their local programming will continue.

If we have a basic criticism of the experiment, it is that local people have not developed much news or comment and opinion programming on matters of local concern. The availability of a mass medium does not mean that it will be used for mass information purposes. This is particularly true in a town where there is little evident social dislocation.

But there is perhaps another reason for this. Radio in North America usually means commercial radio, and that means record-spinning and disc-jockey patter. When you invite people to take part in a radio operation, they think in terms of that model and perform accordingly. Besides, if you have to fill three hours a day, every week day, your information inputs run this thin unless you have a dedicated local information-gathering group behind you.

With this in mind, we are planning a second experiment in an LPRT community. There will be no studio, no high-profile hardware, no expectations about a local radio station. Rather, in response to a community request for local information input into its CBC transmitter, we will provide an animator who will help the community to organize around this need and define and develop the kinds of information input it wishes to communicate. Many of these communities need something as simple as an emergency system of mass communication in times of snow-storm or spring flooding. The local input will all be talk, for fifteen-minute periods at important times in the local daily life. Broadcasting will be effected by means of a rudimentary microphone and amplifier housed in a box attached directly to the transmitter. (A female jack might be added for input from a cassette tape recorder-player.) The cost of this Facility for Local Input (FLIP) is about two hundred dollars, as compared to $3,000-$8,000 for a simple recording studio. The emphasis will be on satisfying the local need for shared information, and not on the need to emulate background music radio.

If this experiment proves useful, then we shall be prepared to encourage community radio groups in other LPRT towns to gain access to our remote transmitters. Right now, we are developing guidelines for community radio operations, in order to ensure that the local committee will be representative and autonomous, that the local programming will be accessible and that our mandate for balanced and useful programming will be maintained.

Further, we hope to be able to provide CBC program people, from various departments within the Corporation, for short-term, on-site training periods, to help these local operations get started, or to help them enrich their programming once started. We suspect this will provide at least as much benefit to our programmers as to the local radio committees.

Canadian Broadcasting Corporation,
P.O. Box 500,
Terminal A,
Toronto 116,
Ontario, Canada.

COMMUNITY RADIO IN BIG TROUT LAKE
by Paulette Jiles

During the long ride on a DC-3 from Sioux Lookout straight north to Big Trout Lake, the passenger can see nothing but lakes and a black spruce forest shaded by stands of poplar and birch. There are no roads and the villages are too small and scattered to be seen from 1,700 feet. It is three hundred miles to the long, white stretch on the horizon which tells you are approaching Big Trout Lake.

The community radio station there was put in as an experiment by the Federal Department of Communications; the Northern Pilot Project.* It is a little FM transmitter with console and two tape recorders, housed in a former teacherage. It broadcasts for fifteen miles, for seven hours a day. The language is Cree, the people are Cree, and Buffy Ste. Marie's records are played until they have more scratches than grooves.

The village of Big Trout Lake is populated by six hundred and fifty Cree, and peripherally by fifty whites. The white people are either working for the government or running businesses. They are there because of Health and Welfare (nurses), Department of Transport (weather station), Ministry of Transport (airstrip), the Hudson's Bay store, a tiny airline operation called Bearskin Air Services with a couple of Cessnas and a crippled Norseman. In addition there is a good-natured fourteen-year veteran of the north who runs a little store and restaurant. The Cree people are there because they have always been there.

This is the country above the Canadian Shield where rivers run north to Hudson's Bay. To the newcomer it seems incredibly empty. As the newcomer becomes less new, he finds it is a jungle of government bureaucracies desperately grasping for water, for mines, for cheap Cree labor and for gold. Indian Affairs employees try to keep their jobs by proposing programs the Indians don't want; the Indians try to fend off the worst programs and suggest more intelligent ones. The radio station was one of the more intelligent ones; a small victory, and perhaps not preferable to a laundromat or to electricity to Indian homes, but a win nevertheless.

The only way you can get around the small settlements in the north is by bush plane; to fly out on the scheduled DC-3's round trip to Sioux is so expensive that only those with government jobs can afford to go down and enjoy the running water, bars, hotels and jumbo shrimp on Sioux's main street. To charter a plane is even more expensive. And yet the officials come in their chartered Otters, twins, singles; their Beechcrafts and Norsemen, the OECA, the MOT, the DOC, AMIK, ICB, the H&W. The little radio station up in Trout, itself without running water, doesn't even bother to broadcast news of their arrival. They come with new down parkas, government-issue five-star Woods down sleeping bags (as if they were planning on spending the night in a snowbank instead of the palatial Indian Affairs cabin, with its running water, showers, electric stove, table lamps, a washer-and-dryer). The radio station just keeps on playing music, local announcements and arguments over whether the chief should quit or not.

"I've got a message here," says a volunteer announcer, "Maria MacKay wants Elija to bring a baby bottle from Jessia's place as quick as possible."

"And here's Cher with 'Half-Breed'..."

"Now we got some local talent, I'll play Eno and Bill for you from this tape—jeez, I can't find it. (Off-mike but still audible) Ron, where's that tape...? On yeah, Eno and Bill singing, 'Me and Bobbie McGee'."

"Here's a song dedicated from Randall Roundsky to Sherry Nanokeesic called 'Sensuous Woman'."

"Maggie Siannawap needs some moosehide, if anyone has some, please call in to the radio station and leave a message."

The bureaucrats fume. They keep saying what a wonderful little radio station it is, but why isn't there any news from the south? Why can't *they* go on the air and explain their programs? Be careful now, and don't let the white people in the village take over, they say, you people are doing a great job with your radio station in your own language. But surely some of that Cree needs to be translated for the benefit of the fifty whites. None of them, of course, bother to learn Cree.

The government supplies any white person coming to the north with Don Mills housing. Running water, electricity, washers and dryers, drapes, rugs, formica, hot water, picture windows. The Indian people live in cabins, they haul their water, they pump up the Colemans every evening. During the winter one woman I know missed her volunteer hour at the radio station, because she was chipping frozen baby turds out of the diapers with a screwdriver. The Bay was out of Pampers again.

When the radio station first went on, the consule arriving on a DC-3 and the white technicians working like mad with their jacks and two-way feeds to meet the deadline, the white people *did* dominate the radio station. They read news from the south when their Toronto Star was delivered. They played classical music, they read magazine articles—anything to connect them across that thousand miles of black spruce, the rock-hard lakes and three foot snows, to the Canadian Imperial centers of Montreal and Toronto. They chatted about the goings-on in the white community, the nurses told everyone to eat vegetables every day (canned, fresh from the Bay, with a month layover in Pickle Crow), somebody read the weather report from the D.O.T. meteorological station.

At first the community people saw the radio station as just another imposition, a useless gadget from down south. Then, the station changed. It's hard to say why.

The older people began to come on the eight-to-nine hour and talk. The volunteer broadcaster, a young man educated in the south, drew them out. They talked about hunting caribou, about a plane crash in Pickle Crow, they recounted stories of the first time they had ever seen a white man, and stories of the oppression of Hudson's Bay from as far back as their grandfathers' time. That hour began it.

Then a big, tall trapper decided, out of the blue, that a noon-hour talk show was needed.

"Come on," he said, "I've been thinking about this." Probably for months. "We need opinions, and gossip, and local talent. Let's get the cassette recorder and tape people." The man was six-foot-four and I followed him all over the village literally at a run, as I am short and the snow was deep. Singlehandedly he turned the noon-hour show into an hour of opinion-broadcasting and local talent and village reporting. We chartered a battered Cessna to Angling Lake to tape the fiddlers there—they played St. Anne's Reel and Hen-and-Chickens for us. We came back to Trout and broadcast that. Things started happening fast. We asked for volunteer reporters. Three young men showed up.

"Where do we start?"

"Why don't we go down to the radio-telephone and call around to the other settlements, see what's happening."

Pretty soon there was a seven-o'clock news report.

"Sam killed a moose in Sachigo today. Muskrat Dam says nothing is happening and everyone's going to bed. George Barben flew into Bearskin Lake today from Weagamow to visit his mother..."

The big flat, country which had seemed so empty when I first came up, when the whites in the village were only broadcasting news and thoughts and ideas from the south, now seemed a beehive of activity; via the HF radio and then broadcast on CFTL, the north was full of people, travelling, hunting, visiting, going down to Sioux for high school, getting married, having babies, going off to their traplines, building houses...

A community radio is for the community; it is their voice, a night watchman, a kind of bell that tells the hours as they once did in medieval villages, an immediate messenger, a mercurial postage stamp that gets your sentences across the settlement even to isolated cabins ten miles away, where people sit beside a woodstove and a kerosene lantern and a little F.M. transistor. They know the voice of every announcer who comes on the air. "That's George." "Oh-oh, there's Dean again, he is going to play 'Don't feel my legs'." When the snow is very deep, and it's fifty below outside, it's good to hear your friends' voices coming over the airways.

Most of the content of the seven hours is music. I've heard people say about this or that community radio, "Oh, they're just record-spinning all the time." Perhaps they are. But when Tom dedicates "Monday, Monday" to Suzanne, a message has gone over the air. Southern radio stations are at a high level of competition for listeners for advertising and hot news. In a large city the listeners need that feeling of contact with a cellular world around them. But in a small village the needs are different. A high-pressure audial attack would have a disastrous effect on a place like Trout Lake.

How does the radio station handle dramatic events? There have been several tragic deaths this winter—two in Fort Severn, one in Trout. A southern commercial station would have blasted them over the air and hyped the bloodiest details. CFTL, in the case of one of the deaths, a heart-attack, remained silent until the elders came to the radio station and presented a long eulogy of the goodness and uprightness of the deceased. The man's wife then came on and sang some hymns. In the case of the deaths in Fort Severn CFTL was silent on the subject for four days. After the story was told in church, the father of one of the victims came to the radio station to eulogize his son. One of the reporters played "Amazing Grace" softly under the father's voice as he spoke. The radio in a crisis becomes a meeting place. People know the details already and, as in a village council, they rise to speak in general terms about death and life, ritually saying farewell.

While I am writing this in May, the radio is in a bit of a slump. Baseball season has hit and every volunteer and announcer is out on the ball field. It's mostly music, and since the Aurora Borealis interferes with the H.F. radio, and this is a hot time for the Northern Lights, the seven o'clock news is scanty. It's break-up time; the ice is beginning to go now and the planes cannot land in the smaller villages which remain isolated. Great strings of snowgeese fly over like daytime stars and the hunters are waiting at the rapids when they drop to rest. But even if it's just Buffy Ste. Marie singing "Native North American Child", that's the radio, it's the town caller, the pulpit for a community voice, and somebody you know is on the air.

For some reason the area north of the Shield is now being inundated by communications technology. Out of Ottawa and Toronto there are plans for microwave systems, more satellites, television—all controlled from the south. Bell is in here. Satellite dishes are going up in villages no larger than 150 people. The technicians arrive with mysterious boxes and hire local people to unload them and tote them around.

I got into a fierce argument this winter with a communications consultant, who had come up to observe the Big Trout radio station. He insisted that technology changes a people, that the old ways up here must go, that the north and the Cree people will never be the same. He argued that radio here must learn to reach outside the village for more news, information, ideas, introducing these people slowly to the drastic jars that technology will bring into their lives.

I, on the other hand, argued that technology does not necessarily change a culture—it's a matter of *power*, and who controls the innovations. Technology has "changed" native people all over the world, because it has always been in the best interest of international capitalism to destroy self-sufficient cultures and make them dependent upon factory plastic and styrofoam food. When the white man first came to this continent, be used the technology of the Indian to invade the vast spaces of Canada. Capitalism came in, loaded in a birchbark canoe; free enterprise on snowshoes; the concept of property and hierarchical command arrived by dogsled. Yet the white man did not change by using Indian technology.

The second point I made was that the Indian people had been changed by technology long before the provincial and federal bush-planes started flying over the Shield. If people here want laundromats instead of another anthropological study to find out what they called their grandmothers, it's because a hundred years ago people here began to wear cotton clothes. You don't scrub out moosehide. They have been living with white man's technology for over a century; but whenever possible, they have used it to further their traditional way of life. They use guns, because it is easier to get a caribou with a 303 than a bow and arrow.

The Plains Indians, to use another example, took one of white man's advantages, the horse, and blossomed into one of the finest mobile societies ever known. They had already developed a profound philosophy of community, of life and the natural world. They fitted the animal into this world; they became the world's finest light cavalry and for a hundred years they incorporated, of their own free will, this accidental gift from the whites into their society. They had guns, but it was not they who wiped out the food supply.

What makes me so angry is that the McLuhanesque communication experts from the south never talk about technology as something that can enhance a traditional way of life. They never mention laundromats, electricity for Indian homes or even public baths. What they push is technology which is centrally controlled by big business or big government: TV, radio, wheaties, water diversion schemes. They call this technology "value free".

CFTL-FM is one bit of technology which is being used by the people in Trout Lake to create a whole new concept of radio. Slow-paced, no advertising, in the native language, a language that has reached a high form of expression and wit because it has until recently been entirely an oral language. Radio where people yawn loudly and complain they're going to close up if the next volunteer doesn't get here. Radio where an older lady crashes in and says she wants to say something about that call-in show that was broadcast yesterday and the mike is handed over to her.

The territory north of the Shield in the Severn River system has long been under the sovereignty of the Cree people. They have a lot to say about the soil, the animals, the river system, the fish. Perhaps they will say it over the radio; perhaps someone will end up listening.

At a bush-camp twenty miles away from the village of Trout Lake we were all resting around the fire. For the first time I wanted to listen to a radio. Five fat geese were piled in a corner, the babies asleep, Northern Lights blazing wildly overhead in the region of the Big Dipper, the woodstove warm and crackling. But I wanted to hear from Trout, and I guess so did everybody else.

Lillian got that look on her face that meant she had an idea. Taking my snowshoes, she unwound a long strip of brass wire that was holding a fracture in the frame. She attached it to her little FM transistor, and stuck the wire out the cabin window as far as it would go. She fiddled around with the knob, grazing 100.3 on the dial. Then, faintly, George Nothing's voice came in:

"Hello, all you crazy people out there in Trout and Angling, going to play this song for you, 'Then came the Whiteman', by..."

Paulette Jiles is a writer and broadcaster living in Toronto, Ontario. She has recently spent eight months in Big Trout Lake as an animator employed by CBC Community Radio.

BEER AND LOAFING IN MIAMI
Alligators and Porpoises Broadcasting in the native sun

By Holt Maness,
Founder, Bascomb Memorial Foundation,
Miami, Florida
(Applicant for FM stations in Miami and Palm Beach)

(*I tried to give you theory in this book. I figure it might help you to know how one person took the theory and ran with it—all the way across the country. I guess I'll have to tell you the depressing part too: Holt made his filing 4 years ago, and has just received his CP. A combination of financial problems, The Moody Creep Bible Institute, and Channel Six has kept this application in the dungeons of the FCC until now. Still, the article is a good case study, and in parts will tell you what not to do as you work toward your broadcast outlet's ultimate licensing.*]

In June, 1971, I left Los Gatos and headed for Miami, Florida, to start an educational FM broadcast station. When I arrived in Miami I had one spare dime for a phone call.

I spent my dime locating a phone that I could use free for the rest of the afternoon. That phone was at the United Methodist Church. The church secretary gave me a cup of coffee, a peanut butter sandwich, and a plate of cold beans. The mission of the UMC in Miami is directed toward the derelicts of the inner city. I took my place in that line...and recalled that in Castro's Cuba the free food lines are long, but that pre-revolutionary Cuba had no lines for free anything. Miami is most like the old Cuba, so the meager generosity of the church was an oasis.

With the Yellow Pages as my major aid, I called every organization of liberal persuasion and carefully followed all suggestions of interested persons. My public list was supplemented by references of friends from northern Florida. The two lists were soon exhausted. I wasn't sure but I thought I had contacted all persons likely to be interested in the project. With the names of all sympathetic persons imprinted in my brain I took my first vacation.

When I returned to Miami the next week to resume my search for prospective board members, I had a fresh list of references gained from a Jacksonville friend. After talking to two persons my old and new lists melted together. I knew the time to form a corporation had come. Nine volunteers met and formed *Potlatch Enterprises, Inc.*

A potlatch is a ritualized festival of competition in giving. The practice was observed by the Kwakiutl Indians of the Pacific Northwest until 1939 when pot-latching was outlawed. The Kwakiutl Indians had been fierce and warlike until the potlatch sapped all their competitive energies. Perhaps the pacific influence of the potlatch was thought to be un-American, or maybe it was anti-capitalistic. No public potlatch has been allowed since 1939. The Miami group set about to employ the concept by giving away radio stations.

Creeping paranoia suggested the possibility that the U.S. drug culture would extract the pot from potlatch and try to make smoke. The name of the corporation was changed to *Bascomb Memorial Broadcasting Foundation, Inc.*

Bascomb members contracted all local tv and radio broadcasters with a proposal for tower sharing. One broadcaster responded by quoting a fee of $1 per foot per month. Since Bascomb's station was to be low power (10 watts) and low finance (10 grand), the price of $800 per month for the necessary height to cover the city was exorbitant. Finally Bascomb secured permission from a local church to use the church property as a proposed site for the FCC application. The site was well located in the center of Miami and thus was ideal for Bascomb's low power operation. The location within the population concentration soon proved to be a curse rather than a blessing.

The first Bascomb application was soon completed and submitted to the FCC, but the application was not accepted for filing. The commission set aside the application because the transmitter site in the center of a residential neighborhood might interfere with the reception of Channel Six TV. No one had warned Bascomb of a classical problem. Channel Six had said nothing when approached for tower sharing. After a year of meeting people and hassling with the tedious bureaucratic paperwork of the FCC, Bascomb was in check.

Representing Bascomb, I went to Washington to discuss the situation with the Commission. The advice given was simple: "Co-locate with Channel Six and no interference will occur. But since the Channel Six tower is twenty-eight miles from the population center, the proposed signal won't reach the city." The Commission laughed at Bascomb's broadcast service to the pigs and chickens of Homestead, Florida. After being assured of cooperation by Channel Six's chief engineer, I took another vacation.

Upon returning to Miami I learned the value of the written word or contract. The owners of Channel Six had decided they wanted no part of Bascomb on their tower...we should go build our own. With a letter promising only ten thousand dollars as our only asset we realized an absurd situation and suspended all activity. The conventions came and brought new friends and new solutions to Bascomb's problem. The heat of the summer passed and so did the hurricanes. The Big Chance came in the early fall.

Late summer had brought a change in the ownership of Channel Six and, said the engineer, a change of heart. Co-location was agreeable. Bascomb quickly complied revisions to the original application. The amendments provided for higher power (13.2 KW) from the Channel Six tower and higher finances with which to accomplish the new proposed station. All seemed well. The Truth came to The Fore. Moody Bible Institute filed an application for the same frequency that Bascomb wanted. Moody's proposed 100KW station at Boynton Beach would interfere with Channel Six just as the original proposal of Bascomb did. Channel Six apparently was using Bascomb as a pawn to defend it from Moody.

Since the early Fall of 1972 the situation has not changed much. The complexity has increased. Presently, Moody has as application for 100 KW from Boynton Beach that is competitive with Bascomb's 13.2 KW application from Miami. Moody will still interfere with Channel Six, although Moody contends that no interference will exist. Channel Six is trying to move out from under the situation by applying for a UHF channel in the area of conflict with Moody. That move will leave Bascomb and Moody to fight for 88.9 mHz. Channel Six's application is opposed by Channel Four TV in Miami. Channel Four is affiliated with the Washington *Post*. A strike application questioning all *Post* affiliates' service of the public interest, convenience, and necessity puts cloudy skies over application. Moody has moved to 88.7 mHz with their application which will allow both groups to go on the air if either will lower power. If Bascomb lowers power their signal will not reach critical areas of support in Northern Miami.

Bascomb has prepared for the possible battle by expanding its statement of program service. The original ten watt application offered minimal seven hour daily broadcast. The broadcast day has been expanded to twenty-four hours. The proposal now includes an expansion of live local programming and more comprehensive national and international service. Affiliation with National Public Radio as proposed will bring selections from that network to Miami and South Florida for the first time. Financial qualifications have been substantially improved by promissory notes designed to qualify Bascomb for matching government funds.

Bascomb is a local group proposing local programming with local control. Moody is Chicago based and proposes basic evangelism and religious 'education.'

The only argument we can see in Moody's favor is a contention that a 100 KW circle of coverage will service a greater population than will Bascomb's 13.2 KW circle. Our answer to that argument results from an experienced consideration of the geography of South Florida.

South Florida appears to be a wide projection from the soft underbelly of the U.S. In fact, the dry and habitable land of South Florida exists only on the Atlantic and Gulf coasts as thin strips bounded by ocean and swamp. Any large circle of coverage generated from a center in one of those population strips will encompass great expanses of the Everglades and the Atlantic Ocean.

Moody's 100 KW transmitter at Boynton Beach will service the Everglades from a point west of Lake Okeechobee and the Atlantic Ocean nearly as far as the Bahamas. Of course that coverage includes the populated area of north of Palm Beach to Northern Miami.

Bascomb proposes two nearly overlapping small circles of coverage—sort of like a figure 8. One circle will be radiated from south of Miami at the Channel Six tower. The other circle radiates from West Palm Beach. The circles nearly overlap at Ft. Lauderdale. A one-watt translator is being considered to fill the gap. Both Miami and Palm Beach applications are currently on file with the FCC. Bascomb's low power figure eight (13.2 KW + 20.0 + KW + .01 KW = 33.21 KW) will serve the populated area from Key Largo, forty miles south of Miami to some point forty miles north of Palm Beach.

Bascomb answers Moody's extravagant plans to serve all the alligators and porpoises of South Florida with a pablum of Evangelistic RF. Bascomb's answer is a synergistic suggestion: let us get more coverage for less power consumption. Bascomb's service to the wildlife of

South Florida, a precious population, will be of secondary nature...measured by our effect on the ecological awareness of our human listeners. Let the Native Sun Shine...
ERP FOR MOODY? QUACK FOR BASCOMB!
Only time and the FCC can resolve this conflict. Suggestions welcome.
Bascomb Memorial Broadcasting Fdtn.
Box 330069
Miami, Florida 33133

NOW IT CAN BE TOLD
(Stuck in here where no one will ever find it.)

[In 1967—I did a frequency search, another in the endless series of frequency searches, we are always looking for black holes in space—and found that there was a frequency available in Washington, DC. A community station in Washington, DC, the great well-head of American power.

I knew that Pacifica had been interested, so I called my friend Stu Cooney who was then Vice-Chairman of Pacifica, told him that I was interested, inquired if they were still so. He told me to come to the next Pacifica board meeting, and they would tell me.

I went, and they were, so I told them that I would show them how to do an application—if they would give us free tapes from their archives for a year or so for KRAB to play on the air. They agreed.

For two weeks or so, Stu and I worked to put together an application. It was not unlike that deadline college newspaper that the five of you have been procrastinating on putting out for the last month or so: When you finally go to work on it, you work day-and-night, sometimes have a ball, and at the worst—get to know the other people involved. You always look back on the experience fondly, one of those pin-points of your life when you were putting your mind together with other minds to exceed all expectations.

Towards the end we got silly—what with lack of sleep and a near conclusion. We were making public-community interviews (we didn't have to: I thought we did: no non-commercial stations are required to do them at this

stage) and writing them up, and one day I made up one I thought would amuse Stu. I stuck it in his pile of real interviews.

WELL, HE NEVER SAW IT: gave the whole stack to the secretary there at Berge Fox, and she typed it up, just like it was real. I never noticed, Stu never noticed—I finished up the engineering, we had one final drink, and I left.

They sent me a copy of the whole application, and I thumbed through it, and I saw, there, as filed the month before with the FCC, my spurious (and rather titillating) interview. Jesus!

I called their lawyer—told him to get it out of there. He didn't want to make too much of a deal out of it, so he resubmitted the whole 50-page interview file to the FCC, citing some "mistakes" in the previous filing.

It worked. No one ever noticed. Which was good: Pacifica was heading towards a hearing in Washington—and that might have been one of the things to shoot them down. But it remains as one of my favorite—albeit secret—interviews. Made up: of course. But that doesn't detract from the flavor of it. Nor the naughtiness.]

George Reticule
1551 West Eastern, N.W.

Digital Programmer

March 24, 1968 — Time: over an hour

Feels the need for more 'intimate programs'. Feels that there is too much censorship of 'rough and ready' stuff on regular stations. Says he goes downtown to the 'peepers' whenever he can. says he would like to see the TV out in the Potomac where he could see 'his old friends' like 'Boobs LaMae' and 'Little Miss Snuggles'. Has been spending a lot of time at home recently 'boning up on the new literature'. Feels there is a place for 'getting to know others better through close radio'; says he feels that Television is a 'peep'. Likes 'slow sexy' music—but not 'too damn much of that classical stuff'. Wants something that he can hang on to. Says radio must keep him 'from getting cold—for as you get older and more wrinkled you get colder'. Says he would like to meet some of our younger volunteers, offers to be volunteer director in charge of 'sub-teens'.

SEX & BROADCASTING

THREE MORE PAPER GAMES

There are a few other things you can do to influence the law and the practice of communications. One is called *Petition for Rulemaking*.

In that, you can ask *anything* of the FCC. You can ask them to turn off all the VHF television channels. You can ask that they open up Long Wave for community stations. You can ask them to make commercial broadcasters donate 10% of their airtime to community groups. You can ask that each year, 2% of the broadcast licenses be revoked—at random—so that some of the rest of us can get a change to be heard on AM.

They won't do it of course: but at least they might accept the Petition, ask for comments on it from other interested parties. And since these things scare the hell out of commercial broadcasters, you might see some scrambling by them and their lawyers to make comments on your Petition.

The form is the same for all *Petitions*—all you have to do is write to the FCC and get a copy of a current one, and use its first-page style. You have to file an original and 14 copies with the Secretary of the FCC.

Then there are comments. Whenever the FCC is Making Inquiry—and they have on-going inquiries at all times—they will invite comments, and give specific deadlines for those comments. Also, you can make comments on all current Petitions for Rulemaking.

All notices of Inquiry are published in the Federal Register—and often in the trade magazines. Inquiries of especial interest also are described in the Washington *Post* or the New York *Times*.

As in the Petition—you must file an original and 14 copies of your comments.

Finally—if the Senate or the House ever decides to hold hearings on a proposed bill, you can ask as a broadcaster, or even as an interested citizen, to appear and make your views known.

Here are some examples of a Comment I made on the Fairness doctrine two years ago; and, more recently, a filing I made with the Senate Subcommittee on Communications, on the subject of 5 year license renewals.

Both, I may say, have subsequently been lost in the giant paper whirl-bag called Government Documents Filed Forever. Where they rest, I guess, to this day.

5 University Avenue
Los Gatos, California

Mr. Ben F. Waple
Secretary
Federal Communications Commission
Washington, D.C. 20554

Dear Ben:

We are happy to transmit herewith an original and 14 copies of comments in response to Part II of Notice of Inquiry in relation to the Fairness Doctrine, Federal Communications Commission Docket Number 19260.

If there are any questions concerning this, kindly communicate with us at the above address.

Love and Kisses,
The KTAO Bunch

Before the
Federal Communications Commission
Washington, D.C. 20554

In the Matter of

The Handling of Public Issues
Under the Fairness Doctrine and
the Public Interest Standards
of the Commucations Act

Docket
No.
19260

To: The Commission *en banc*

Comments in Response to Part II
of Notice of Inquiry

1. The Fairness Doctrine, as it stands now, is a fair and a viable document. I think it has been partially successful in giving the general public access to the aether which is and should be their property.

2. In addition, the Fairness Doctrine hangs as a gentle and subtle threat to commercial broadcasters—a threat that says that they must learn to share their enormously profitable broadcast frequencies in order to accommodate the public weal.

3. Any lack of success of the Fairness Doctrine as it stands now is a result of the piggishness of commercial broadcasters. They have been given licenses to coin gold—and still, in many cases, they are reluctant to share even a small portion of this mine with the public.

4. The work of people such as ourselves—and, we should hope, the Federal Communications Commission—should be to prove to broadcasters that they do not have to be so greedy; that there is enough wealth and prosperity implicit in the ownership (or stewardship) of a broadcast license to allow more time to be used for the discussion of important public issues.

5. I should hope therefore that the Commission will continue to enforce the Fairness Doctrine as it stands now—without imposing some mathematical or exact formula basis to its usage.

6. Implicit in this whole filing is a belief that American broadcasters have continually and willfully squandered a vital natural resource—namely, the aether. They have done this not because they are bad men, or because they are unfeeling—but because of the coming of several individuals with an overweening necessity to make their frequencies into some monumental financial jackpot. This has made them lose sight of the wonder and beauty of true community television and radio.

7. As part of this squandering is the hopeless situation—caused by dreadful fly-by-night montebanks—of jacking up the cash flow, and profitability, and consequently, the purchase price of existing stations. This story of the waste of a vital natural resource called the spectrum is painfully outlined in Eric Barouw's superb three volume *History of Broadcasting in the United States.*

8. Unfortunately—the college and school broadcasters have been of little help. Frequencies reserved for educational purposes have been used for teaching in the narrowest (and dullest) sense—or as the stamping ground for unimaginative bureaucrats in school 'communications' departments. Only a visit to England or Canada—and a week with their radio and television outlets—will show what we have pitifully missed in the area of adult education in the broadest sense, the Greek sense.

9. We feel the Fairness Doctrine can be expanded slightly to encourage a fuller public participation in broadcast outlets. We think it would be sensible of the FCC, for instance, to require existing licensees to explicate—once each day—the rights of individuals to respond to questions of controversy under the Fairness Doctrine. We could envision, for instance, one minute spots each day (with a required number in prime time) which would outline in brief form the content of the Fairness Doctrine—and the fact of the station's availability to the public as a forum. In the best tradition of commercial broadcasting, which has done so much to bring the 60 second spot to its present artform—we think broadcasters could put their own imaginations to work on making this daily message interesting, exciting, and (hopefully) thus be able to interest many members of the public in using their facilities for the discussion of public issues. We feel that broadcasters have always thought that the public misunderstands their role with relation to public issues—and we are sure that each and every AM, FM, and TV licensee will welcome the opportunity to explicate to the public—daily—their willingness to serve them in a fair and friendly way.

10. However—some broadcasters—it is possible, may use this opportunity to eliminate what few controversial programs they may now be airing, in order to preclude any use of their frequencies for Fairness purposes. For that reason, we should like to suggest a positive bribe to put more controversial programs on the air, in prime time. We think that a financial incentive may be the only method to get broadcast executives to open their airtime to the public.

11. In this document, we are emphasizing the need to get the public to the microphone and the camera. Like those who value this country, and its noble experimentation with democracy, we fear for its future. We fear for its future if the poor and the dispossessed continue to be displaced from the public consciousness as represented by the media. We fear the rage of the angry and the Chicano and the Black if they do not get some satisfaction that at least the FCC, or a few broadcasters, or *someone,* is working to get them heard, to get their faces on television, to get their voices on radio. For the unheard and the unseen are coming to be well aware of the great golden profits to be made from American radio and television—and we fear that they are coming to feel it is at *their* expense; that they, the poor and the old and the uneducated, are being bought and sold like slaves, through the terrible rating game. We fear for our future if we continue to displace them from the radio and television stations—or at the worst, only [to allow them to be seen on news programs] when they go on a rampage or a march. In other words, we are making these suggestions because we fear for our country, and for our future as broadcasters. When the bombs start coming down, they will not differentiate between us, and George B. Storer, and Gordon McLendon.

12. The power to tax is the power to destroy—but it is also the power to persuade, to encourage. The FCC has now the power to persuade—through its annual license fees. We

should like to see a form of rebate of these fees to those broadcasters who choose to open their time to any group wishing to speak on any controversial topic.

13. We should like to see something called *Rebate Time* come into existence. Broadcasters would invite groups to apply for half-hour or hour programs on their stations. These groups would be given time on a strict first-come, first-served basis. There would be no restrictions placed on them by broadcasters, outside the usual standards of obscenity and libel. The groups would oe given time at no cost—and the programs (discussions, panels, lectures, dialogue, drama) would be presented live.

14. The broadcasters would bill the FCC for this time at their standard rates. Thus, if a television station in a major market gave an hour of *Rebate Time* to the Urban League, and the prime time cost over that station were $2500 an hour, then the FCC would be billed $2500. Only that sum would be deducted from the station's annual license fee—and the amount of the annual rebate would be no greater than what the station would have paid to the Commission annually.

15. In this way the FCC would be actively encouraging open discussion and dissent in a positive manner; and the communities would benefit by having their frequencies opened (at last) to the now-frustrated minorities; free speech would be given a worthwhile workout; and broadcasters—who universally loathe having to tithe *anything* to the FCC—would have a good excuse to discount and possibly eliminate this government license fee. And those of us who have been deploring American radio and television for so long would have a chance to see and hear some fascinating on-the-air material: things we would never get to see or hear in any other way.

16. Broadcasting is magic. It is a wonderful tool for enlightenment and self-knowledge. But—like organized Christian religion—it is in the hands of men, not gods. Those who run American broadcasting do not seem to be generous enough—nor wise enough—to recognize the magic of radio and television in relation to the needs of the public. It is not enough for them to take from the great aether goose a constant series of golden eggs. No: they have to milk it of its blood as well.

17. The power to tax is the power to persuade. The FCC has that generous power right now. No broadcaster—enjoying the rich benefits of *giving away airtime,* and for that, receiving a rebate from the government (and, we may say, the plaudits of the public as well)—is going to berate the Commission for interference, or censorship.

What with the broadcasters' recurring fears of the FCC, we should think that they would delight in the chance to kill two troublesome birds with one stone. We—and the public—could ask no less of them.

Respectfully submitted,

Reginald A. Fessenden
Radio Station KTAO

December 20, 1971

(*The Radio Times* #102)

Statement of Lorenzo W. Milam
Prepared for Presentation Before
The Senate Subcommittee on Communications
Meeting July 31, 1974
On the Question of Five-Year Licenses
For American Radio and Television
Stations
(HR 12993)

Instead of the ten minutes that I have been allocated—I would like to ask for a full day to make my presentation.

If you granted me the time, I would say nothing. Rather, I would bring with me a radio, and a television set. For eight hours, I would ask that we listen and watch a sampling of the seventy radio and television stations that serve this area.

We would then have a fair idea of American broadcast policy—not as the National Association of Broadcasters tries to make it appear, but as it really is. We would have, after that eight hours of drudgery, an excellent idea of the system of American broadcasting that so many want to reward with a five year sinecure.

I am always amazed at the way that people do not listen. I have been a broadcaster since 1952, and for the last fifteen years, I have helped organize, build, and manage ten radio stations. I listen to radio constantly—and not necessarily my own stations. I am always fascinated by the changes and delusions of the broadcast industry.

What I read and what I hear always seem to be at such extraordinary variance. When I read the testimony given before the House on this bill, I even wonder if we are talking about the same industry.

The members of the NAB and the broadcasters speak of freedom and "public affairs" and license security as if it were not one of the most lucrative industries in the world. They hold up their cup of burning gold as if it were the grail—ignoring the tragic waste of the aether used for profit, and profit alone.

The capacity of radio and television to enlighten and ennoble has been proved in England and Japan and Italy and New Zealand and South Africa and Canada. Especially in Canada, where the system of great broadcasting is not divorced from the opportunity to make money.

Those of us who despair for American radio and television do not fault the profit motive. The balance sheet is an effective tool for gaining work and some degree of efficiency.

No, what we are opposed to is excess profits—that 60 or 70% return on invested dollar, annually—which is, according to official FCC figures, the norm for the industry.

It is greed that has done us in. And it is that greed that you will reward with half a decade of license security if this bill passes.

There must be some way to get American broadcasters to yield up five or ten percent of their goldmine return in order to improve our broadcasting. The citizens' groups have tried to do it by threatening the media's own oil depletion allowances, the government granted licenses. The industry has responded by demanding even more security in the form of this hokey bill.

Is there any way that I can convince you that the Federal Communications Commission is *not* threatening our security? The FCC has always been, and always will be—an instrument for conservation. Let me try to explain:

As we all know, American radio and television stations can be bought and sold on the open market like so many pigs. And, like goats or pigs, the price paid depends on the value of the item to the buyer.

Each year, approximately three percent of the existing broadcast operations are traded. And this three per cent tells us what shrewd investors think the broadcast business is worth.

In 1968, there were 345 radio and television licenses traded, approved for sale by the FCC. The total price paid for them was $152,455,412. The average price per unit was $450,000. Investors thought that a broadcasting unit was worth slightly less than a half-million dollars.

In 1972—after the WHDH decision, long after various citizens' groups had come to be involved in the transfer and relicensing process, 276 radio and television stations were sold for a total of $268,330,537. The average price per unit was $972,212. Investors, highly informed and selective investors, valued radio and television station units at almost a million dollars.

To phrase it another way: in 1968, when broadcasters were relatively secure in their operating licenses, when the citizens' groups were only beginning to feel their power, when the concept of "strike" applications was relatively unknown, and "petitions to deny" were unheard of—the average broadcast property was worth one-half of what it is worth now (or rather, what it was worth in 1972, the last full year we have figures for.)

Or, to phrase it still another way: the marketplace has taken the protests of citizens' groups into account by doubling the worth of the average station license, and the National Association of Broadcasters and all those other troglodytes who are fighting for a five-year license bill are definitely and distinctly putting you on.

298

Broadcasters have little or nothing to fear from those who challenge their licenses. The FCC has always been and will always continue to be an instrument for the conservation of property; a body which is designed, for good and for ill, to protect broadcasters, large and small, from outside forces.

Between the years 1948 and 1952, a total of 14 broadcast station licenses were revoked, or license renewals were denied. This amounted to about two-tenths of one percent of the total number of stations on the air.

Between the years 1968 and 1972—a time of greatly increased activity by minorities and other petitioners, the total number of licenses revoked and renewals denied rose to twenty-four —or slightly more than three-tenths of one percent of all licensed stations.

Thus, in a time of greater citizen participation and a rumored "more involved" FCC, the rate of license loss jumped one-tenth of one percent over a similar period twenty years ago.

The National Association of Broadcasters and those who petition you for redress against the rising hordes of minority citizen activists are definitely putting you on—and it would be a great pity for the American media consumers if you were to listen to them.

The delusional argument of the NAB is that the radio and television stations of the United States need five-year licenses in order to be more secure, in order to make long-range capital plans, in order to devote more time to public service programs.

For the worlds "security" and "stability" and "Public Service," I ask that you substitute the words "capital gains," and "profitability," and "increased annual gross." The security of

American radio and television station owners is as great as ever, and their net profitability is even greater. (I am not talking about income, only—I am talking about the hidden benefits of fast equipment depreciation schedules, and what the FCC calls "payments to partners, proprietors, and stockholders." I am talking about the immense benefits of trade outs, and the inevitable price rise of station licenses.)

Profitablility is not the subject of this hearing—but that gross profitability of radio and television is here with us in this room, a spectre that hangs on our every word. Because excess profit is the name of American media—and every Black or Latino or Indian who asks for time, and is denied, is being denied in the name of gross profit.

They have heard, all of us have heard, of the pot of gold that waits at the end of the aether. And they wonder why this incredible richness is being used as a continual block to using the media for their ideas. And they wonder why they cannot get themselves and their brothers into ownership, and management? Why are they being left out?

The only route for having his voice heard for some poor, or Black or other minority person is through the FCC license renewal process. It may be a tedious and rickety process—a bizarre entryway—but it is the only way that works, the only way to get rich and powerful and disinterested broadcasters to listen.

Most of you only have a vague idea, like most citizens in this country, of the way that we radio and television station owners work to protect ourselves and our million-dollar licenses. And in particular, you have no idea how much my fellow broadcasters scramble to hire some Black or Latino in that difficult period immediately before license renewal.

I know: I hear the stories, listen to the gossip, talk to my fellow radio station owners. In California, in the last half-year, you cannot guess the number of minority persons who have been hired on because this year is license renewal year.

You won't know, and most broadcasters won't tell you—about how many public affairs and "community" broadcasts are spouting forth, all of a sudden, on a hundred background music stations. Media owners know—even if few others do—that tomorrow, August 1, 1974, is FCC license renewal deadline. And they'll do anything, *anything* (even good programming for a change) to avoid threats to that precious license.

It is a silly system. If you and I could work it any other way, I am sure that we would. To see to it that broadcasters filled their prime-time hours with deep, thoughtful, interesting discussions, and interviews, and talks.

If there were any other way, I am sure than we would not use the license renewal process to encourage broadcasters to be color blind, to hire any and all to their news, and sales, and top management positions, without regards to race, or color.

If there were any other way: but there isn't. I am convinced—I have struggled with the problem for fifteen years. For fifteen years I have tried to convince my fellow broadcasters to listen to the voices of the dispossessed. I have asked that they read the terrible writing on the wall, the words *Let Me In*. I have said that if they ignore, and continue to ignore those words, that all of us, good and bad alike, are going to suffer.

It may be too late to change the structure of American broadcasting. But it is not too late to change American broadcasters. I think we can drag them, kicking and screaming, into the 20th Century, away from the robber baron traditions of the past. They will despise the process — but I think that all of us will be the better for it.

What I am proposing is not a five year, nor even a three year, license period. Rather, I would like to see broadcasters subject to a greatly simplified, almost automatic *pro forma* one year broadcast license period.

I am proposing a new form for license renewal: it would be short — one page, to be exact. I include my proposed form at the end of this statement.

The National Association of Broadcasters and *Broadcasting Magazine* are most usually a gold-mine of misinformation. However, in one lone regard — they are correct: the present FCC license renewal forms are abortions: thick, fat, engorged, self-serving, useless monstrosities.

The present renewal process embodies almost two hundred community interviews, which are ignored by all, and read by no one except some harassed GS-11 at the FCC who can do little or nothing for the community problems in Moline, Ill. My proposed new form would discard this verbiage in favor of a few simple direct questions, addressed to the owner of the license, to be answered in no more than five pages.

The form would be filled out by the broadcaster each year, and filed with the FCC. Announcements of license renewal would be made over the air, and if there were no protests, the station would be relicensed for another year, automatically.

We know that perfection does not lie on this earth — at least, not yet. There are and would be some broadcasters who carry their public trust so poorly that they deserve, and would deserve, the loss of license. Or as Father Drinan has stated, in a different context: "It is inconceivable that 100% of American radio and television stations are so good that they deserve automatic license renewal." There are and always will be some terrible miscreants who should lose their government licenses to coin money. And I have two specific suggestions as to how this could be accomplished.

You will notice on my proposed license renewal form the following questions:

"For the sample broadcast week list:

1. *The number of commercial spot annoucements carried;*
2. *The number of commercial programs carried;*
3. *The number of community-news-public affairs programs carried, and their duration;*
4. *The number of community-news-public affairs programs carried in prime evening time* (for FM or television), *or in 'drive time'* (for AM);
5. *In relation to your weekly gross,*
   a. *What was the total cost of these programs, and*
   b. *How much revenue did you sacrifice to carry them?"*

In these five questions, I am trying to get at the heart of the economic determinism of contemporary broadcasting; the fact that wretched radio and television does not come about because media executives are bad people; but rather, that they are under constant pressure to deliver high profitability. And what this question series would do is tie, directly, profit to relative amount of "community-news-public affairs" programming.

301

I am of two minds about what the FCC should do with this valuable information. One possibility is that each year, the AM, and FM, and TV stations in each ADI Market* which show the lowest ratio of public affairs programming to profitability should automatically lose their FCC license. Like that: without question.

Another, and far more litigious solution—thus likely to be closer to the hearts of broadcasters and their omnipresent not to say omnivorous communications attorneys—would be the setting for hearing of the renewals of those broadcasters in the bottom 5% of their market, in reference to public service versus profit considerations.

When I talk about hearings, I am not talking about the present cumbersome, expensive, paper-clogged, lawyer-crammed, three, or five, or seven-year hearings (with appeals) that presently engage the full toils of the FCC judicial process.

No: I would rather think in terms of an expedited hearing; one that would take place at the studios of the licensee in question.

Broadcasters in this country continue to try to isolate their listeners from the facts of FCC licensing procedure. They manage to infer that the renewal process is an inviolate process: one of interest alone to the broadcaster and the FCC. This proceeding that you and I are involved in today is a further attempt to extend that process of exclusivity.

I would like to see that change. I would like to see the public air, the medium itself, involved in the actions of license renewal.

I would like to see the airing of a new program each year. The stars would be those radio or television stations that showed the lowest ratio of public affairs programming, and the highest relative profitability. The program would be carried by the station itself, and it would be called "License Jeopardy."

On a day arranged by the FCC, the Hearing Examiner (now called "Administrative Law Judge") would arrive at the studios of the starring radio or television station. On hand would be the station management, and owners. Listeners would be there as well: they would have been advised of the program over that very station, and by all the other radio and television stations in the same community. They would have been advised that they were welcome, and would be welcome to question the station owners, and staff: to praise or blame, fully and freely, live, on the air.

The FCC Hearing Examiner would be there to bring system and order to the program: to direct the questions of the complainants, to encourage discussion, to phrase questions ignored by the live or telephone call-in participants. He would be free to suggest, not require, some alternatives for the licensees.

"License Jeopardy" would continue through the day—and into the next if the Examiner felt that it were necessary for a full airing of the problem. And the Hearing Examiner would have full discretion as to the ultimate resolution for the offending station: license renewal, revocation—or probation for another year.

If the Examiner, in his best judgement, decided that the license should be revoked, it would revert to the FCC, to be issued to another applicant or applicants who would demonstrate,

---

*An ADI market is a ratings device dreamed up by a television and radio numbers racket called ARB. ADI means "Area of Dominant Influence." Broadcasters accord the same reverence to the concept of the 208 national ADIs that, say, religionists do to the Shrine of Fatima.

through tapes or actual programs, their willingness to make better use of the frequency than the offending licensee.

To some, this might come as a radical solution to a difficult problem. Broadcasters might protest the actual use of the revocation process as contemplated in the Communications Act of 1934, and the loss of their golden goose.

Communications attorneys, whose proliferation habits are second only to the *lagomorphs* of Australia, will certainly deplore this reduction in their utility—a debatable utility at best—and the consequent reduction in their income which is and always has been, after all, a tax on the listener and viewer not unlike the tax of so-called "free" advertising.

But balancing this, and I think overwhelmingly so, would be the fact that the staff of the FCC, and the listeners, and the broadcasters would be using the medium, the aether—to inform and involve us all in the license renewal process.

It would be a rare utilization of the McLuhanesque concept of radio or television being used as its own perpetuating force: a hearing on license renewal, on the very air that will be affected, is a diverting use of the power of broadcasting; probably far more diverting than the present insanity that troubles our aether, and our minds.

I started off my testimony by suggesting that you set aside a day so that we could listen to eight hours of American radio, and watch eight hours of American television. All this was predicated on the wish that we could actually have you vote on a subject that you would have been exposed to, mercilessly.

Rather than sit there and have twelve dozen "industry" representatives tell you about their plight of instability and poverty—we could listen to and *see* the abject poverty of rich and meaningful programming, get some feel for those $75,000-a-minute television advertisements, or the 18-hour, $100 a shot radio spots that darken our air and mock our hopes for an enlightened democracy.

We won't do it: spend that eight hours. We don't have time, I guess: and I know, in any case, we would be bored silly with the repetitiveness of a media which does not have to be repetitive, and so unresponsive to the needs of our world.

At this point I can merely ask that you consider the words of someone, like myself, who has seen commercial radio and television from the inside, who knows that broadcasters are not in trouble, do not need your assistance to protect them from the angry crowds on the streets.

I ask, rather, that you understand that the crowds are on the streets because they have been denied access so long—that they know the only way to get their voices heard is through rhetoric and violence that will get them (and their message) on the Seven O'Clock News.

What a silly way to express the voices of a dissident democracy. Let us hope that more, not less, responsive licensing processes will make broadcasters aware of the fact that they are one of the few remaining hopes for an enlightened populace, who ask nothing more than to be heard, without having to pay so dearly for that right.

I thank you.

SEX & BROADCASTING

*Proposed Federal Communications Commission Form
For Renewal of AM, FM, and TV Stations*

(Response is to consist of this page and no more than
five additional pages of double-spaced typed material)

Call letters of radio or television station:
Community of Service: ADI:
Power/Frequency/Class of Service:
Station Manager or Person Responsible for day-to-day Decisions:
Owner:
Gross Income of Station for Past 12 Months:
Net Profit of Station for Past 12 Months:***
Estimated Sales Price of Station (if it were to be sold today):*****
Owners' Capital Gains (if it were to be sold at that price):

Notes: ***In response to net profit question, please indicate amounts of income you are deducting as depreciation (and your depreciation schedule) as well as salaries for managing partners and owners. Include, as well, figures for all trade-outs and other management benefits and include all advertising amounts for businesses in which owners, management, or staff have any interests.

Note: *****The estimated sales price is included so that the FCC annual license fee can be tied to that—rather than the present "Highest One Minute Spot Rate." The latter lends itself to constant changes and interpretations, whereas the former tests the economic acuity, pride, and fallibility of the facility owners.

303

In 1970—the FCC got a hard on about naughty broadcasting—and instead of going after the smarmy sniggering on the Jack Parr Show, the real obscenity of American communications, they decided to go after KRAB, a small, obscure FM station which had been struggling along in Seattle for 8 years. · Instead of renewing KRAB's license, the FCC decided, because of what they thought were dirty programs, to give the station a one-year license renewal. Mike Bader, the attorney for The Jack Straw Memorial Foundation, felt that the FCC was picking on KRAB for all the wrong reasons, and the Board agreed to fight it.

They did—for three days of hearings, and seven tons of briefs, rebriefs, unbriefs, overbriefs, and other legal legerdemain. The whole was held in Seattle, before Hearing Examiner Ernest Nash—and his final decision (the only real decision to ever come out of the FCC on the question of "obscenity") is excerpted here.

Before the
Federal Communications Commission
Washington, D.C. 20554

In re Application of

The Jack Straw Memorial Foundation

For Renewal of the License of
Station KRAB-FM
Seattle, Washington

Docket No. 18943
File No. BRH-1430
File No. BRSCA—801

Appearances

Michael H. Bader, Esq., (Haley Bader & Potts), on behalf of The Jack Straw Memorial Foundation; and Walter C. Miller, Esq., on behalf of the Chief, Broadcast Bureau, Federal Communications Commission.

INITIAL DECISION OF HEARING EXAMINER ERNEST NASH
Issued: March 22, 1971; Released: March 25, 1971

Introduction

1. KRAB-FM is a non-commercial educational broadcast station operating on 107.7 mHz, Channel 299 at Seattle, Washington. It is licensed the The Jack Straw Memorial Foundation. An application for renewal of KRAB's license was filed by the licensee on November 4, 1968.

2. In a letter to the licensee dated January 21, 1970, the Commission granted a short-term renewal of KRAB's license. In its letter, the Commission said that it had received complaints from the public that profane, indecent, or obscene language had been broadcast during the past license period. Referring to the station's stated policy against broadcasting obscene and libelous material, the Commission concluded that in broadcasting a program presented by Reverend Paul Sawyer, KRAB had violated its own programming policy. Commissioners Cox and Johnson issued statements dissenting from the views expressed in the Commission's letter and from the action granting the applicant a short-term rather than full term renewal (21 FCC 2d 833).

3. KRAB filed a petition on March 20, 1970, asking that the Commission reconsider its action and grant a full 3 year renewal of its license. In response to this petition the Commission reconsidered its earlier action to the extent of offering the applicant a hearing as to whether or not it was entitled to a full-term rather than a short-term renewal (FCC 70-655, July 7, 1970). KRAB accepted the Commission's offer of a hearing and its application for renewal was thereupon designated for hearing upon the following issues: (FCC 70-873, August 19, 1970).

> "(1) to determine whether KRAB-FM has exercised proper licensee responsibility in effectuating its policy regarding the suitabilitly of material for broadcast; and

*"(2) Whether in the light of issue (1), the public interest would be served by a one year or a full three-year renewal of the license of KRAB-FM."**

4. In its Order of Designation, the Commission also directed that the hearing examine into KRAB's handling of the Reverend Paul Sawyer broadcast, which took place in August 1967, and programs broadcast on March 9 and 10, 1969, which involved discussions with members of the San Francisco Mime Theatre. The Broadcast Bureau was also directed to give timely notice to the applicant if it intended to rely upon any other broadcasts relevant to the issues designated for the hearing.

5. On September 9, 1970, KRAB filed a Motion to Clarify and Enlarge Issues. In that Motion the applicant, among other things, requested the addition of a meritorious program issue. This request was granted and the following issue was added to the proceeding:

"to determine whether the programming of KRAB-FM has been meritorious, particularly with regard to public service programs." (26 FCC 2d 97)

A prehearing conference was held in Washington, D.C., on September 23, 1970.* At that conference, the Broadcast Bureau gave written notice that it intended to rely upon a number of other programs during the course of the hearing. In this notification, the Broadcast Bureau detailed the alleged obscenities which had been broadcast and gave the names of the complainants who had brought attention to these programs. As listed in the Broadcast Bureau's notification, the following programs were added to those specified by the Commission: Two programs in which the principal speaker was the Reverend James Bevel and which were broadcast during December 1967; a program with Dave Wertz broadcast 9:30 to 10:30 p.m., October 1, 1968; a program entitled "Murder at Kent State" broadcast 5:30 to 6:00 p.m., August 10, 1970.

*After the prehearing conference the Hearing Examiner originally designated had to withdraw from the proceeding. Examiner Ernest Nash was designated, with the consent of all parties, by Order of the Chief Hearing Examiner. (FCC 70M-1418, October 15, 1970)

6. Hearings were held in Seattle, Washington, on November 12, 13, and 16, 1970, and the record was closed on November 16, 1970.

*Commissioner Bartley dissented to the form of the hearing. He would have had the hearing deal with the issue of whether the license should be renewed at all. Commissioner Johnson concurred in the result.

Findings of Fact

1. The Jack Straw Memorial Foundation is a non-profit educational corporation organized under the laws of the State of Washington. It is the licensee of KRAB in Seattle and KBOO in Portland, Oregon. A Board of Trustees consisting of 11 members, nine of whom live in the Seattle area, are responsible for the formulation of the policies under which KRAB is operated. These policies have taken the form of written resolutions, oral understandings, or statements published in the KRAB program guides.

2. KRAB operates as a "free forum broadcast station" designed to encourage free and complete public expression. Its basic policies regarding program suitability were originally formulated by Lorenzo W. Milam who was the founder of KRAB and owned the station until he transferred it to the Jack Straw Memorial Foundation. These policies were largely oral understandings until they were reduced to writing and formally adopted by the Board of Trustees after the Commission had raised questions regarding the suitability of the content of certain programs which KRAB had broadcast.

3. KRAB is listener supported. It receives its funds in the form of contributions from listeners and has been the recipient of grants from various foundations. It operates on an annual budget of about $14,000. Most of the regular employees of the station receive little or no pay for their work. A good deal of the work needed to run the station is performed by volunteers from among its listening audience.

4. KRAB's policies as to determining whether or not a program is suitable for broadcast were related to the Commission in a transmittal made November 21, 1967, as follows:

> "The station will not avoid programs because of their unusualness or outspokenness. The primary criteria of broadcast standard is fairness: that the station should provide a great deal of time to speakers, writers, and thinkers from a wide variety of viewpoints. It is crucial that their material be well thought-out, meaningful, and insightful; there should be no sensationalism for its own sake.
>
> "In the case of material which raises questions as to its merit for broadcast because of some social, moral, aesthetic, or scatological outspokenness, the material shall be referred by the Program Director to the Station Manager for audition and judgement as to whether it should be broadcast entire, elided, or not at all.
>
> "If the program inspires concern on the behalf of the Station Manager as to its appropriateness for broadcast, the program shall be auditioned and passed on by the Board of Directors meeting as a whole, or by those directors appointed by the board to judge the material.
>
> "This simple procedure has worked well in the past

with, perhaps, one or two programs a month being eliminated by the Station Manager or Board or a Committee of the Board for obscenity, obscurantism, sensationalism, or simple boorishness. It relies on the judgement and good taste of the station staff, integrated with that of the Board—both with respect to programs presented and those referred to higher authority."

These standards or policies are the same ones which are in effect now and which were in effect during the broadcast of the programs which resulted in this proceeding.

5. [Here omitted are a list of the Trustees]

6. Greg Palmer has been station manager since January 1970. Preceding him in that post were Lorenzo Milam, who founded the station and served as manager until March 1968; Chuck Reinsch, who served as manager for a short period beginning in March 1968; and Gary Margason, who succeeded Reinsch and was manager until Palmer took over the job.

7. Each of the employees who auditions programs uses his or her judgement as to whether or not a program is suitable for broadcast. Each has developed expertise in certain fields and reviews programs in his fields of expertness. Besides obscenity, the auditioners look for such other matters which would affect suitability for broadcast as advocacy of law violation, boorishness, and obscurantism. If any of the four employees should have a question regarding the suitability of any material for broadcast it is discussed with Palmer, the station manager, and any problems Palmer may have as to suitability for broadcast he discusses with the Board of Trustees.

8. Palmer has a general knowledge of what is to be considered obscene so far as broadcasting is concerned. In his view, a stricter standard should apply to broadcasting than is applicable to literature and other media. He relates his standard of what constitutes obscenity to the standards of the community rather than his personal view as to what may be considered obscene. Particular standards are developed in discussions among the employees responsible for auditioning programs and among the Board of Trustees.

9. As a matter of station policy, if anyone of four particular words or their derivations should be used in a program proposed for broadcast, reference must be made to the station manager for his decision as to whether these words are to be deleted from the program before broadcast. According to Palmer, in editing programs for suitability, 99 per cent of the time when something is deleted for obscenity, it is apt to be one of these words or one of their derivations. It is not the policy of KRAB to exclude these words from all programs broadcast regardless of the context in which the words are used. Palmer has never been confronted with a situation in which an entire program had to be rejected because it was obscene. From time to time it has been found that certain words or expressions had to be deleted before broadcast in particular cases.

10. Palmer keeps abreast of current decisions and pronouncements of the Federal Communications Commission. He receives such material from the station's communictions counsel. He is the principal liaison between the personnel who operate the station and the Board of Trustees.

11. A number of programs were edited and changes were made for suitability before broadcast. Palmer deleted two words from a tape entitled "The Army on Trial" because he felt that in the context in which they were used they were obscene. "Running the Bulls in Blue," a documentary produced by KRAB, was edited but mainly because of obscurities, such as crowds moving from one place to another, rather than for the deletion of obscenities. A commentary by Selma Waldman dealing with the Women's Liberation Movement contained some talk about the words men use to describe women. After a discussion with Selma Waldman, Palmer, apparently with her agreement, deleted two of the words from the broadcast. Following is a list of some other programs which were edited to remove obscenities before being broadcast over KRAB.
   Comedy of Lenny Bruce
   Women of the Seventies: Rights, Roles and Risks, local panel discussion
   William Kunstler Speaking at the University of Washington
   Stanley Crouch: Ain't no Ambulances for no Nigguhs Tonight, Flying Dutchman recording
   A Night in Santa Rita, Flying Dutchman recording
   Pregnancy: Love it or Leave it, local panel discussion
   Son of Earth Day, tape from Pacifica
   The New York Panther 21 Manifesto, from Radio Free People
   Vamping on the Panthers, local documentary

12. From time to time Palmer has referred questions regarding the suitability of programming to the Board of Trustees. For example, it become known that a group called the Seattle Liberation Front would hold a demonstration. He knew where it was going to take place and it was conveyed to Palmer the anticipation of a possibility that there would be violence. It was thought that this demonstration was important enough that the program dealing with it be broadcast live, if possible. This matter was discussed with the Board of Trustees at a meeting because of the expectation that a program dealing with the demonstration might result in the broadcast of words or expressions inconsistent with the station's standards of suitability. After consideration and discussion it was decided to broadcast the program without editing even though the tape recorders might very well pick up words or expression not considered suitable for broadcast under the station's usual standards. There was no evidence presented at the hearing to indicate that the program did involve broadcast of any offensive words.

13. Another example of a matter of obscenity discussed with the Board of Trustees by Palmer was a proposal to remove one of the taboo words from the station's list of four. This was proposed because Friede had heard the particular word in a broadcast of a national educational television program. There was some feeling that there was no longer a need to continue the taboo against this particular word. This matter was tabled by the Board of Trustees after discussion. KRAB continues to have four taboo words. Broadcast of any of these may be permitted only after special consideration by the station manager or by the station manager in consultation with the Board of Trustees.

14. Two programs were specified for particular consideration in this proceeding by the Commission in the Order of Designation and consistent with that Order the Broadcast Bureau designated three more programs for special consideration. These programs were the broadcast by the Reverend Paul Sawyer; the interview with a member of the San Francisco Mime Theatre; the talk given by the Reverend James Bevel; the record entitled "Murder at Kent State"; and the bluegrass program hosted by Dave Wertz. These programs were alleged to have violated the station's policies in that obscenities were permitted to be broadcast. These programs and the circumstances under which they were broadcast were as follows:

Paul Sawyer Broadcast

15. At the time of the broadcast mentioned by the commission's Order, Reverend Paul Sawyer was Minister of the Lake Forest park Unitarian Church located in a suburb just north of Seattle. Lorenzo Milam was manager of KRAB at that time. He had come to know Sawyer through a mutual interest in sound and sound techniques. Sawyer had been a participant in some programs on KRAB and hosted a regular program dealing with sound effects.

16. Milam found out that Sawyer had been preparing a tape recorded autobiography. By the time Milam found out about it, the autobiography was about 30 hours in length. Milam listened to portions of this tape, thought it was interesting, and thought that it would be worth broadcasting on KRAB as an "autobiographic marathon." Nancy Keith and one or two other employees at KRAB listened to parts of the Sawyer autobiography. Neither Nancy Keith nor Milam recalled hearing any objectionable language in the portions of the tape which they heard. In discussing the Sawyer autobiography, some of the station personnel expressed a view that it should not be broadcast because it was dull. Nevertheless, the decision was made to go ahead with the broadcast.

17. Broadcast of the taped autobiography took place on August 5, 1967. Miss Keith was on duty at the station. Sawyer was there to handle the playing of the tape because of problems with the quality of the sound. Milam was at home. At about 9:00 a.m., he turned on the radio to listen to KRAB while eating his breakfast.

18. As Milam describes it, soon after he started listening to the autobiography, he heard a word which frightened him. Apparently, the autobiography included some descriptions of Sawyer's intimate relations with his wife. Obscenity frightens Milam, and he recognizes obscene words by the emotional response he has toward hearing them, characterized by sweating and coldness of his hands. Milam could not remember the exact word or words spoken which caused this emotional response while he was listening to the Sawyer tape. He conceded that the actual words were probably those related by the Broadcast Bureau in their Bill of Particulars or their equivalent.

19. Milam called the station and talked to Miss Keith. Miss Keith had already heard what had upset Milam and she was also quite upset. Milam told Miss Keith to talk to Sawyer. She did and Sawyer apologized saying that there would be nothing more like that on the tape.

Broadcast of the tape continued but more language frightening to Milam and upsetting to Miss Keith came out. Milam called the station again and talked to Sawyer. He told Sawyer that he was threatening the station's license and he didn't want him messing around like that. Seemingly, Sawyer didn't have the same concern over the use of obscene words that Milam had, but he did give assurance that nothing else obscene was on the tape.

20. Sawyer was permitted to continue to broadcast the autobiography, but obscene words continued. This time, Milam got into his car and drove to the studio. He took the Sawyer program off the air and substituted a program of Indian music in its place.

San Francisco Mime Theatre

21. A group known as the San Francisco Mime Theatre presented some performances in Seattle about the middle of February 1969. P.J. Doyle, of the Adult Education Department of the Seattle Public Library System, attended these performances and was favorably impressed with the group. Doyle, however, was annoyed with what we considered to be an excessive use of the four-letter Anglo-Saxon verb denoting the act of sexual intercourse. Doyle's job with the library system calls upon him to use radio. He is not a professional librarian, having been a book dealer prior to coming with the library system about five or six years ago. Doyle broadcasts a regular weekly program over KRAB dealing with new book acquisitions or with books in the library's collection which have a bearing on outstanding current events. Doyle's programs are productions of the Seattle Public Library.

22. He made arrangements to interview an actor, Joseph Lamuto, who was a member of the San Francisco Mime Theatre company. Lamuto and Doyle met at the KRAB studio where the interview was taped. Although station personnel were present during the interview, it was not supervised or auditioned, as such, by a member of the station's staff.

23. Generally, the interview dealt with the Mime company's performance, but a small portion of the interview, a transcript of which is included in the record, consisted of a discussion of the consequences of using the four-letter Anglo-Saxon verb previously described. In the transcript of the interview, Doyle assures Lamuto that it is all right to say the verb and it is used about four or five times. Among other things, Lamuto illustrated his argument that people react to this word by referring to a poem by Lawrence Ferlinghetti in which this verb is used about 20 or 25 times in such a way that it loses its usual affect, according to Lamuto. In this short discussion, Lamuto and Doyle dealt with the word and its use. It was not a discussion of the act it described and, the word was not used as epithet or expletive.

24. This interview was broadcast about two weeks after it had been taped. Prior to broadcast the tape was not auditioned by KRAB personnel nor was it edited by KRAB. Doyle had had discussions with Miss Keith regarding the standards of suitability which KRAB applied to its programs. He was aware of these standards. He considered his interview with Lamuto to have been a serious discussion about the use of the English language which had backfired. He was concerned that a program produced by the city library should get KRAB

into trouble. Books which contain "four-letter words" are on the open shelves in the Seattle Public Library. Included among these are the poems of Lawrence Ferlinghetti. These books may be taken out by any holder of an adult card and an adult card may be obtained by anyone 12 years of age or older. Doyle still broadcasts his weekly program over KRAB. When he comes to words such as the one that caused problems in his interview, he substitutes a "blank" and feels foolish for having to do so.

Dave Wertz

25. Dave Wertz describes himself as an amateur expert on bluegrass music. He had a program on KRAB which consisted of bluegrass music and pertinent accompanying commentary. Wertz tried to imitate the style of such well known programs of bluegrass as Nashville's Grand Ole Opry and Richmond's Old Dominion Barn Dance. Between broadcasts of music selections, Dave Wertz would tell what he called "corn country jokes."

26. When Wertz came to work for KRAB, it was made clear to him that he was not to use any obscenity on the air. His type of jokes do not contemplate the use of obscene words. He had no recollection of what he may have said on his broadcast of October 1968, but he had been told that someone had called to complain about the program. He may have told a few of his country stories. An example which he gave is the one about "the hillbilly whose bathroom caught on fire but fortunately the flames didn't reach the house."

Murder at Kent State

27. "Murder at Kent State" is a recording produced under the Flying Dutchman label. KRAB played three records under this label during 1970. Two of these records, after preview, were edited for obscenity and a number of words were removed before the records were broadcast. "Murder at Kent State" was also previewed before broadcast, but the same or similar words spoken on this record were not deleted prior to broadcast.

28. "Murder at Kent State" is a reading of a series of articles by Pete Hamill which appeared in the New York Post. These articles describe the incident which occurred at Kent State University in which a number of students were killed during a confrontation with the Ohio National Guard.

29. Before it was broadcast, this record was auditioned by the station manager, the program director and some of the Trustees of KRAB. The record took a total of 46 minutes to play and included about a half-dozen obscenities including an epithet directed at the Vice President of the United States.

30. KRAB played the record on August 10, 1970, several months after the incident which it describes had occurred. In playing the record without deleting the obscene or indecent language, KRAB's management was moved by the consideration that any editing would

adversely affect the emotional impact of the record. It was thought that the record was news worthy and important, particularly to the university community at the University of Washington, which was a considerable proportion of KRAB's regular audience. Since the University station had returned to broadcasting programming which did not include matters of current relevance to the student body and faculty, KRAB felt it had an obligation to fill a void by giving the University community a program such as that represented by the recording "Murder at Kent State."

Reverend James Bevel

31. On December 9, 1967, KRAB broadcast the tape recording of a talk by Reverend James Bevel given at the University of California, Berkeley. This tape had come to KRAB from the Pacifica Foundtion and the box in which it had been forwarded indicated that some deletions from the tape had been made. There were complaints to the FCC about this broadcast and the station was visited by an inspector from the Commission. This inspector asked for and was given the tape for copying. The tape was returned to KRAB. A previously scheduled broadcast of the tape for December 26, 1967 was cancelled. At that time, Lorenzo Milam, who was station manager, was out of town. Before the broadcast of December 9, 1967, the tape had been auditioned by an employee of KRAB.

32. A meeting of the Board of Trustees was held on January 2, 1968, to discuss what to do about rebroadcast of Bevel's talk. It was the unanimous decision of the Board to rebroadcast Bevel's talk, but to preface the rebroadcast with a statement by Milam describing the events which had taken place since the tape was played on December 9, 1967.

33. A transcript of the tape and a transcript of Milam's introduction were received at he hearing. An offer that the tape of these talks also be included with the record was rejected, but the Examiner did listen to the tape. Bevel's talk is largely a rambling discourse directed at what is apparently a predominately white student audience whom he considers to consist of radicals. Bevel uses certain expressions which may be described as well-known slang or vulgar references to virility; or common blasphemies or abstruse expressions which sound like they ought to be somebody's obscenity. There are all listed by the Broadcast Bureau under the heading of alleged obscenities including a reference to academic "pimps to freak you off". This last quoted set of words, if it is an obscenity, is a contribution to the Examiner's education in an area where he had thought life had foreclosed all possibility of novelty.

34. Milam in his introductory remarks unleashed a somewhat candid though not entirely novel evaluation of a broadcaster's feelings toward the FCC. Probably, the best way to make findings as to the tenor of the talks given by Milam and Bevel would be to quote a representative portion of each presentation. Milam's introductory statement was much shorter than Bevel's talk, but it does give an insight into the licensee's attitude which motivated it in broadcasting Bevel's talk without deletions for obscenity. Milam had the following to say:

"The FCC has responsibilities to exercise care in the power of licensing of broadcast stations. The Communications Act of 1934 specifically states that the FCC shall in no way indulge in censorship of programs. The creators of that government body were wisely concerned that freedom of speech through broadcasting should in no way be curtailed. This is where the issue has been joined with Reverend James Bevel and KRAB and the local official of the FCC. In the month that we've had to stew over this event we've come to feel that this confiscation of tape was a case pure and simple, of censorship. Censorship of the cruelest form, for it created in us a deep sense of fear over the future of KRAB, the disposition of our valuable license and all the deep questions of qovernment control. We've decided to rebroadcast the James Bevel tape. We've done so fully aware of the dangers to our permit, our broadcast license. We, and now I'm speaking for the board of Directors of the Jack Straw Memorial Foundation which is the parent corporation of KRAB, have met and discussed at great length the possible consequences of the act of rebroadcasting of this material and we've decided that we must replay it.

"The FCC in it's rules has very wisely demanded that broadcast licenses should have full responsibility for the material they broadcast. They, the broadcasters alone must act on behalf of the public interest, convenience and necessity. No one else can be responsible and if the broadcaster fails in this duty he's subject to the revocation of his license. We here at KRAB feel that we would be sabotaging the public interest, convenience and necessity if we didn't play the James Bevel tape at this time. For despite his strong language, a language that is an integral part of his message, James Bevel is trying to tell us something important, trying to express a crucial view of Negro-white relations in this country. KRAB has always been a forum for the dispossessed, we've opened this frequency at 107.7 megacycles in Seattle to hundreds of different viewpoints about hundreds of different subjects. We've done this not because we agree with any one speaker, we couldn't conceivably do so, but because the miracle of free speech in this country lends itself to knowledge and understanding of so many disparate viewpoints, even those which may be offensive to us. For by understanding the hundred voices of antagonism one can and does become an active, knowing and thinking part of the democratic system. James Bevel doesn't speak for KRAB, no-one does really, but James Bevel is a representative of an important and sometimes frightening new force in America. By failing to play this talk KRAB would be doing a disservice to its listeners denying them a knowledge of the important forces around them. We would be saying, in effect, that the license of KRAB is more important than freedom of speech and freedom of knowledge. We simply cannot as responsible broadcasters ignore this duty; we'd be foolish not to play the words of James Bevel."

35. Bevel spoke for about an hour. His choice of language was not such as one would expect to hear from a pulpit. His ideas were expressed in a stream of consciousness form with little attention to the niceties of rhetorical organization. To get an idea of what Bevel spoke about and how he expressed himself, it is best that we let his own words speak for him. A fair sample of what he had to say and the choice of language he made is the following:

"Man is a love animal and love is an energy just like oxygen that man needs in order to act rational and when a man can inhale and breathe into his body love energy he acts rational, natural and truthful, that's why you hear the brothers saying 'acting natural'. To be natural is to consume love energy that is present in the universe and there's only one thing that can stop man from consuming and acting rational and that is if man begins to fear anything he lose the capacity to love, himself, that's the nature of the problem. Lot of folks want to argue with the chancellor, and a lot of folks want to argue with LBJ, and a lot of folks here want to argue with their mommas and their daddys and very few people here are prepared to say that the reason that the administration function as it does is that it's fearful, and very few of us here are prepared to say the reason we are here today is that we afraid that if we don't pick up a piece of paper we can't have protein. Most of us can't say that, but the realities are that we are here not because we are wise and not because we are in the pursuit of education but because we afraid not to be here. Fear, fear is a disease it's a sickness, for fear does not allow man to perceive the universe as it is because it locks man out of himself, and in the past, if you ever studied literature. A lot of you jive folks studied literature and didn't even know what the hell you was doing. In class if we study literature of the past, men who live at another period when the energy was in another form, you read a story in the old testament about adam being locked out of the Garden of Eden, man being locked out of himself, because he feared something, and when man is locked out of himself, and when he begin to fear he acts the same way, he starts hating folks. You see I get tired of walking round this country listening to city jitterbug fascists, who call themselves radicals pretending that they're any different from the Administration, when they know damn well they're driven by fear just like the Administration, but you see fear makes man hate, what it does is make man project his contempt for himself on to other folks, and why we pretending that its the Administration that is holding up our freedom, and like we want to pretend like Reagan is holding up our freedom, and we want to pretend like johnson is holding up our freedom, and the realities are that we hold our own freedom because we afraid to pick up our own nuts* and say I'm a man here in the universe and I ain't going no goddam place, that's why we don't have freedom!"

*There was no claim of obscenity regarding this usage.

PROGRAMMING AND PROGRAM POLICY

36. KRAB does not avoid programs because they are unusual or outspoken. Its musical programs cover a broad range from jazz to classical. Its policy in music programming is to avoid music which is broadcast by other stations in the area. Programs of oriental, preclassical western and other types of unusual music are broadcasted. KRAB also programs jazz, blues, rock, bluegrass, renaissance and baroque as well as music from foreign countries such as Japan, Norway, Sweden, New Zealand, Korea and others.

37. KRAB broadcasts a substantial number of political programs and discussion programs not ordinarily heard on radio. In a recent primary election, more than 20 candidates were each given a half hour of time to present their views in their own way. Some candidates spoke for the half hour, others received calls from listeners, and other were interviewed. A recent referendum dealing with the State's abortion laws lead KRAB to broadcast a two hour discussion moderated by a member of the staff with panelists representing both sides of the question. Religious programs have included interviews with clergymen who "speak in

tongues"; the "1970 Annual Gymanfa Ganu", a Welch religious program; an interview with the Hare Krishna sect; as well as interviews with religious personages and presentations of religious programs not ordinarily heard in the Seattle area.

38. KRAB submitted 31 pages listing by title and participants its public service programs of note. Ordinalily, such lists do not tell us much about a station's programming. In the instance, however, some idea of the range of subject matter and variety of personages heard over KRAB are apparent. The following is a selection taken from KRAB's public service programming exhibit:

PICKETING IN BELLINGHAM. In March, 40 picketers protesting the war in Viet Nam were arrested and booked for disturbing the peace. A program of interviews and comments.

SHOULD COMMUNISTS BE EXPELLED FROM UNIVERSITY FACULTIES? Debate between Fred Schwartz of Christian Anti-Communist Crusade and Otis Hood, Chairman of Communist Party of Massachusetts.

HAS THE COURT USURPED THE POWERS OF CONGRESS, Robert M. Hutchins.

THE CHANGING MEANING OF THE ORGANIZATION, Dr. Harry Levinson, address on industrial management and the psychological meaning of the organization of work.

EUTHANASIA, local, panel discussion.

THE WILL OF ZEUS, Stringfellow Barr discusses his book and compares the political problems of classical Greece and contemporary America.

GOLD AND THE GOLD SITUATION, panel discussion, Dr. Ernest Patty, former Pres., University of Alaska, and Pres. and Mgr. of some gold mining operations, Edward McMillan, from NB of C, Dr. Frederick B. Exner from KRAB, and John McFalls, stock consultant, local.

PEACE KEEPING UNDER THE RULE OF LAW, panel discussion on national sovereignty and the world community, Justice Earl Warren, Kenzo Takayanagi, Chairman of the Japanese Cabinet Commission on the constitution, Senator J. William Fulbright, others.

JACQUES COUSTEAU, producer of World without Sun, lecture in Washington, D.C., on exploitation by man of natural resources.

PRODUCTION VS. REPRODUTION, the Population Problem in the U.S. and in Calif., panel discussion, Marriner Eccles, moderator, Alice Leopold, Lewis Heilbron, Dr. Karl Brandt, others.

TRIP TO DJAKARTA, Beverly Axelrod on her meeting with Vietnamese women.

ACADEMIC FREEDOM, Arthur Flemming, President of Univ. of Ore. and former Sec. of Health, Ed., and Welfare, address, at EWSC at Cheney.

YOUNG AMERICANS FOR FREEDOM, Jack Cox, YAF State chairman for Calif., address on government errors.

A PEEK AT PIKE, documentary on Pike Street Market.

EMMETT MCLOUCHLIN: CATHOLICISM AND FREE MASONRY IN AMERICA.

POLITICAL CONDITIONS IN SOUTH AFRICA, sociology prof. Pierre van den Bergh.

TRAFFIC IN NARCOTICS, Detective Chet Sprinkle of Seattle Police Narcotics Bureau.

DR. GATCH AND THE DIET OF WORMS, Southern physician on poverty conditions.

39. As earlier stated, KRAB receives its funds from its listeners and from various foundations. It has received money for various purposes from such organizations as the Jaffe Foundation of Philadelphia, which is administered by Ambassador Walter annenberg; the James E. Merrill Trust of New York; the Gerber Foundation; and others, including:

> "*$10,000 from Seattle's PONCHO.* PONCHO is a fund raising organization for the arts in Seattle. It made a lump sum allocation to KRAB to extend its operations. PONCHO representatives made the grant to KRAB because it 'performed' a very valuable and unique func-

tion as an open forum radio station through which arts needs, as well as all kinds of other social needs, local, regional needs, could be explored and examined."

"$7,500 from the Corporation for Public Broadcasting. KRAB was one of 73 out of the more than 400 noncommercial radio stations in the United States to receive CPB awards. The funds were used principally for program improvement, i.e., morning show, program director, two correspondents, and general efforts to encourage news programs, documentaries, and similar programs.

"$1,200 from the Washington State Arts Commission. Part of a matching fund award from the National Endowment for the Arts."

40. There were about 25 witnesses who were neither employees nor trustees of KRAB. They appeared to testify to the usefulness and excellence of KRAB's programming. These witnesses were either regular listeners to KRAB or individuals who had used the broadcast facilities either as participants in programs or on behalf of public institutions which they represented. A few witnesses had heard some of the five programs which were given particular consideration in these proceedings. There were no Witnesses who appeared to support a view adverse to the station. Some of the testimony given in support of the station's usefulness to the community is given in the succeeding paragraphs.
[*Partially omitted here are a list of station supporters.*]

41. Robert W. Means is an air traffic controller who has resided in the Seattle area since he matriculated at the University of Washington in 1930. He and his wife listen to KRAB regularly. He also tapes programs for broadcast over the station. Events that he has taped have included a Ralph Nader press conference, a speech by John Howard Griffin, a speech by the former Chief of the Cuban Air Force, and a community meeting where people discussed their concern about nuclear waste materials. Means had the following to say about KRAB:

"One value of KRAB to the community is the broadcasting of such diverse material as this: some of it inspirational, some of it frightening; all of it relevant in some way to the enormous, complicated, unresolved problems of our challenged society. It has been a major but important function of the station to open up for public discussion such edgy questions as abortion reform, treatment of criminals and the insane, white-collar crime, changing forms of religious experience, and the like.

"Of equal or greater value to me are the programs of ethnic music, book and movie reviews, commentaries from abroad or from a foreign point of view, far-ranging expeditions into American folk music, systematic explorations of the classical music catalog, and the many and varied humor programs. KRAB has been of intellectual experience. This has not been accomplished by being comfortable, complacent and conformist."

42. John Stewart Edwards is a native of New Zealand and a professor of zoology at the University of Washington. He is a frequent listener to KRAB and finds that for him and his colleagues it forms a significant part of their intellectual input. He has, from time to time, heard "four-letter words" used on KRAB, but has never been offended by any such language. In that connection he expressed the following point of view:

"I have heard them from time to time, yes, and if you are interested in my response to it, I would say that the response to the words as used more as epithets or as what everybody knows is used in common, everyday life, these words I find less offensive than say the kind of innuendoes, for example, on the Johnny Carson Show or on some of the popular television programs. In fact, last evening there was just such an example of what I considered an obscene innuendo on the Johnny Carson Show. I have never heard anything of that type on KRAB, although specific Anglo-Saxon four-letter words have been used mainly as epithets, which one finds used in journalism. Any reader of the New Republic or Harper's magazine will find these frequently, and I would imagine that the average listener to KRAB is more like the reader of a journal that carries these words without question these days."

43. Robert J. Block is an investment banker and real estate broker. He is a listener and has been an unsuccessful candidate for office. As a candidate, he had been offered and has used the facilities as KRAB. He has also participated in discussion programs on the station. He has never heard any program on KRAB which offended him. As he put it:

"I think that their programming has been stimulating and highly useful in the rather plastic society in which we live. Nothing they have ever done has certainly offended my sensibilities."

44. Fred Cordova is the director of public information for Seattle University. Seattle University is a private institution conducted by the Jesuit Fathers. It has an enrollment of over 3,000 students. Cordova has been an occasional listener to KRAB. KRAB has given coverage

to campus events at Seattle University such as appearances by Mortimer Adler and Barry Goldwater, a symposium for Filipino American youth and a symposium on Indian problems. He gave the following evaluation of the usefulness of KRAB to Seattle:

> "For a station like KRAB I think it is quite a necessity, if I must do a little bit of editorializing here, I think it is quite a necessity here in our city. Our media, especially in radio, is quite commercial, regardless of whether it is AM or FM, and KRAB is the only station that I know of here in the Pacific Northwest where it deviates from a normal type of programming, radio type of scheduling, and it allows, I think, good free thought on controversial as well as other urban issues that have to be aired."

45. Elsie B. Martinez is a postal clerk and a music school graduate. She listens to KRAB and has never heard anything offensive broadcast. She said,

> "I can turn on KRAB and I get all sorts of ethnic music, I get Bach and Scarlatti, harpsichord, get all sorts of beautiful classical music, and then I get the latest rock and roll and everything. It is very pleasant to listen to."

46. Other witnesses appeared and gave testimony supporting the view detailed above. All witnesses gave strong support for the useful and unique qualities of the station's programming.

CONCLUSIONS

1. KRAB is non-commercial, listener supported and it broadcasts a variety of programs of outstanding quality. Its programming is of a type not usually heard on radio and its appeal is directed to an audience of people with a high degree of intellectual curiosity. KRAB provides its audience with a broadcast service which is attractive and uniquely appealing. As a matter of policy, KRAB is committed to providing the Seattle area with unusual, stimulating and extraordinary programs. KRAB's programming is meritorious and the station does render an outstanding broadcast service to the area which it serves.

2. KRAB is directed by a Board of Trustees who are above average in their educational backgrounds and who represent a variety of tastes. This group is responsible for setting station policy and for exercising overall supervision over programming. It is actively involved in carrying out its duties. In order to bring its audience the type of unusual programming that its policies call for, KRAB experiments with the unique and gives time to an extraordinary variety of programs. In doing so, KRAB sometimes falls short of the expectations of its management, its audience or the licensing authority to which it is accountable for its franchise. Thus it is that this proceeding, to determine whether or not KRAB's license should be renewed for a short

term of one year or a full three-year period, came about. A few of KRAB's programs involve the broadcast of words or expressions described as obscene.

3. It is not KRAB's policy to use obscene or indecent language in its broadcasts for the sensational or shock effect that such language might have. This licensee eschews obscenity, profanity, and indecency. Its procedures for clearing programs for broadcast are designed to avoid material which would give offense to the community. This proceeding was instituted because KRAB did broadcast some programs which did give offense to some members of the community in which its programs are heard. We are directed, therefore, to determine whether in broadcasting certain programs specified by the Commission and the Broadcast Bureau, KRAB violated its own standards. This determination must be made, however, in the content of standards laid down by the Commission.

4. Our most current applicable source as to the Commission's policy regarding broadcast of such offensive material is the analysis in the *Notice of Apparent Liability, In re WUHY-FM*, FCC 70-346, April 3, 1970. In its discussion in that notice, the Commission renewed its committment to the right of licensees,

> "...to prevent provocative or unpopular programming which may offend some listeners. *In re-Renewal of Pacifica,* 36 FCC 147, 149 (1964). It would markedly disserve the public interest, were the airwaves restricted only to inoffensive, bland material. Cf. *Red Lion Broadcasting Co., Inc. v. F.C.C.,* 395 U.S. 367 (1969)."

5. Taking up the matter of obscene language, the Commission did prescribe standards to guide in determining the permissible and impermissible areas. In setting these guide lines, the Commission did recognize the difficulties which arise in trying to steer a course between the censorship which the law forbids the Commission to exercise and the indecent or obscene language which the law forbids the licensee to broadcast. After relating some of the obscene and offensive language which had been broadcast by WUHY, the Commission observed that:

> "8. ...these expressions are patently offensive to millions of listeners. *And here it is crucial to bear in mind the difference between radio and other media.* Unlike a book which requires the deliberate act of purchasing and reading (or a motion picture where admission to public exhibition must be actively sought), broadcasting is disseminated generally to the public [Section 3(o) of the Communications Act, 47 U.S.C. 153(o)] under circumstances where reception requires no activity of this nature. Thus, it comes directly into the home and frequently without any advance warning of its content. Millions daily turn the dial from station to station. While particular stations or programs are oriented to specific audiences, the fact is that by its very nature, thousands

of others not within the 'intended' audience may also see or hear portions of the broadcast. Further, in that audience are very large numbers of children. Were this type of programming (e.g., the WUHY interview with the above described language) to become widespread, it would drastically affect the use of radio by millions of people. No one could ever know, in home or car listening, when he or his children would encounter what he would regard as the most vile expressions serving no purpose but to shock, to pander to sensationalism. Very substantial numbers would either curtail using radio or would restrict their use to but a few channels or frequencies, abandoning the present practice of turning the dial to find some appealing program. In light of the foregoing considerations we note also that it is not a question of what a majority of licensees might do but whether such material is broadcast to a significant extent by any significant number of broadcasters. In short, in our judgment, increased use along the lines of this WUHY broadcast might well correspondingly diminish the use for millions of people. It is one thing to say, as we properly did in *Pacifica,* that no segment, however large its size, may rule out the presentation of unpopular views or of language in a work of art which offends some people; and it is quite another thing to say that WUHY has the right to broadcast an interview in which Mr. Garcia begins many sentences with,____, an expression which conveys no thought, has no redeeming social value, and in the context of broadcasting, drastically curtails the usefulness of the medium for millions of people."
(Footnotes omitted.)

6. Going on to the standards to be followed, the Commission concluded that for broadcasting,

"10. ...we believe that the statutory term, 'indecent', should be applicable, and that, in the broadcast field, the standard for its applicability should be that the material broadcast is (a) patently offensive by contemporary community standards; and (b) is utterly without redeeming social value. The Court has made clear that different rules are appropriate for different media of expression in view of their varying natures. 'Each method tends to present its own peculiar problems.' *Burstyn* v. *Wilson,*

343 U.S. 495, 502-503 (1951). We have set forth in par. 8, *supra,* the reasons for applicability of the above standard in defining what is indecent in the broadcast field. We think that the factors set out in par. 8 are cogent, powerful considerations for the different standard in this markedly different field."

7. A person could be a regular listener to KRAB and not hear any obscene or indecent language broadcast. The most that can be said is that a regular listener may occasionally hear some four-letter Anglo-Saxon sexual or scatological terms. KRAB is not a station that presents smut regularly or frequently. There were only five programs broadcast over a period of three years which led to the controversy which resulted in this proceeding. We will consider each of these programs:

8. There is no evidence that Dave Wertz used obscene language either in the particular program specified in the Bill of Particulars or in any other of his shows. He broadcasts bluegrass music and tells the kind of stories that are associated with that type of entertainment. We conclude that telling "corny jokes" entails risk and may give some offense, but we can not conclude that Dave Wertz broadcasts anything contrary to the policies of the licensee or the Commission.

9. P.J. Doyle conducted an interview which he thought to be a serious discussion of language usage. His program was produced and presented under the auspices of the Seattle Public Library. This discussion involved the use of the one word most likely to offend if heard over the air or anywhere else.

10. Doyle's broadcast took place without prior audition by the licensee's staff or management. Doyle now knows better and he is careful about the language which is used on his program. In this instance, the question we must resolve is whether or not the licensee failed to exercise proper care by not having auditioned this program in advance of its broadcast. There is nothing in the record to show that Doyle's prior programs gave any indication that preauditioning of his program was necessary in order to avoid broadcast of material which might be offensive or otherwise in bad taste. In addition, we must bear in mind the auspices under which this program came to the station. It was, after all, produced and sponsored by the Seattle Public Library. We concluded that the licensee acted with reasonable diligence in its handling of this program. It is clear that Doyle is now aware of his responsibilities and that the material which he now broadcasts does not fall short either by the station's standards or the standards which the Commission would have its licensee observe.

11. Reverend Paul Sawyer was known to Milam and the licensee had had some experience with Sawyer as a performer prior to broadcast of the taped autobiography which caused problems. Whether a station should broadcast anybody's autobiography for 30 hours is not our concern. What did happen was that such an "autobiography marathon" was begun. Sawyer was not known to be a person who used obscene language. Part of the material which he planned to broadcast was auditioned and nothing heard in these auditions was obscene.

When it became apparent during the actual broadcast that Sawyer's autobiography did include words or expressions which were unsuitable, his broadcast was taken off the air. We conclude that the worst that can be said regarding this incident was that it was an error in judgment which was expeditiously corrected.

12. "Murder at Kent State" and the James Bevel broadcasts bring us head on to the issue of whether a licensee may under any circumstances broadcast (a) material known to be obscene or offensive; or (b) material not considered offensive or obscene by the licensee but which might be so considered by others. In the case of the "Murder at Kent State" record, the language used included words which the licensee did consider obscene and ordinarily would not permit to be broadcast. In this case, after careful consideration, the licensee's Trustees and managerial employees decided that in their judgment use of the particular language was necessary under the circumstances involved. This is a matter of judgment which we concluded the Commission has left to licensee determination. In this case, language was not broadcast for shock or sensationalism, but rather for the purpose of presenting a vivid and accurate account of a disastrous incident in our recent history. We conclude that in this exercise of judgment, the licensee conformed to the standards prescribed by the Commission as well as its own policies regarding suitability.

13. It is too bad that Reverend James Bevel did not take a little time to organize his material. He had some very interesting and provocative ideas which some people may have lost. Reading his entire text without being forewarned to expect "dirty words" one could possibly miss some of them altogether, as indeed happened with "nuts". Bevel is an emotional and colorful speaker. But, Bevel's language was not anything like that used by Garcia and Crazy Max in the program that brought a $100.00 sanction upon WUHY-FM. Bevel's talk really comes within the scope of the concern with which the Commission was dealing in its letter of January 21, 1970, to Mr. Oliver R. Grace (FCC 70-94) rather than the more provocative WUHY-FM program. In its letter to Grace, the Commission said:

> "The charge that the broadcast programs are 'vulgar' or presented without 'due regard for sensitivity, intelligence, and taste', is not properly cognizable by this Government agency, in light of the proscription against censorship. You will agree that there can be no Governmental arbiter of taste in the broadcast field. See *Banzhaf v. FCC*, 405 F 2d 1082, (CADC), certoirari denied 395 U.S. 973, cf. *Hannegan v. Esquire Magazines*, 327 U.S. 146 (1946)."

14. In concluding that some of the language used by the Reverend Bevel was vulgar rather than obscene, we are unavoidably treading into an area of often stormy controversy over our changing mores. There was no real effort made to produce evidence as to the extent to which anyone in Seattle was offended by anything heard on KRAB. Neither was there any particular effort made to show that the words designated as obscene by the Broadcast Bureau were not offensive to the community. KRAB under its own policy would ordinarily avoid giving offense by avoiding the use of such language.

15. There is really no quarrel by KRAB with the standard set by the Commission that broadcasters should avoid language that is patently offensive by contemporary community standards and utterly without redeeming social value. We can not avoid the difficult result that what particular language may be unacceptable for broadcast is not susceptible to being reduced to an immutable, time resistant glossary.

16. All but one of the "obscene" words listed by the Broadcast Bureau are now to be found in Webster's New International Dictionary, 3rd Ed., 1961, G & C Merriam Co. Every one of these words, with one exception, is characterized as vulgar rather than obscene by the scholars who produced the dictionary. Our times are indeed changing. Consider what Mr. Clive Barnes, the drama critic of the New York Times recently said:

> "Incidentally, have you noticed how the currency of swear-words, those honestly shocking oaths only to be emitted in times of intense stress, have become hopelessly devalued. A new Broadway play quite casually ran the whole lexicon, and no one seemed to notice. We appear to have overcome obscenity by incorporating it into polite conversation."*

17. Our recent history has been embellished by this event. "Love Story", Erich Segal's long continuing best selling novel, is now a very well attended motion picture. Our most prominent citizen saw the movie and following is a portion of a press report of what he had to say after that event:

> Chatting informally this morning with newsmen about his State of the Union message, President Nixon said he had seen the movie in Camp David recently, had enjoyed it and, the President added, 'I recommend it.'
>
> "However, he said, he was mildly upset at the film's profanity."
>
> "He said his wife and two daughters, Tricia and Julie, had read the book and felt the 'shock of the dialogue they put in the girl's mouth.' "
>
> "'I wasn't shocked,' the President said, 'I know these words, I know they use them. It's the 'in' thing to do.' "
>
> "However, Mr. Nixon said, the dialogue 'detracted from a great performance' by Ali MacGraw, who plays the female lead."
>
> "Discoursing briefly on profanity, Mr. Nixon said that

*New York Times, February 28, 1971.

swearing 'has its place, but if it is used it should be used to punctuate.' If profanity is overused, he said, 'what you remember is the profanity and not the point.'"†

"Love Story" includes virtually every word cited by the Broadcast Bureau as obscene.

18. We can not emphasize too strongly that while KRAB did broadcast a few programs that included some language offensive to some people, they did not do so with any intent to give offense, to pander, to sensationalize, to shock, or to break down community standards. KRAB should be given credit for a real desire not to debase community standards of taste and decency. In considering their policies and their programming as an entirety, the licensee of KRAB seeks and most often attains those standards of taste and decency in programming that we should like to see reflected more often in our broadcast media.

19. We conclude that KRAB's programming, in total, is outstanding and meritorious. We conclude that the few instances in which KRAB did broadcast obscene language, either willing or unwittingly, do not justify denying grant of a full term, three-year renewal of its license.

20. Accordingly, IT IS ORDERED that unless an appeal to the Commission from this Initial Decision is taken by a party, or the Commission reviews the Initial Decision on its own motion in accordance with the provisions of Section 1.276 of the Rules, the application of The Jack Straw Memorial Foundation, for renewal of license of station KRAB-FM IS GRANTED.

Ernest Nash
Hearing Examiner
Federal Communications Commission

†New York Times, January 23, 1971.

RUNNING FOR POLITICAL OFFICE

Now what the hell does running for town council in Los Gatos have to do with you starting a radio station?

Nothing. It's just that I felt the whole campaign was so inspired, and so ignored in the press [the population of Los Gatos is 25,000] that I wanted to be sure our campaign platform got read somewhere else.

Because what it is—is a universal platform. To reverse the deadly progress which has turned Santa Clara Valley from a flower-filled orchard-filled paradise, into the Orange County of the Bay Area.

We lost—getting no more than 20% of the vote. We don't even think of it as a moral victory: our candidacy was so wierd that some other good guys got eliminated, and the builder-contractor combine took over as it has for most of the other towns of Santa Clara County.

As with so many campaigns of this type, the only victories we achieved were in the press. The best news articles appeared in the LOS GATOS TIMES OBSERVER which, under normal conditions, is about as Dengue-filled a newspaper as could be expected in such a place. Our press releases (a situation where we write news stories for them) gave them a chance to provide the following droll introduction to our campaign. The full platform follows.

CANDIDATES PROMISE TO ERECT
ANTI-SMOG FANS

"We would like to extend Billy Jones Railroad through Los Gatos for rapid transit. BART will be made compatible with Billy Jones," said Cese McGowan, the latest of several new faces in the Los Gatos political arena last Friday.

She was one of four Los Gatans—all associated with radio station KTAO—who dropped by the Civic Center to take out nomination papers for town positions to be decided in the April 11 election.

Miss McGowan and station owner Lorenzo Milam intend to file for town council positions; Merrill Kelly will run for the town clerk's post; and Telemachos Athanasios Greanias is eying the town treasurer's position.

All four will run individual campaigns, Miss McGowan said, but all four intend to run on the same platform.

Planks in the platform, in addition to the Billy Jones extension, include items such as a plan to erect a series of large fans "to blow smog from San Jose, San Francisco and Oakland back to those communities".

"We plan to transform the Los Gatos Chamber of Commerce into the Milpitas Chamber of Commerce," Miss McGowan said. "This way the Chamber can attract industry and people to Milpitas, rather than to Los Gatos."

Miss McGowan said another plank in the campaign platform involves the hiring of a "town fool to stick potatoes up the exhaust of all cars belonging to Triple A, Bank of America, Standard Oil, and the Embassy of the Union of South Africa."

"We also plan to pelt Governor Reagan with marshmallows if he ever comes to Los Gatos," she added.

The four candidates met Friday for a lunch or organic sandwiches and carrot juice and to discuss campaign strategy, Miss McGowan said. All four definitely plan to file before February 3 and all will remain in Los Gatos for the campaign.

This last was a reference to a campaign a couple of years ago in the state of Washington when a Lorenzo Milam-sponsored candidate, Richard Greene, announced his candidacy for Washington Land Commissioner and then immediately left for Hawaii.

Miss McGowan said this would not be happening in the Los Gatos election. "I have to go to San Francisco one night," she said, "but I'll be right back."

POLITICO QUIZZUM

Highway 17, running like a river of smog through the heart of Los Gatos, should be rechristened "The Red River" in honor of its petrochemical exhausts. The new town song will thus become "Red River Valley."

The Los Gatos Chamber of commerce shall be rechristened "The Milpitas Chamber of Commerce." The $25,000 annual dole from the Los Gatos Hotel-Motel Tax to the Chamber will continue in order to attract new businesses, new inhabitants, and new industry to Milpitas.

There shall be constructed in Los Gatos a plant which will manufacture air so that we can breathe. In addition, a series of large electric fans will be raised along the northern and eastern perimeters of the town which will blow the foul airs and miasmas back to San Jose where they belong.

Los Gatos long hairs should be hired under the Federal Food Parity Program to dispose of surplus potatoes in the following way: they will be stuck up the exhaust pipes of automobiles which are owned by ecologically unsound companies like PG&E, General Motors, Standard Oil, the AAA—and, as well, any consulate cars of The Union of South Africa.

Los Gatos shall secede from Santa Clara County and apply for permission to become an extension (or Westerly Spur) of Alpine County. Any takeover attempt by gays shall be resisted stwenously.

Surplus marshmallows should be disposed of in the following ways: if Gov. Reagan chooses to visit the town, he will be pelted with marshmallows. If Max Rafferty chooses to visit Los Gatos, he will be pelted with chocolate-covered marshmallows. If General Westmoreland chooses to visit us, he will be pelted with toasted marshmallows.

The Los Gatos Town Hall—now a nesting place for obtuse bureaucracy—should be turned into a 24-hour child daycare center. All officials and bureaucrats should double as babysitters.

The Billy Jones Railroad will be extended to cover the entire area of Los Gatos. This shall be called 'rapid transit.' BART will be invited to make its lines and coaches compatible.

The Western California Telephone Company will be given a subsidy to improve its service. The first subsidy will consist of five thousand feet of string, and a hundred tin cans.

All dogs running loose in the downtown Los Gatos area should be fitted with corks.

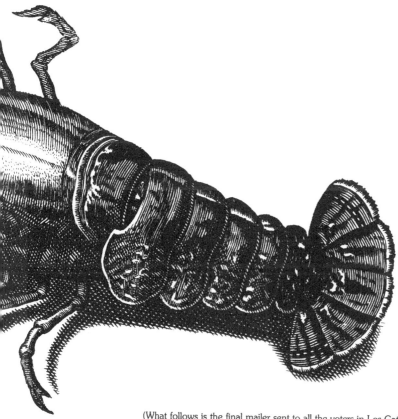

(What follows is the final mailer sent to all the voters in Los Gatos to convince them that our silly campaign was not all that silly.)

332

ONE: THE AUTOMOBILE

The automobile is our enemy. We have been subsidizing to for over fifty years—and now it is threatening to rise up and destroy us.

We have no idea what its petrochemical exhausts are doing to our lungs, and to the lungs of our children. We can only read the figures which tell us that the smog visible to us represents *only* two percent of the total airborn pollutants; that the amount of lead in trees and bushes next to freeways has quadrupled in the last decade. In these figures we can begin to read our own death notices—and know that it is time to do something, *anything*.

Since there is so little happening on a state or national level—we must begin work on a local level to defuse the automobile, before it kills us.

If there were a raw sewer running through the heart of Los Gatos, there would be meetings, protest, outrage. Highway 17 is such a sewer: but because we cannot *see* the harmful wastes, we pretend that they aren't there.

We have to begin to heed the warnings of our scientists. We have no idea what we are doing to our bodies with this accretion of automobile wastes in our lungs. We must make those who own and use automobiles feel some sense of the danger and greed involved—and fear for what they are doing to all our children's lives.

. . .

We will use the full force of the town government to *force* the State Division of Highways to put a lid on Highway 17. Not only will we make our views known in Sacramento on town policy—but we will begin to use local policy to harass noisy, smog-producing trucks and busses which use the Freeway. Just because it is state right-of-way shall no longer be an excuse for permitting this grossness in the heart of a once-beautiful town.

Locally, we begin an active street-narrowing project. We will invite townspeople to plant trees and bushes on sidewalks and asphalt parking areas. Local businessmen will be given tax credit for such expenditures, and local home owners will be given town sponsored prizes for new planting. We will do everything possible to arrest the proliferation of ecologically destructive paving-over of our precious earth.

We will begin to experiment with novel forms of public transportation. We will purchase special town bicycles which will be provided free for all townspeople and visitors to get from here to there. We will import Pedi-cars from Hong Kong, and invite people to taxi persons for small, fixed fees to every part of the town limits.

We will begin—more and more—to ban automobiles from our streets. Merchants will be invited to carry their stores *outside*: to display their wares on the streets and sidewalks. The pedestrian will come to be far more important that the car.

We will give a town subsidy to the Billy Jones Wildcat Railroad, so it can extend its lines throughout the town for public transit. The old southern Pacific right-of-way, now wasted on off-street parking, will be converted to bicycle paths and horseback trails. We will encourage and alternative form of transportation—and begin a war on the automobile mentality which thinks nothing of taking a 5000 pound machine a half-mile to buy a loaf of Wonder Bread at the Alpha Beta. We must eradicate this destructive mental set.

The Los Gatos Planning Department, under the guise of 'preserving' the town, makes horrendous requirements for off-street parking. All applications for home and apartment improvements end up with the property owner required to pave over another 500-1000 feet of land for the ubiquitous auto. We must purge the town bureaucracy of such thinking which may have been proper in 1935 or 1947 — but which is ruining us now.

When you build for progress — it comes, and with a terrible vengence. Each concession to the automobile brings demands for more. We must bring the town government into the realities of the 1970's. And we must do this by example. Each official government car must be pollution-free: and bureaucrats will be encouraged to ride bicycles, horses, and scooters.

We will join with the town of Santa Cruz to reactivate the steam passenger train which once ran between the two towns. Together, we will float a bond issue to build the railroad, buy right-of-way, and import old passenger cars from other countries. With this antique form of transportation, we will pull railroad buffs and tourists to the area, and give them a scenic and memorable chance to recreate the past.

We will put an end to all town engineering projects which encourage and ease the use of automobiles. There will be no more one-way streets. Stop lights will be dismantled and sold to San Jose — and all replaced with the relatively efficient (and inexpensive) four way stop signs. Any spending which benefits the automobile, and makes its use easier — will be halted. Completely.

 TWO: PROGRESS

The Los Gatos Planning Commission holds long, boring (and expensive) hearings on all applications to build, change, and modify. These are supposed to maintain the 'character' of Los Gatos — but somehow, they have been transmuted over the years so that it is impossible to preserve and use the older buildings in the town. Through an archaic Fire Code (designed for San Francisco, not Los Gatos), and through foolish parking requirements — it becomes cheaper and easier to rip down old homes and buildings, and put up plastic and terra-cotta stamp-out buildings in their place.

St. Lukes church asked the town to let them use their old parking lot for a combination free tennis court for the young and Sunday parking area when they remodelled. Instead, they were required to asphalt over the whole area, including some grassy plots, for their once-weekly services; and our government refused to let them do something creative 'because it wasn't in the code.' This anti-people thinking in our town must be stopped if we are to survive.

Some real estate dealers and speculators make a considerable profit from our land and our air, and they are assisted in this by federal and state tax laws which give them considerable subsidies. It becomes an easy road to bucks to buy up an old and scenic farm in the hills, divide it into tiny lots, and jam in 40 or 100 instant-plastic fall-apart homes. The people who do this are called 'developers,' but they are really destructors. We are the ones who must live forever with the strains put on our resources and utilities.

Our governments pride themselves on their policies of encouraging growth, and homebuilding. But this policy has gone insanely beyond all hope and good-sense: individuals feel helpless before the monster called 'Progress.' We are being forced to contribute — through our taxes — to the tearing down of the past, the destruction of fine Victorian homes, the constant plowing under of our precious orchards. We must learn to care for the past, and we have to learn to care for our air and land and water.

We must rid ourselves — immediately — of the Fire Code which makes it impossible for us to use the great old buildings of the past for creative stores, offices, and homes. We must encourage the use of our heritage and—at the same time—put and end to the tearing down of our old buildings. We should go even further: that is to encourage property owners to bring into this area the homes which are being destroyed in the more obnoxious growth-oriented towns in the area like San Jose or Santa Clara. We should see Los Gatos as a repository of all that was fine and precious in the past of the once gracious Santa Clara County.

We will evolve a system of taxing 'developers' and 'speculators' who try to venture into this town—make it very unprofitable for them to disturb the fabric of our lives here. In addition—we will make it a policy of the Town Government that local banks (who profit so from our business—even though they are owned by large, outside corporations) will grant loans so that groups or individuals can get 70-90% financing of purchase of older homes. We will make preservation of the homes a matter of town, financial, and aesthetic policy—and will do everything (taxation, laws, zoning) to make this part of our future here.

Persons who take older homes and build them up, and care for them, will be given tax-advantages by the town. Although we cannot end the onerous (and inequitable) property tax system *at this point*—we can offer rebates to those who spend money out of their own pockets to preserve the past.

In addition, we must put the brakes on population growth of Los Gatos. It may be profitable for the merchants and the Chamber of Commerce to foresee 50,000 crammed into this town by 1984—but for the rest of us, this is an unnecessary burden to the schools, and the utilities, and the scope of the town.

One way to put an end to this growth is to put an end to the annual dole of $25,000 (that comes out of our pockets through the hotel-motel tax) to the Los Gatos Chamber of Commerce. Nothing is more repulsive to those of us who value the smallness of this community—than to know that the Chamber is getting almost $500 a week—in order to encourage more residents. The profit that comes to these business people comes directly out of *our* pockets—through the *expense* of growth. [If you have any doubt about how much it will cost *you* to have 25,000 more people here—look at the recent study commissioned by the City of Palo Alto, which established, beyond a doubt, that the best way to serve the community was to forbid the development of further sprawl in the adjoining mountains. "The expense to the community services leads us to the conclusion that the development should be avoided at all costs..." was the apparent conclusion of the extensive report.]

We must do everything possible to enlighten the businessmen-developer-bankers that 'Progress' (viz., profit) to them means deterioration and destruction to the rest of us. More population means more money for the banks—but more taxes for the rest of us.

To help change their way of thinking—the Town government will impose a 1% annual tax on gross of those businesses which profit through the destruction of the ecology. This 1% tax—to replace the annual business tax now imposed—will be levied against all new and used car dealers in the town, as well as gas stations, mortgage companies, real estate dealers, sporting goods stores (which sell such things as snowmobiles or outboard motors), large shopping center complexes—and all garden and hardware stores which sell destructive insecticides. In addition—all those businesses which have contributed monies to defeating ecological bills and initiatives in the past 10 years will be taxed (which would include our local branches of Wells-Fargo, Shell Oil, PG&E, and Bank of America.)

All these taxes will be forgiven those business which prove that they have used a like amount of money to *support* such legislation—or who spend such monies for the planting of trees and grassy areas and bushes in the Los Gatos area.

In addition, the Los Gatos Chamber of Commerce will be levied a special tax of $25,000 for being allowed to do business in the town.

THREE: TAXES

Nothing is more cruel—especially to the old and the disabled—than the county property taxes. It is a regressive tax, which gets more burdensome each year, and it falls most heavily on those who can least afford it: those on pension of social security. This tax goes to subsidize the speculator-builder, the quick-buck artists who are out to ruin our land and our lungs.

The county government is grossly unresponsive to our needs: as always, distant and incomprehensible to those of us who need its assistance. It is designed to embroil the old and the untutored and the young in a terrible Kafka-like machinery. There is no humanity, no caring in this (supposedly) efficient bureaucracy.

Los Gatos will offer a tax rebate on the home property levy—to all those people within the town limits who are on social security, unemployment, pensions, food stamps, or welfare. This tax rebate will be drawn from the 1% ecological tax. All the citizens of Los Gatos who receive these rebates will be encouraged to join in weekly work projects sponsored by the town to clean up areas of the town which have been defiled by the tourists. Areas will be set aside for volunteers to plant trees and flowers—and the town department of parks will assign volunteer coordinators to assist in these projects.

The old people of the town—now surely the most underused and over-ignored segment of our town population (relegated as they are to 'Old Peoples' Homes' and 'Nursing Homes' and 'Senior Citizens Recreation' Centers)—will be drawn back in to the mainstream of the town life. They will be encouraged to work in a volunteer capacity with the young and the poor. In Sweden, the old and retired are offered a chance to live-in at charity child-care centers, and homes for mentally disturbed and physically handicapped children. Even if it means bussing the elderly to East San Jose, we must do everything possible to make them feel a part of the continuous development of civilization: and make it possible for them to work with the young and the underprivileged. It has been found that the older people who have an active and useful life live longer, and are more healthy—than those who are just stuck away in some 'home', ostensibly out of everyone's way.

Because the County of Santa Clara is so insensitive to needs of the disadvantaged, and serves merely as a bureaucracy to tax the underrepresented without mercy, we should begin working towards getting Los Gatos out of the clutches of Santa Clara County. We will begin the long proceedings of setting up a separate "Los Gatos County" and will invite the neighboring towns to join in our secession. Only with our *own* county government, with Los Gatos as its County Seat, can we preserve the beauty and heritage of this area. Only in independence can we defy the rapacious builders and developers who wallow in the bones of Santa Clara County.

Our taxing machinery will be used to create an ecologically sound community of people and lifestyle. We will use the taxing powers to punish the malefactors of our community. For instance, two local newspapers, *The San Jose Mercury-News* and *The Times-Observer*—have been profoundly unaware of the juggernaut progress which threatens our health and our lives.

Both newspapers have encouraged the gross and destructive paving over of the area (the *Mercury-News,* for instance, gave glowing publicity to the building of the Eastridge Shopping Center. It ignored the laying low of thousands of stately trees, refused to acknowledge the disastrous ecological consequences to the area—all because it was 'business' and they knew that Eastridge meant more gross profits for them. Money to them means far more than the coarse ravagement of a gentle land.)

The *Mercury-News* and the *Times-Observer* are allowed to use our tax-built sidewalks to hawk their awful dishwater, even as the rest of us are forbidden by the town to set out potted plants or trees on these areas because of 'ordinances' and 'laws.'

These machines should be removed at once: they are an eyesore and a blemish. However, in order to encourage life on the streets, we will encourage these and other newspapers to set up wooden stands with real live newsboys selling their wares. (Which might make some inroads on the terrible lack of employment for the very young.)

The town taxing machinery will be used—for a change—to encourage the maintenance of all orchards still remaining within the town limits. Subsidies will be offered to property owners who resist the temptation to sell off these parcels of land. We must as well set up emergency ordinances to preserve the few orchards remaining in the town.

FOUR: STRUCTURE OF TOWN GOVERNMENT

Somewhere along the way, they decided that the best form of government would be that which was efficient (meaning cold) and bureaucratic. It was felt that the only way to assure perpetuation of the forces of order were through an isolated 'Town Manager' system—so that we could never really *find* the power (and the responsibility) in our local government. It was felt that if a bureaucrat did not smile, or show joy, or show life—then we could rest assured that our community was safe from the taint of humanity which might think of people before laws. It was felt that a cold and harsh aspect (which thrives in the present architecture of the Los Gatos Town Hall) would make sure that we were all being treated equally—and that no-one was getting a bare glimpse of warmth which should be human interaction. This system has created a whole terrifying breed of non-humans who call themselves 'professionals' or 'city managers' or 'bureaucrats.' Some of us who have studied the history of Nazi Germany realize all too well that this efficiency in the face of all human appeals was the be-all and end-all of men like Eichmann. 'He does his job,' is the theme of the bureaucrat. Not: "He cares."

We will begin charter revision activities to rid ourselves of the Town Manager. This form of government may have been effective 50 years ago when we all feared the East Coast type of government—but in these enlightened times, we do not need this cold and bureaucratic protection.

Instead, we will look towards the implementing of a full time involved town council—an elective body which will have great power and great responsibility.

People will be hired for jobs in town government not through some imagined absolute cold civil service guideline (which seems to give rise to nothing but pretty tyrants who use their power to lord over the rest of us) — but rather through standards of humanity and warmth which will make us feel that it is our government which cares for us: does not see us as an impediment to the shuffling of papers; but rather, a part of the joy and humanity of our days.

The beautiful turn-of-the-century town hall was ripped down by all those who thought Progress should be the point of our lives — and replaced with the present garish, and icy, and efficient Town Hall which is nothing more than a spawning ground for further exercises in bureaucracy (not to say, costing us enormous sums in maintenance of this bureaucracy).

We must seek some creative use for this complex. For instance, it would be a perfect site for a 24 hour free public child-care center for the working mother. It could be used for art exhibits, parties, festivities, dances for the young: anything which will emphasize that it is *our* town hall, and not the sole property of a dry and desiccated bureacracy, who have hung up their *humanness,* and turned themselves into unresponsive mannikins.

FIVE: THE POLICE DEPARTMENT

It is as inappropriate to a peaceful community such as this one to see our local police patrolling the streets with shotguns strapped to their dashboards as it would be if they were wearing death's head on their visors. There is — built into this Police Department (as in all) a para-military efficiency — which tells us that police cannot be allowed to be human because they are militant defenders of The Law. This distance and efficiency makes them an object of fear and hatred — rather than of respect and affection.

None of us doubt the need for a police force — men are brutes, and greedy brutes at that. Still — there is something terribly wrong with a militant force which will send a 16 year old kid into the homosexual nightmare of Santa Clara County Juvenile Detention Center for a couple of marijuana cigarettes — and, at the same time — allow some land developer who rapes the earth, burns down the trees, destroys the wildlife — to go scotfree.

Because the police department seems to be unswervingly dedicated to building walls between the members of the department and the younger, long-haired members of our community — the society is being ripped apart. The whole fabric of democracy must stand or fall on the premise that good will and humanity — rather than militancy and hardnosed law — should be the way that we perpetuate our rights, and freedoms, and happiness.

We must do everything possible to humanize the police: to make them feel that they are a needed and necessary part of the community — not an isolated force of righteousness and toughness. We must use every method available to the town to end the forced moral and mental attitudes which split this (and all) police departments from the rest of us. We must urge our police to become humans again.

We will end the requirement that local police carry guns with them at all times. A gun exposed in a holster is like a uniform to some: a signal of hostility, a red flag. To some people — especially the oppressed and the young — a gun, a badge, a uniform is a signal to rage, and attack. We must displace this excuse to violence.

For it is archaic and unnecessary that we have clung to the ancient symbols of law and power, A Show of Force. Like a fluorescent rotating cross on a church, it is an unnecessary and garish symbol out of another age.

We will encourage our police *not* to be police when they are off duty. They will be encouraged to leave the official sets and attitudes at the police station, and join the community of men when off and away from their place of work.

We will try to integrate the police and the non-police duties. They will be urged to take 3 or 6 months leaves of absence, to work in other branches of the town government. We will encourage the police to diversify their talents, to see the world through (for instance) the eyes of the firefighter. We will try to change the *set* of the minds of the full-time, 24-hour-a-day, week-in week-out policeman.

We will restructure the pay and advancement of the police—so that there is *continuing* and *long-range* growth in their pay. This will put an end to the present system which offers no chance for advancement after four or five years, leading to despair and stagnation. We will do everything possible to make the pay and standards of the Los Gatos police the highest in the country—to attract the best people to work here.

We will introduce ideas which should humanize the life of the police. For instance, we will create the post of a counselor. This person will have the same powers as the policeman, but he will never carry a gun, nor wear a uniform.

The counselor will be called in to mediate disputes between people—non-violent controversies between neighbors, within families. The counselor will thus avoid the either-or situation which arises when one 'calls the police.' He will mediate between contending parties, in the mode of a classical man of wisdom.

Problems which, in the past, have had to be solved by hauling people into our overcrowded jails, jamming the courts—will be solved, instead, through an informal intense brain-session with the contending parties.

The counselor will be a peacekeeper, the neighborhood policeman from fifty years ago. He will be a sage and generous member of the community who will have the power of the executive, with wisdom of the judiciary, and the interest of a friend, neighbor, and advisor. Understanding and good will must take precedence over militant law enforcement.

For one of the weaknesses of our country (long since abandoned by the English, who invented it) is the strict division between legislative, executive, and judicial. Such arbitrary, single-cell classification of the duties and responsibilities of men can only create schizophrenia. We have forced on our police an attitude of Grown-Up Boy Scout which—because of its enforced iciness—tears us all apart in its unreality.

It is important that we create less militant symbols for the police to work with. There will be a new set of horse trails interweaving the town, and we will set up a large mounted police force. We will also seek to recreate the 'walking of the beat' where the policemen can become part of their home areas.

In any event, we must get the police out of their super-charged, super-polluting white Dodges. Expenditures of town funds for such cars is a contradiction of what we need in human priorities.

We must discourage the destructive Civil War now taking place between the young and the police. Although Los Gatos police have tried (within their restricted world) to be fair—they have also joined in the past with the far less human Santa Clara County Sheriff's Department to use an 'informer' system in order to entrap young marijuana users and sellers. The paying of sixty or seventy dollars to an impressionable youth is poor schooling in the ethics of American law enforcement. It breeds vituperation among the young, and for none of them does it increase the necessary respect the law must have in order to properly function.

In addition, we must encourage the police to bring young volunteers into the corps—young people who will patrol their own parties and gatherings to warn their peers as to the actual content of the law, without the radical tool of arrest and detention.

When men are forced into their unreal sets for too long—they come to believe them implicitly. Someone, somewhere tried to make the force of law a cold and militant machine for controlling men; and now we are faced with the foolish need to defend and continue these unreal sets. We can afford to no longer; we must put an end to them.

SIX: THE PHONE CO., PG&E, THE BANKS

Unfortunately, the Los Gatos Telephone Company was taken over recently by General Telephone. Thus we have a government-protected oligopoly which pulls officers and upper management from other places in their octopus network, and ignores the needs and interests of this community.

Profit becomes the be-all and end-all of the telephone company. And these profits do not accrue to the immediate benefit of this community—but rather, to large stockholders and upper management far away.

This and all telephone companies have been crudely insensitive to the needs and interests of the poor and the unemployed and the aged. They have never given concessions to those on welfare, or pensions. Their cold and plastic office workers always say, "Regulations don't permit..." and "We aren't allowed..." and never tell us that they, as a powerful national semi-monopoly have profound influence on the national and state utilities regulatory commissions. They write the rules, and then tell us that they have to obey the rules.

The service of Western California Telephone is a monstrous joke: or haven't you noticed how many of your calls go dead; haven't you noticed that you may try to call Saratoga and end up getting Milpitas; haven't you noticed that the quality of calls to San Jose is slightly worse than an overseas toll call to Manila.

Western California Telephone Company and Pacific Gas and Electric—ignoring their powerful monopoly position—insist on advertising heavily with newspapers, in magazines, on unsightly billboards, on radio and television. They always claim that this is public service...but they don't tell us that if they cancelled all advertising, our rates would go down. Not only would we gain the direct savings of this money now wasted—but we wouldn't see the existing facilities (electronic and power-production) strained to care for the unreal demands created by the ads.

Pacific Gas & Electric has shown itself to be truly hostile to campaigns to protect the environment. Not only do they use possibly unsafe defoliants on our trees to 'protect' the lines—but they actively contribute money, their government-protected income (our money), to defeat state and local legislation such as the recent anti-high rise initiative in San Francisco.

In other words—our state government makes their profit-protected monopoly possible: and not satisfied with that power, they take our money to defeat necessary environmentally-protective legislative and initiative measures.

Finally, there are the banks. What can we do with them? So grossly insensitive to the needs of the community. Refusing to offer attractive mortgages to those of us who want to buy older homes—to preserve them forever.

Banks will lend small amounts of money 'on the land,' but always say (of older homes and stores) "The building isn't worth it." Our heritage isn't worth it? What they are telling us is that they will loan us money to plow under the great old buildings in the area—or, perhaps, to tear down some gracious orchard, but they will do nothing to help us save the past. They won't see that if we protect and honor that past, it will pay off handsomely, especially in a town which caters to tourists in search of all the glories of yesterday, running away from the plasticity of today.

Banks—along with PG&E—use their large resources to contribute to the defeat of ecologically sound legislation and initiatives. They seem to have lost touch with the reality of our day-to-day lives, and the threats to our future health and safety. They are—as with most American large corporations—run by middle managers to whom profits are all-important; and for whom human values are programmed out so they can be 'efficient.' They forget us as people—and rather see us as objects to be bought and sold with billboards and media ads.

• • •

We will begin the lengthy process of making the telephone company a municipal utility—owned and operated by the people of Los Gatos. We will apply to the state utilities commission to reduce the rates, both for local services and long distance calls. The municipal telephone company will be operated as a not-for-profit corporation. We will strive to improve the system, the quality of the service, and see to it that there are special free telephones provided for those people on welfare, pensions, retirement, food stamps, social security, ADC, and the poor and unemployed.

In the same way, we will work towards setting up a local power company to buy electricity and gas in bulk from PG&E, and distribute these cheaply to the townspeople, with special concessions to the old and disabled. As with Palo Alto (which has had this system for 70 or so years) we will be able to end the high rates that a profit-making monopoly must demand, with their extensive advertising and high executive salaries.

Through the 1% annual gross tax on banks, we must move these organizations to become more aware of the ecological consequences of their actions—and perhaps persuade them to make provisions for loans on older buildings and structures. The power to tax is the power to destroy—but it is also the power to convince. We must use every legal means to convince banks and financial institutions to work with the town to preserve the past and defeat the future.

SEVEN: GENERAL QUALITY OF LIFE

Some time ago—God knows when—it was decided that American town governments should be efficient and cold; that the purpose of towns was to provide acres of cement for automobiles; that street life should be suppressed by jack-in-the-box plastic buildings; that town services should be performed by "Experts" with cold and analytical minds; that town governments should be dedicated to protecting the profits of real estate dealers, businessmen, respectable bankers, and speculators; that the old should be labelled 'Senior Citizens,' and safely shunted off to 'Retirement Communities' and conveniently forgotten; that the young should be kept off the labor market until age 25; that public property should be considered the private property of bureaucrats; that amenities in towns should be limited to patches called 'parks' with no real human use of interactions; that all zoning should be strict and exclusive—giving us tightly compartmentalized 'living' and 'working' and 'shopping' areas: as if we and our lives and our souls could be so easily divided into such convenient parts.

Compartmentalization has severely damaged the joy and hopefulness of all of us: governments have become antagonistic machines, protecting the prosperous, enmeshing the poor and the confused, disenchanting the young. We must start work to end this strict division of life and government and pleasure and work and feeling. We must begin to reconstruct the life of the town, the life of all of us.

The streets of Spain and Mexico and Italy give us a great feeling for the life of the communities, the souls of the people. The sidewalks and alleys and streets are filled—day and night—with pedestrians and peddlers and street artists and musicians. There are donkeys and carts and horses and bicycles; there are endless coffee shops and bars and vendors and awnings and color and life, spilling out into the 'traffic' area—turning the towns and cities into living organisms, into vital conjunctions of humans, all interacting with each other.

We must reinstate the street life in the community of Los Gatos. Ordinances and long-winded hearings which are presently required for those who want to use the sidewalks and streets for commerce and coffee houses and bars and art shows will be repealed. Merchants will be encouraged to spill their wares out on the streets. All restrictions against street musicians, and street artists, and street peddlers will be repealed.

The town government will embark on a vigorous campaign to encourage all—the young, the old, retired, the families, the couples, the kids—to participate in town street life at all hours of the day and night. Streets (much like Portobello Road in London) will be set aside and blocked off to automobiles so that the street merchants can sell clothes and jewelry and trinkets. Downtown streets will be closed off to vehicular traffic at regular intervals so that fairs and festivals can be held. All ordinances against selling and consuming foods, coffees, and spirits in public will be repealed. Pedestrians will be made masters over the cars.

The downtown area will be honeycombed with bicycle paths and horse trails and walkways. Public buildings—schools, the Town Hall, Fire Houses, town maintenance shops—will be opened to the townspeople without question for gatherings and shows and festivities.

Those who contribute their taxes to participatory democracy will come to have a sense that the town is theirs, for a change, for them to enjoy. Police will have a positive duty to get rid of their mace and hide away their guns and give us a feeling of warmth and human interchange so that visitors will marvel over the social and political integration of the community.

And finally—and perhaps best of all—the town will hire a full-time fool to show us our sins of delusion and pretension: an adult-child in costume who will gently mock the pomposity of government, businessmen, hippies, policemen, educators, media executives, councilmen, planners, dentists, contractors, clerks, and laundrymen. The town fool will be paid generously to make us aware of all our foolishness: a good fool, a fool who will do as fools have done through the generations. And that is—to point out that, perhaps, the rest of us are somewhat foolish; and that he himself is, perhaps, the only wiseman in the community.

NOTE

This mailer was prepared by The Clean Slate (also called The Meal Ticket) and concerns the municipal elections which were to be held in Los Gatos on the 11th of April, 1972.

342

It was put together with all-volunteer labor, and the four members of The Slate: Cese McGowan and Lorenzo W. Milam (Town Council), Telemachos Athanasios Greanias (Town Treasurer), and Robert Lee Gray (Write-In, Town Clerk).

You will notice that we avoided the usual rhetorical flyer—the boldly printed sheet devoting two sentences to each Important Issue, with a picture of the candidate (many teeth) with wife (more teeth) and three or five pre-set children (most teeth). This is called 'humanizing' the candidate in Public Relations terms.

Well—we are all humans. Our spouses and our children can smile with the best of them. But we tend to feel that an in-depth explication of the problems facing this community was worth far more than a thousand canned photographs.

Our candidacy for town office started in a rather jocular fashion some months ago. We had some funny banquets and testimonials and such. We even invented a mythical enemy: a man named Mr. Big, leader of The Boys of North Santa Cruz Avenue. It was all very amusing until someone (is there *really* a Mr. Big?) got the courts to knock one member of The Slate off the ballot. Apparently, it was felt that the decision of a single judge was superior to the desires of 10,000 voters. That move took Merrill Kelly from the ballot, and disqualified him for office of clerk.

We wanted the voters of Los Gatos to have a clear choice in the election. So we then asked our campaign manager—Robert Lee Gray—to run as write-in for Town Clerk. This will mean a bit more trouble for you : if you want to register your vote *against* the juggernaut called Progress, not only will you have to vote for McGowan, Milam, and Greanias—you will have to memorize the name *Robert Lee Gray*, write it in to the proper place, and rubberstamp an X next to it.

We should like to thank the incumbents, the other candidates, the legal staff down at Town Hall, and the system of American Democracy for our enlightenment in the process of elective office over the last two months. We should also like to thank Oscar Wilde, whose words—uttered some 75 years ago—have become of theme of our campaign. He said:

"*The trouble with participatory democracy is that it takes too many evenings.*"

343

GROWING PAIN

when i was very little, my father bought a shotgun and some shells and took me hunting for the first time. he got me up early one cold morning and drove to the country and made me follow close behind while he searched the ponds and lakes for ducks. every time he found one of the green-headed birds he blasted it out of the water and after he had shot about a dozen ducks, he would then drag me and the dead ducks around to his friends, neighbors and relatives, showing off the beautiful ducks he had killed. it always amazed me that he killed only the beautiful birds, never the ugly mudhens. and to tell you the truth, i never really cared for this sort of thing, yet he made me do it because i was little. but i am grown now and i know better.

—Statement of Merrill Kelly of qualifications for election to office of Los Gatos Town Clerk.

A FINAL (AND SAD) STORY ABOUT KDNA AND RADIO POLITICS
(Which I may no longer believe, even)

[*The following was printed at the end of the Second Edition of* Sex and Broadcasting. *Like most authors, I have the debilitating habit of re-reading my own books from time to time, to be sure, really sure that I wrote what I thought I wrote, or to see how I have changed, or to get some idea of what all that tomfoolery meant back then, when it was so important.*

But we change and grow (both little and tall), and the ideas and feelings that prompted this final work have changed too. The page never varies: the me that wrote it then has grown a bit more skeptical, a lot more dis-believing in some of the final process of community radio.

Interior politics—the very thing I wrote this book to protect you from—have taken their toll. I keep thinking of what I was, when I wrote this: wondering where those particular sunrises have squatted off to, and why.

KDNA is no longer. We sold it to get money to start other community stations. When we did that—I thought it would be a simple process: sacrifice one to create dozens of others.

But it didn't work out so simply. The good that was KDNA—the greatness, really—got sold down the river. I lost some of my virginity in the process.

The blacks and the minorities that were so isolated from the change we all want when I first wrote this book have become more alive to their possibilities: there are some minorities who have the ability to split the boards and staff of community stations and make eunuchs of all in the process.

FM has grown up—is now where AM radio was at the end of WW II. Community radio has become crowded, and there are controversies between stations and between staffs of the stations which are no longer rewarding or interesting.

And the person who elects to be the leader of such an operation or several such operations is asking for endless shit, no matter how he tries to isolate himself from it. People care too goddamn much.

Which is, I guess, what we were working for all along. But I never knew it would work out...exactly like this. When we set off on that trip into the mountains, I never guessed that we would end up down at the bottom of this rock quarry, with the slag-heaps all around us, and the trees denuded and ugly in the cold harsh 5 o'clock sunset.

I am keeping this essay in the book, and at the end of the book, because it represents me 4 years ago, at a time when I thought we had no where to go but way up there. Then, in one of those funny mirror moebius strip journeys, I found out that we went down or over, all the while I thought we were going even further up.

Too bad. It was good times. It's not that I don't care any more about Poor Peoples Radio and Telling the Message and getting station #23 on the air. I still do—care, that is.

It's just that the next dozen or so have to be done another way. And I am not sure exactly how that should be. I don't know yet: and won't—at least not until I have retreated to a Hot Springs and in myself for the next two years, to smell the air—real air, not aetherized air—for awhile. And figure out where it all went wrong, if indeed it did go wrong; and figure out how it could be done better in the future.

Meanwhile—here is my favorite essay...a mirror of me gone past. I stick it in here at the end because this is where it all started, and ended. All together. Where we belong.]

I was on the air, playing some Gagaku Court Music of Japan. A listener came in. He was wearing a cape, and a magic pointy hat. He have me an eggplant.

345 345

"Here," he said: "Take it. Do you know that this eggplant contains all the vibrations of the Universe?"

I didn't, but I took it anyway. I set it on the Gates Studio Consolette to keep me company as I was playing the Japanese Court Music.

Now I had never been exposed to the Vibrations of the Universe before. At least I didn't think I had. But since I was an old radio man, I figured I could hear them by simply holding the eggplant up to my ear.

I wasn't sure of what to listen for. Once I had read in a book that the Universe turned on slots of gold. I thought that the noise of this particular eternity might be a soft, golden moan.

I put the eggplant up to my ear and as far as I could tell, there was no especial sound. At least *I* couldn't hear it. All I could hear was the Gagaku Court song "Ichi Uta."

After the selection was over, I told the KDNA audience about this adventure. I said: "Here. Maybe my hearing is bad. Or maybe I am just insensitive. Maybe you can hear the sound of the Universe better than I can." So I held the eggplant up to the microphone so that they could hear what I couldn't hear.

And as I did that, the tape machine (which I had neglected to shut off) came on with another selection of Japanese Court Music "Suruga Uta."

At KDNA, the radio station and the volunteer living quarters are all in the same building. When you get tired of playing radio downstairs, you can go up to the 2nd floor and borrow one of the beds for awhile and wake up at 3am and hear Gabriel playing James Brown at the 120 db level...and it creeps up from downstairs through the heat ducts like fog, mixing with your dreams, and outside the Blacks come and go all night at the Rex Hotel&Bar next door.

KDNA is set in the St. Louis ghetto. The streets are filled with pimps and whores and angry Blacks and drunks and kids in rags playing and forty-five year old bohemians and junkies: right outside the window you can see them laughing and talking and running and falling down.

(The neighbors are convinced that KDNA is a whore house in disguise. Two drunks came in once, wanted to see the girls. "No," Jeremy said: "It's a radio station. Look." And he took them back to see the records going around, the sound coming out. They weren't convinced. So he went on the air, said that there were two visitors who believed themselves to be in a house of prostitution. He asked that some listeners call up to explain that it was the Sound of Music, not young ladies, what made that particular operation go around. They did, but there is no record that the visitors were at all convinced. Prejudices die hard in the ghetto.)

The last time I was at KDNA, I was drafted to do all the shit work on preparing their application for renewal of broadcast license since I am such a good bureaucrat. Besides, I am the only one connected with that operation who can spell.

One of the questions on the form really got to me. It said *State in Exhibit Number* (blank) *the methods used by the applicant to ascertain the needs and interests of the public served by the station...*

"Jesus Christ, F.C.C.," I thought: "The needs of St. Louis come right in the goddamn door and steal the tape recorders and tools for another fix. And when they aren't doing that, they are following the female volunteers down the street, trying to get them into the alleys so they can expose them to other community needs." It's just like asking a concentration camp inmate fresh out of Auschwitz whether he was *sure* the Germans were prejudiced against Jews.

As far as I know, KDNA is the only ghetto radio (or television) station in the country. I must say it adds a certain *verite* to the programming. Like the staff has to go no more than 20 feet to see a city dying because of ruinous speculation, and a petty, bickering city government, and prejudice. But living and working in the ghetto makes the staff tough, less than willing to be open and free as we are (I think) in the garden paradise flower of Los Gatos. They have lost too much equipment to the sticky fingered junkies; the women have been too brutalized on the empty streets. There are two locks on the front door; a peep-hole and speaker, and when you knock after six in the evening, a 1984 voice says 'Who's there?" *That Orwellian voice is one of us.*

Across the street, the Apollo Broadcasting Company bought up a two story building and ran an automated FM station there until the ghetto kids burned it up for the third time so that they got sick and tired of it all and moved downtown to the marble-and-glass section of St. Louis where the needs of the people weren't so *pressing* and *real* for god's sakes. KDNA staff shrugged their shoulders and boarded up all their street level windows and bought seven more fire extinguishers.

When I was there, I realized that KDNA sees so many community needs that they might well become one. Radicalization, I guess they call it. What happens if you are a young white college dropout hippie type, into experimenting with good radio—and you find yourself in the middle of the dingy community of the poor—with all its accumulated grievances: it's enough to make you think, "As part of the most gorgeously rich country in the world, why does this city look something like 1944 London?" Are all Blacks just slobs who tear down neighborhoods and rip up buildings? Or is it something more subtle than that? *Like the destruction of self-belief of a whole culture—born of slavery, prejudice, five centuries of intolerance?*

So you find your ideas becoming a bit more militant. And then there's the St. Louis Police. Doing their bit. To radicalize.

It's not enough for the staff to be living pisspoor, getting $10 a month for expenses, giving up every possible freedom of affluence to this crazy station—then there's the force of order and law.

It is a balmy night in September. Late at night. 1:18 AM, to be exact. And they come bludgeoning in the door, dragging the staff out of bed, claiming to have found a couple of baggies of you-know-what (rabbit-and-hat style) in some previously empty cabinets.

KDNA, a station which has programmed some material which is very critical of our foreign policy, state and city governmental corruption. A station which is striving to exercise those all-too-rare freedoms of speech. Is that why you are here, officer? I would like to explain to you officer, but there you are, waving the two black eyes of a sawed-off shotgun at me, warning me not to turn on the microphone, warning me not to tell the listeners about what is happening to me, and my heart, and the future of free speech radio in St. Louis. I worry about your finger, officer, and blood *my blood* and flesh *my flesh* all over this quiet control room. Are you trying to radicalize me officer? Are you trying to give me some feeling for arbitrary, untrammeled power, police power? Is that the reason you are here, officer?

I started out by telling you that the theme of *Sex and Broadcasting* would be the potent need, we all have, to communicate. I got sidetracked a bit on the FCC and all those Civil Service fiefdoms which don't give a good goddamn about you and me, and the sweet tangled parts of our personalities. I told you how they live by rules and laws—which is supposed to give us a great sense of security, but for those of us who have read some about Eichman and Hoess, we get a bit ancy about the 'true' bureaucrat. (One of the great lessons of Nazism is that paper-pushers have a tremendous capacity for evil, and a tremendous ability to isolate themselves from that evil.)

But there is another vital part of listener-involved community alive free form radio which I have not communicated. And, since I am damn sick and tired of writing and rewriting this book, I'll have to stuff it here at the end, like some shirt-sleeve dragging out of the side of the suitcase.

What it is is that I have yet to tell you about the *fears*. That big choo-choo train, streaming down the line, a gold letter plate screwed in the boiler marked

PARANOIA

Some community radio station people have good reason to wake up and sweat at night. The staff at KPFT Houston kept thinking *Why are we being so silly? Why do we have all these imaginary fears?* Then these imaginary fears turned into two very unimaginable blasts in the transmitter shack. And KDNA fears as I have mentioned came in the door with Remington 12 Gauge Shotguns.

And KPFA. There's my next book. Growing up inside me right now, a sprout and shoot which will render me sleepless for the next two years *as I have been for the last two weeks getting you out of my system*

Old KPFA. The first. In a long line. KPFA. On the air in 1949. Off the air in 1950. On the air in 1951. And, in 1948, 1949, 1950, 1951, 1952 et al: that was when the staff tore each other apart. Each day. Hourly. Endlessly.

There was good reason for the KPFA people to be scared for the first 10 years of the station's life. The United States was locked in The Fear of the Red Beast. It was a rumor that John C. Doerfer, recently of Wisconsin, friend of you-know-who, Chairman of the FCC, had vowed that he would get that son-of-a-bitching station off the air.

Fortunately, the country grew up some, and Mr. Doerfer was caught with his hand in a cookie-jar known as the George B. Storer Company company yacht. (Regulators should be social virgins, and not fishing cronies with those they are supposed to be regulating. Doerfer, surely one of the worse FCC chairmen, left office shortly after the news got out. He is now working for—ready?—George B. Storer.)

But listen: the fears of the community radio people, I am loath to tell you, come as strongly from within as without. It works like this: people like you and me who are involved with strange and honest broadcast operations have a looseness in the brain-pan. We (you and I, love) operate best through tension, insane schemes, and bizarre fears. We seem to create nests of slander, inwit, neurotic outrage, and mental dyspepsia.

I tell you all this not to cover you and me and the existing community stations with calumny. But rather, to suggest that as you move towards getting your operation on the air, you should

also set about defusing the madness inherent in the people who will come to be volunteers or staff for you.

See: commercial radio stations have a built-in defusing process which is *make-money*. You don't have that. What you have is a group of dedicated sincere people who want to Do Good and Right. And they are all crackers. Aren't we?

Choose your fellow workers carefully and well. Get people who are stable and loving and involved, but get people who have a life outside the station. Because they can drive you (and it) balmy.

Listen: the reason KRAB was such a benign operation through its first five years was not just because Seattle is such a benign city where the outrage of free speech has been tolerated up through the ages. Nor is it because for the first five years we were convinced that no one ever listened to us: what with our two hour concerts of Korean Temple Bells and weekend extravaganzas of the music of Dahomey. No—it was because Nancy and Gary and Jeremy and James and I were careful to people the station with richly self-contained individuals. Good people, who loved listener-supported community radio, and what it could do for our minds; but, individuals who valued life outside the station.

It was not just that we took a couple of gallons of Mountain Red to the board meetings; it wasn't that we practiced an anarchistically politically detached wryness in our daily lives: it was, most of all, that we had a loud early warning system which went off whenever 'political' types came in the door. And I ain't talking about communists or John Birch Society members.

You will have hundreds of volunteers. They, and your board and staff, should be apolitical. Apolitical in the most inner sense. Apolitical in that you can only survive through openness, warmth, and a militant avoidance of rumor. You must be a lightning-rod.

I can tell you all this, but I am not so sure that I can *really* tell you, unless I were to meet you face-to-face, belly-to-belly. It's all tied up with Existential mental sets, whatever the hell those are, and the willingless to confront the peas hiding under the mattress. You must you should please try to defuse (read *diffuse*) intrigue and rumor. You will survive longer than—if not as brilliantly as—all those committed, secretive, artistic, tortured political boobies who try to convince us that their creativity is tied to their troublesome difference.

There is a story I have to tell you about all this. It is right where it should be: at the end of this book, at the beginning of all others. I told it to Larry Q. Lee, the resident wag of Pacifica, right after he called me "The Peter Pan of Listener Supported Radio." If I can tell him, I can certainly tell you:

It concerns Lew Hill. He started the whole listener-supported radio station thing in this country. He is the Father of Us All. Or was, until, in 1957, on a windblown hill outside of Berkeley, in a 1953 Dodge, with the windows rolled up, with the motor running, with a green water-the-lawn garden hose snaking from exhaust pipe to left side vent window, he did himself in.

Why did you do it, Lew?

And Gertrude Chiarito the Great Earth Mother of KPFA for 18 years from the miniscule beginnings until the harpies ran her away in 1966, dreamed a dream. A dream, a horrible dream, which tells you and me all about the crazy madmaking tear-apart weirdness which is the community of community radio:

It is late at night. It has been raining very hard. It is after midnight. There is a banging at the door of my apartment. I put on my robe and run downstairs, open the door. There is Lew Hill. His clothes are all sopping wet. The rain is pouring down his face.

"Why, Lew," I say: "You can't be here. You're dead." YOU CAN'T BE HERE. YOU'RE DEAD.

"No," he says. "I'm not dead." *I'M NOT DEAD.* "It's just a rumor, spread by my enemies," he says. *IT'S JUST A RUMOR. SPREAD BY MY ENEMIES. SO THEY CAN TAKE THE RADIO STATION AWAY FROM ME.*

A NOTE ON THE THIRD EDITION

Except for the first three pages of text, the whole of *Sex and Broadcasting* has been elaborately revised. It is now about ten times longer, with about one-third of the bombast of the first edition.

The book came out in June of 1971 because I got goddamned sick and tired of writing up single-space five page letters for all those people wanting to set up alternative, community radio stations. I put the whole thing together in three days. It showed.

But it sold. We exhaused the first printing in less than three months, with the kind help of *The Whole Earth Catalogue, The Village Voice, Source, Liberation News Service,* and *Vocations for Social Change.*

(This is in unpleasant contrast to my first book, *The Myrkin Papers,* which took five years to write, three months and $6000 to print, and which sold some 150 copies. If you are willing to pay $5 for *that* turkey, I still have a few dozen cartons under my bed and behind the door.)

That so many people are interested in the art of radio and radio station building is good and hopeful. What is tragic is that so many of my correspondents come from Philadelphia, Boston, Chicago, Cleveland, Detroit and other large cities where there is no good human alive community radio, and where there is nothing in the way of available frequency space. Something must be done: Perhaps the citizens should have access to the over 60% of the spectrum wasted on the military. Perhaps we can convert the unused Long Wave band (100-500 kHz) to a community radio system.

Community radio. A fine thought! Studios, turn-tables, tape recorders, available to anyone who pays a low hourly rate: common carrier—with no censorship in any form. A solution which has turned up in Micronesia—of all places—where anyone can go on the air after midnight and broadcast music, and talk, and interviews. Anyone who has the desire is free to communicate: and there is none of the false mystique imposed by broadcasters.

The need to communicate is as strong as the need to eat, or sleep, or love. If there is any theme to this book—it is that we have come to a time when so many of the other needs have been met, and the driving force to be heard must be met in the same way. That's one theme, anyway. The other is the absolutely mad-making will-o'-the-wisp called The FCC.

As far as sex goes, I would supposed you have been hard-pressed to find much raw naked diddling in this book. It's not that I am shy about offering you some passionate act of congress in the transmitting room, as the tube-blowers rage overhead. No: it's just that I really wanted you to buy this book, and my Great Aunt Beulah convinced me that a book with the word *Sex* in its title would double its sales, and quadruple its readership.

Lorenzo W. Milam
Los Gatos, California
October, 1974